# Lecture Notes in Computer Scie

T0250679

Commenced Publication in 1973
Founding and Former Series Editors:
Gerhard Goos, Juris Hartmanis, and Jan van Leeuwen

## Editorial Board

Victor G. Ganzha   Ernst W. Mayr
Evgenii V. Vorozhtsov (Eds.)

# Computer Algebra in Scientific Computing

9th International Workshop, CASC 2006
Chişinău, Moldova, September 11-15, 2006
Proceedings

 Springer

Volume Editors

Victor G. Ganzha
Ernst W. Mayr
Technische Universität München
Institut für Informatik
Garching, Germany
E-mail: {ganzha,mayr}@in.tum.de

Evgenii V. Vorozhtsov
Russian Academy of Sciences
Institute of Theoretical and Applied Mechanics
Novosibirsk, Russia
E-mail: vorozh@itam.nsc.ru

Library of Congress Control Number: 2006932483

CR Subject Classification (1998): I.1, F.2.1-2, G.1, I.3.5, I.2

LNCS Sublibrary: SL 1 – Theoretical Computer Science and General Issues

ISSN        0302-9743
ISBN-10     3-540-45182-X Springer Berlin Heidelberg New York
ISBN-13     978-3-540-45182-2 Springer Berlin Heidelberg New York

Springer is a part of Springer Science+Business Media

springer.com

© Springer-Verlag Berlin Heidelberg 2006
Printed in Germany

Typesetting: Camera-ready by author, data conversion by Scientific Publishing Services, Chennai, India
Printed on acid-free paper        SPIN: 11870814        06/3142        5 4 3 2 1 0

# Preface

This volume contains revised versions of the papers submitted to the workshop by the participants and accepted by the program committee after a thorough reviewing process. The collection of papers included in the proceedings covers not only various expanding applications of computer algebra to scientific computing but also the computer algebra systems themselves and the CA algorithms.

The eight earlier CASC conferences, CASC 1998, CASC 1999, CASC 2000, CASC 2001, CASC 2002, CASC 2003, CASC 2004, and CASC 2005 were held, respectively, in St. Petersburg, Russia, in Munich, Germany, in Samarkand, Uzbekistan, in Konstanz, Germany, in Crimea, Ukraine, in Passau, Germany, in St. Petersburg, Russia, and in Kalamata, Greece, and they proved to be successful.

It was E.A. Grebenikow (Computing Center of the Russian Academy of Sciences, Moscow) who drew our attention to the group of mathematicians and computer scientists at the Academy of Sciences of Moldova conducting research in the field of computer algebra. We were impressed that this group not only is concerned with applications of CA methods to problems of scientific computing but also carries out research on the fundamental principles underlying the current computer algebra systems themselves, see also their papers in the present proceedings volume. It was therefore decided to organize the 9th workshop on Computer Algebra in Scientific Computing, CASC 2006, in Chişinău, the capital of Moldova. We hope that this will foster new and closer interactions between the Moldova CA group and other research groups working in the field of computer algebra.

The papers collected in the present volume are devoted both to the topics that have already become traditional for the CASC workshops, and to several new topics. Among the traditional topics, there are the studies in Gröbner bases, polynomial algebra, homological algebra, quantifier elimination, the applications of computer algebra systems in the field of the solution of differential equations, celestial mechanics, Newton polyhedra, mathematical physics, nuclear physics, and fluid dynamics.

Two papers present the results in a new topic, which was addressed for the first time during the CASC 2005 workshop: the application of computer algebra techniques in the field of nanosciences and nanotechnology. Another novel theme is the application of CA methods to cellular automata with symmetrical local rules.

In addition to the accepted submissions, this volume also includes two invited papers. The paper by T. Sturm (University of Passau) addresses various aspects of the computer-algebra-based computer logic system REDLOG, which focuses on real quantifier elimination algorithms. An immense potential of quantifier elimination techniques for the integers is pointed out. Another new REDLOG domain is queues over arbitrary basic domains. Both have promising applications in practical computer science, viz. automatic loop parallelization and software security.

The invited talk by S.M. Watt (University of Western Ontario) is devoted to algorithms for symbolic polynomials where the exponents are not known in advance, such as $x^{2n} - 1$. The case is considered where multivariate polynomials can have exponents which are themselves integer-valued multivariate polynomials, and algorithms are presented to compute their GCD and factorization. Additionally, the case of symbolic exponents on rational coefficients (e.g., $4^{n^2+n} - 81$) is treated, and it is shown how to avoid integer factorization.

Our particular thanks are due to the members of the CASC 2006 local organizing committee at the Technical University of Moldova: V. Dorogan (Chair), M. Izman, and V. Dragan, who have ably handled local arrangements in Chişinău. We are grateful to W. Meixner for his technical help in the preparation of the camera ready manuscript for this volume.

July 2006                                                          V.G. Ganzha
                                                                   E.W. Mayr
                                                                   E.V. Vorozhtsov

# Organization

CASC 2006 was organized jointly by the Department of Informatics at the Technische Universität München, Germany, and the Technical University of Moldova, Chișinău, Moldova.

## Workshop General Chairs

Vladimir Gerdt (JINR, Dubna)          Ernst W. Mayr (TU München)

## Program Committee

Alkis Akritas (Volos)
Gerd Baumann (Cairo)
Hans-Joachim Bungartz (Munich)
Andreas Dolzmann (Passau)
Victor Edneral (Moscow)
M'hammed El Kahoui (Marrakech)
Ioannis Emiris (Athens)
Victor Ganzha (Munich, co-chair)
Evgenii Grebenikov (Moscow)
Jaime Gutierrez (Santander)
Ilias Kotsireas (Waterloo)
Robert Kragler (Weingarten)

Richard Liska (Prague)
Bernard Mourrain (Sophia Antipolis)
Eugenio Roanes-Lozano (Madrid)
Yosuke Sato (Tokyo)
Werner Seiler (Heidelberg)
Stanly Steinberg (Albuquerque)
Serguei Tsarev (Krasnoyarsk)
Evgenii Vorozhtsov (Novosibirsk, co-chair)
Michael N. Vrahatis (Patras)
Andreas Weber (Bonn)

## Local Organizing Committee

Valerian Dorogan (Chișinău, chair)     Mihai Izman (Chișinău)
Vladimir Dragan (Chișinău)

## General Organizing Committee

Werner Meixner (Munich, chair)         Annelies Schmidt (Munich, secretary)

## Electronic Address

WWW site: http://wwwmayr.in.tum.de/CASC2006

# Table of Contents

# Comparison Maps for
# Relatively Free Resolutions*

V. Álvarez, J.A. Armario, M.D. Frau, and P. Real

Depto. Matemática Aplicada I, E.T.S.I. Informática, Universidad de Sevilla,
Avda. Reina Mercedes, S.N. 41012 Sevilla (Spain)
{valvarez, armario, mdfrau, real}@us.es
http://www.us.es/gtocoma

**Abstract.** Let $\Lambda$ be a commutative ring, $A$ an augmented differential graded algebra over $\Lambda$ (briefly, DGA-algebra) and $X$ be a relatively free resolution of $\Lambda$ over $A$. The standard bar resolution of $\Lambda$ over $A$, denoted by $B(A)$, provides an example of a resolution of this kind. The comparison theorem gives inductive formulae $f\colon B(A) \to X$ and $g\colon X \to B(A)$ termed comparison maps. In case that $fg = 1_X$ and $A$ is connected, we show that $X$ is endowed a $A_\infty$-tensor product structure. In case that $A$ is in addition commutative then $(X, \mu_X)$ is shown to be a commutative DGA-algebra with the product $\mu_X = f * (g \otimes g)$ ($*$ is the shuffle product in $B(A)$). Furthermore, $f$ and $g$ are algebra maps. We give an example in order to illustrate the main results of this paper.

## 1 Introduction

Calculations in homological algebra are commonly expressed in terms of resolutions. It is not unusual that these resolutions are embedded in the bar construction (or some other standard resolution) in a special way. When this occurs, they are said to *split off* of the standard resolution (see [14]).

A classic example is the Koszul resolution $K = A \otimes E_\Lambda[u_1, \ldots, u_n]$ related to the ideal $I = (x_1, \ldots, x_n)$ in the polynomial ring $A = \Lambda[x_1, \ldots, x_n]$; as $A$ is an augmented algebra over $\Lambda$, the *bar resolution* $B(A)$ for $\Lambda$ over $A$ [5,16] can be constructed. $K$ is also a resolution of $\Lambda$ over $A$ and by the comparison theorem [16], there is a chain homotopy equivalence $B(A) \leftarrow K$. In this case, an explicit contraction (special homotopy equivalence) $B(A) \Rightarrow K$ exists [17]. This contraction makes that the Koszul resolution splits off of the bar resolution. Using this contraction and some homological perturbation tools, perturbations of this resolution can be computed and the perturbed resolutions can be used to make complete effective calculations where previously only partial or indirect results were obtainable. This idea has been exploited in a series of papers by Lambe [12,13,14] and provides an algorithm for computing resolutions *which split off of the bar construction*. This algorithm has been extended to a more

---

* This work was partially supported by the PAICYT research project FQM–296 from Junta de Andalucía (Spain).

V.G. Ganzha, E.W. Mayr, and E.V. Vorozhtsov (Eds.): CASC 2006, LNCS 4194, pp. 1–22, 2006.

general context in [11]. We point out that the notion of contraction is essential in order to find effective algorithms in homological algebra using the set of techniques provided by homological perturbation theory, since the input data of our algorithm have to be codified in this form.

The well-known comparison theorem in homological algebra states that any two projective resolutions are chain homotopy equivalent. For relatively free resolutions $Y = A \otimes \bar{Y}$ and $X = A \otimes \bar{X}$ of $\Lambda$ over $A$ with explicit contracting homotopies, there are recursive procedures for obtaining explicit equivalences $f: Y \to X$ and $g: X \to Y$ where the explicit contracting homotopies play a principal rule. In addition, there are inductive procedures for obtaining explicit chain homotopies of $fg$ and the identity and with $gf$ and the identity. Generally, the maps defined in this way do not form a contraction. Here, assuming that $Y = B(A)$, we give a necessary and sufficient condition for determining when these maps form a contraction, which seems to be new.

In the special case that $A$ is connected, $Y$ is the bar resolution of $\Lambda$ over $A$, and the above maps form a contraction from $B(A)$ to $X$ (i.e., $X$ is a resolution which splits off of the bar resolution), we define a degree minus one map $\tau: \bar{X} \to A$ which is an $A_\infty$-twisting cochain, so that $(X, \tau, \{\Delta_i\}_{i \geq 0})$ becomes an $A_\infty$-twisted tensor product where $\{\Delta_i\}_{i \geq 0}$ is the $A_\infty$-coalgebra structure of $\bar{X}$ transferred from $\bar{B}(A)$ by means of 'tensor trick' [8] (see Algorithm 1). This provides an elegant codification of the differential of the complex $X$ in terms of the $A_\infty$-twisting cochain and the $A_\infty$-coalgebra structure of $\bar{X}$. Furthermore, assuming in addition that $A$ is commutative (but not necessarily connected) and that the contracting homotopy of $X$ is a quasi algebra homotopy (see [19]), we prove that the morphism $\mu_X = f * (g \otimes g)$ ($*$ is the shuffle product in $B(A)$) endows $X$ a commutative algebra structure, for which $f$ and $g$ are algebra maps (see Theorem 7) and give a method for computing new resolutions taking advantage of this algebra structure (see Algorithm 2). In the example given in this paper, we compute a resolution $\widetilde{X}$ of $\mathbf{Z}_p$ over $\Gamma(w, 2n)$ using an initial resolution which splits off of the standard resolution, $B(A) \Rightarrow X$, and perturbing this contraction. The contraction $B(A) \Rightarrow X$ has been computed by means of the comparison theorem. We prove that $X$ is a DGA-algebra as well as $\widetilde{X}$. A computational advantage is deduced from this fact, since it is only necessary to compute the perturbed differential on the generators of $\widetilde{X}$ as an algebra, better than on the whole set of generators as a module. This type of computational advantage is our main motivation for studying the algebra structures underlying the resolutions.

We organize the paper as follows. In section 2 we give the necessary definitions and notations for defining the comparison maps when $Y$ is the bar resolution and $X$ is a contractile relatively free resolution. We also give a necessary and sufficient condition for guaranteeing that a contraction arises. Section 3 is devoted to study the $A_\infty$-structure inherent in $X$, when $X$ is a relatively free resolution over a connected DGA-algebra $A$. In section 4 we analyse the multiplicative behaviour of the comparison maps, assuming that $A$ is a commutative DGA-algebra and that the contracting homotopy of $X$ is a quasi algebra homotopy. Finally, we give an example in order to illustrate the main results of the paper.

## 2    The Canonical Comparison Contraction – A Necessary and Sufficient Condition

We will quickly review some basic notions of Homological Algebra. More details can be found in [16]. Let $\Lambda$ be a commutative ring with $1 \neq 0$, and $A$ an augmented differential graded algebra over $\Lambda$, briefly termed DGA-algebra. The differential, product, augmentation and coaugmentation of $A$ will be denoted respectively by $d_A$, $\mu_A$, $\epsilon_A$, and $\eta_A$. Nevertheless, sometimes, we will write them simply by $d$, $\mu$, $\epsilon$, and $\eta$ when no confusion can arise. In what follows, the Koszul sign conventions will be used. A morphism $\rho : A_* \to A_{*-1}$ is called *derivation* if it is compatible with the algebra structures on $A$. The degree of an element $a \in A$ is denoted by $|a|$. Let us recall that if $B$ is also a DGA-algebra, then $A \otimes B$ is canonically endowed an algebra structure by means of the morphism $\mu_{A \otimes B} = (\mu_A \otimes \mu_B)(1_A \otimes T \otimes 1_B)$, where $T(b \otimes a) = (-1)^{|b||a|} a \otimes b$. If the DG-algebra $A$ is connected, that is $A_0 = \Lambda$ and $d_1 : A_1 \to A_0$ is zero, then there is a canonical augmentation $\epsilon_A = 1_\Lambda : A_0 \to \Lambda$.

Let $n$ be a positive integer. The exterior algebra with one generator $u$ in degree $2n - 1$, the polynomial algebra with one generator $v$ in degree $2n$, and the divided power algebra with one "generator" $w$ in degree $2n$ are denoted by $E(u, 2n-1)$, $P(v, 2n)$, and $\Gamma(w, 2n)$, respectively.

We need here the *reduced bar construction* $\bar{B}(A)$ of a DGA-algebra $A$ (see [16]). Recall that it is defined as the connected DGA-coalgebra, $\bar{B}(A) = T^c(S(\bar{A}))$, where $T^c(\ )$ is the tensor coalgebra, $S(\ )$ is the suspension functor, and $\bar{A} = \mathrm{Ker}\ \epsilon_A$ is the augmentation ideal of $A$. The element of $\bar{B}_0(A)$ corresponding to the identity element of $\Lambda$ is denoted by $[\ ]$ and the element $S\bar{a}_1 \otimes \cdots \otimes S\bar{a}_n$ of $\bar{B}(A)$ is denoted by $[a_1|\cdots|a_n]$. The tensor and simplicial degrees of the element $[a_1|\cdots|a_n]$ are $|[a_1|\cdots|a_n]|_t = \sum |a_i|$ and $|[a_1|\cdots|a_n]|_s = n$, respectively; its total degree is the sum of its tensor and simplicial degree. The tensor and simplicial differential are defined by:

$$d_t([a_1|\cdots|a_n]) = -\sum_i (-1)^{e_i - 1}[a_1|\cdots|d_A(a_i)|\cdots|a_n],$$

and

$$d_s([a_1|\cdots|a_n]) = \sum_i (-1)^{e_i}[a_1|\cdots|\mu_A(a_i \otimes a_{i+1})|\cdots|a_n]$$

where

$$e_i = i + |a_1| + \cdots + |a_i|.$$

If the product of $A$ is commutative, a product $*$ (called shuffle product) can be defined on $\bar{B}(A)$. In this way, the reduced bar construction has a commutative Hopf algebra structure.

We will use here the structure of *twisted tensor product*. Let $A$ be a DG-algebra and $C$ a DG-coalgebra. It is well known that $\tau : C_* \to A_{*-1}$ is a *twisting cochain* if and only if $d^\tau = d_A \otimes 1 + 1 \otimes d_C + \tau \cap$ is a differential on $A \times C$ (see [4]), where the morphism $\tau \cap$ is defined by:

$$\tau \cap = (\mu_A \otimes 1)(1 \otimes \tau \otimes 1)(1 \otimes \Delta_C). \tag{1}$$

The DG-module $(A \otimes C, d^\tau)$ is called *the twisted tensor product (or TTP) of A and C along $\tau$*. We will also use the notation of $A \otimes_\tau C$ for such DG-module.

A *relatively free resolution of $\Lambda$ over $A$* is a pair $(X, \epsilon)$ where $X$ is a graded differential $A$-module of the form $X = A \otimes_\Lambda \bar{X}$ with $\bar{X}$ a DG-$\Lambda$-module and $\epsilon: X \to \Lambda$ a morphism of graded differential $A$-modules which is a weak equivalence, thereby, the homology of $X$ is zero except in degree 0 where it is $\Lambda$. We will call the complex $(\bar{X}, d_{\bar{X}})$ the reduced complex, and it is always obtained in the form $(\bar{X}, d_{\bar{X}}) = (\Lambda \otimes_A X, 1_\Lambda \otimes_A d_X)$, by means of the the classical 'neglect' functor on the category of all $A$-modules to the category of all $\Lambda$-modules. It is standard terminology to call the elements of $\bar{X}$ reduced elements. Given a morphism $\psi: X \to Y$ the notation $\psi|_{\bar{X}}(\bar{x})$ means $\psi(1 \otimes \bar{x})$ where 1 is the unit in $A$ and $\bar{x} \in \bar{X}$. We follow these conventions throughout the paper.

A resolution $\epsilon: (X, d) \to \Lambda$ is called *contractile* if there exists a '*contracting homotopy*', i.e., a family of $\Lambda$-module morphisms, $h_{-1}: \Lambda \to X_0$, $h_n: X_n \to X_{n+1}$, such that $1 = d_{n+1}h_n + h_{n-1}d_n$, $\forall n \geq 0$, where $d_0 = \epsilon$ and $h_{-1} = \eta$. Besides, it may always be assumed to hold that $h^2 = 0$ (see [3]).

Throughout this paper, $(X, h, d)$ will denote a contractile relatively free resolution $(X, d)$ with contracting homotopy $h$.

An important example of relatively free and contractile resolution of $\Lambda$ over $A$ is the *bar resolution* $(B(A), s, d)$ (or $B(A)$) [16, 13]. More specifically, $B(A)$ is the *twisted tensor product* $A \otimes_\theta \bar{B}(A)$, where the twisting cochain $\theta$ is given by

$$\theta([a_1| \cdots |a_n]) = \begin{cases} a_1 & n = 1 \\ 0 & \text{otherwise} \end{cases} \tag{2}$$

where the weak equivalence $\epsilon_{B(A)}: B(A) \to \Lambda$ is the canonical augmentation of $B(A)$ (in fact, it is a homotopy equivalence) and the contracting homotopy $s$ is given by

$$s: B(A) \to B(A) \quad \text{where} \quad s(a \otimes [a_1| \cdots |a_n]) = [a|a_1| \cdots |a_n].$$

From now on, we will use $s$ for denoting the above homotopy.

A *contraction* (see [6], [9]) is a data set $c: \{N, M, f, g, \phi\}$ where $f: N \to M$ and $g: M \to N$ are morphisms of DG-modules (called, respectively, *projection* and *inclusion*) and $\phi: N \to N$ is a morphism of graded modules of degree $+1$ (called *homotopy operator*). These data are required to satisfy the rules: **(c1)** $fg = 1_M$, **(c2)** $\phi d_N + d_N\phi + gf = 1_N$ **(c3)** $\phi\phi = 0$, **(c4)** $\phi g = 0$ and **(c5)** $f\phi = 0$. These three last are called *side conditions* [15]. In fact, these may always be assumed to hold, since the homotopy $\phi$ can be altered to satisfy these conditions [7, 14]. We will also denote a contraction $c$ by $(f, g, \phi): N \Rightarrow M$.

For instance, the bar resolution $B(A)$ of a DG-algebra $A$ gives the following contraction:

$$C_{B(A)} : \{B(A), \Lambda, \epsilon_{B(A)}, \eta_{B(A)}, s\} \tag{3}$$

where $\eta_{B(A)} : \Lambda \to B(A)$ is the canonical coaugmentation of $B(A)$.

By the comparison theorem for resolutions [16], given any relatively free resolution $X = (A \otimes \bar{X}, d_X) \xrightarrow{\epsilon} \Lambda$ of $\Lambda$ over $A$, there is an $A$-lineal morphism (a *comparison map*) $g \colon X \to B(A)$ inductively defined:

$$g_0|_{\bar{x}_0} = \eta_{B(A)} \epsilon_X|_{\bar{x}_0}, \qquad g_{n+1}|_{\bar{x}_{n+1}} = s g_n d_{n+1}|_{\bar{x}_{n+1}}. \tag{4}$$

This map is a homotopy equivalence between $B(A)$ and $X$. Moreover, if the resolution $X$ is contractile with contracting homotopy $t$, $(X, t, d)$, then comparison theorem provides an analogous inductive definition for the $A$-lineal morphism $f \colon B(A) \to X$:

$$f_0|_{B(A)_0} = \eta_X \epsilon_{B(A)}|_{B(A)_0}, \qquad f_{n+1}|_{B(A)_{n+1}} = t f_n d_{n+1}|_{B(A)_{n+1}}. \tag{5}$$

Both of the compositions $fg$, $gf$ of theses comparison maps are homotopic to the corresponding identity maps. Inductive formulae for the associated homotopies are also available:

$$\phi \colon B(A)_* \to B(A)_{*+1}$$

defined on reduced elements and then extended $A$-linearly,

$$\phi|_{B(A)} = (-s g f - s \phi d)|_{B(A)}, \tag{6}$$

and

$$\kappa \colon X_* \to X_{*+1}$$

where

$$\kappa = t(1 - gf).$$

Let us observe that in general $\kappa$ is not $A$-lineal. These formulae are crucial for the work in [13, 14]. Let us observe that the morphisms $g$ and $\phi$ satisfy

$$g(\bar{X}) \subseteq \bar{B}(A), \qquad \phi(\bar{B}(A)) \subseteq \bar{B}(A), \tag{7}$$

but $f$ and $\kappa$ do not satisfy the analogous condition.

Generally, $fg$ is different to $1_X$, but sometimes a contraction arises, which we call '*the canonical comparison contraction*'. A necessary condition for guaranteeing that $fg = 1_X$ is given in [11]. We next give a necessary and sufficient condition for this purpose.

**Theorem 1.** *The data set* $\mathcal{C} \colon \{B(A), (X, t, d), f, g, \phi\}$ *is a contraction if and only if* $dt|_{\bar{X}} = 0$.

PROOF. First we assume that $dt|_{\bar{X}} = 0$. Taking into account that $t$ is a contracting homotopy of $X$ to $\Lambda$, it holds that

$$1|_{\bar{X}} = (dt + td)|_{\bar{X}} = td|_{\bar{X}}$$

Now we will show that $fg = 1_X$, the proof is by induction. We have $f_0 g_0 = 1$ by construction and for $n > 0$, on reduced elements,

$$f_n g_n = t f_{n-1} d_n g_n = t f_{n-1} g_{n-1} d_n = t d_n = 1$$

where the first and third equality comes from (7) and induction hypothesis on $t$, respectively. By $A$-linearity the proof is extended to elements of $X$. It is readily checked that the side conditions hold.

Reciprocally, now, let us assume that $fg = 1_X$. Working in a similar way as above, we have, on reduced elements for $n \geq 0$,

$$1 = f_n g_n = t f_{n-1} d_n g_n = t f_{n-1} g_{n-1} d_n = t d_n$$

Hence, $td|_{\bar{X}} = 1|_{\bar{X}}$ and $dt|_{\bar{X}} = 0$.

$\square$

*Remark 1.* For the remainder of this paper, we will assume that $\mathcal{C}$ is a contraction. In this situation, $\mathcal{C}$ is called *the canonical comparison contraction* for $(X, t, d)$.

A resolution $X$ *splits off of the bar construction* (see [14]) if there is a contraction (called comparison contraction) from $B(A)$ to $X$. Note that this contraction can be different from the canonical one.

With this definition at hand, we can state the following proposition:

**Proposition 1.** *Let $(X, t, d)$ be a contractile relatively free resolution. If $dt|_{\bar{X}} = 0$, then $X$ splits off of the bar construction.*

In the sequel proposition we analyze the contracting homotopy $t$ of $(X, t, d)$.

**Proposition 2.** *Let $(X, t, d)$ be a contractile relatively free resolution which splits off of the bar construction by means of the canonical comparison contraction. Then $t = fsg$.*

PROOF. First, due to the fact that $s \colon B(A) \to \bar{B}(A)$, we can use the inductive definition of $f$ in this composition $fs$, thus

$$fsg = (tfd)sg$$

since $sd + ds = 1$,

$$tf(ds)g = tf(1 - sd)g = tfg - tfsdg = tfg.$$

The last identity results from the fact that $tfs = t(tfd)s = 0$ (because $t^2 = 0$). Finally,

$$tfg = t$$

since $fg = 1_X$.

$\square$

*Remark 2.* If we consider $\epsilon_Y \colon (Y, t_Y, d) \to \Lambda$ any contractile relatively free resolution of $\Lambda$ over $A$, instead of $B(A)$, the comparison theorem for resolutions provides similar formulae (comparison maps):

$$f \colon Y \to X, \quad g \colon X \to Y, \quad \phi \colon Y_* \to Y_{*+1}, \quad \kappa \colon X_* \to X_{*+1}$$

where under the hypothesis that $t_Y(Y) \subset \bar{Y}$, it is possible to get the result analogous to Theorem 1.

# 3    Differential Structures in the Comparison of Resolutions

Up to now little has been said about the nature of the differential $d_X$ in $X$ for an arbitrary contractile relatively free resolution with contracting homotopy $t$. Here we show that $d_X$ can be rewritten in terms of an $A_\infty$-twisting cochain $\tau : \bar{X} \to A$ and the $A_\infty$-coalgebra structure of $\bar{X}$ transferred from the coalgebra structure of $\bar{B}(A)$. To this end, we prove two previous results (Theorems 4 and 5) which claim that working with resolutions (*à la Cartan*) is equivalent to work with reduced complexes (*à la Eilenberg–Mac Lane*) from a homogical point of view. We describe a method for passing from one way to the other.

Now, we recall the concept of a perturbation datum. Let $N$ be a graded module and let $f : N \to N$ be a morphism of graded modules. The morphism $f$ is *pointwise nilpotent* if for all $x \in N$ ($x \neq 0$), a positive integer $n$ exists (in general, the number $n$ depends on the element $x$) such that $f^n(x) = 0$. A *perturbation of a DG-module $N$* is a morphism of graded modules $\delta : N \to N$ of degree $-1$, such that $(d_N + \delta)^2 = 0$ and $\delta_1 = 0$. A *perturbation datum of the contraction* $c : \{N, M, f, g, \phi\}$ is a perturbation $\delta$ of the DGA-module $N$ verifying that the composition $\phi\delta$ is pointwise nilpotent.

We now introduce the main tool in Homological Perturbation Theory: the Basic Perturbation Lemma ([4, 7, 8, 3, 19]).

**Theorem 2.** *(BPL)*
*Let $c : \{N, M, f, g, \phi\}$ be a contraction and $\delta : N \to N$ be a perturbation datum of $c$. Then, a new contraction*

$$c_\delta : \{(N, d_N + \delta), (M, d_M + d_\delta), \; f_\delta, \; g_\delta, \; \phi_\delta\}$$

*is defined by the formulas:* $d_\delta = f\delta\Sigma_c^\delta g;\; f_\delta = f(1 - \delta\Sigma_c^\delta\phi);\; g_\delta = \Sigma_c^\delta g;\; \phi_\delta = \Sigma_c^\delta\phi;$
*where*

$$\Sigma_c^\delta = \sum_{i \geq 0} (-1)^i (\phi\delta)^i = 1 - \phi\delta + \phi\delta\phi\delta - \cdots + (-1)^i(\phi\delta)^i + \cdots.$$

Let us note that $\Sigma_c^\delta(x)$ is a finite sum for each $x \in N$ because of the pointwise nilpotency of the composition $\phi\delta$. Moreover, it is obvious that the morphism $d_\delta$ is a perturbation of the DG-module $(M, d_M)$.

The transference of the algebra structure up to homology equivalence has been considered in [8, 9, 19]. Next, we review several notions.

**Definition 1.** *[19]* Let $A$ and $A'$ be two DG-algebras and $c : \{A, A', f, g, \phi\}$ be a contraction. The projection $f$ is *a quasi algebra projection* if the following conditions hold:

$$f\mu_A(\phi \otimes \phi) = 0, \qquad f\mu_A(\phi \otimes g) = 0, \qquad f\mu_A(g \otimes \phi) = 0.$$

The homotopy operator $\phi$ is *a quasi algebra homotopy* if the following conditions hold:

$$\phi\mu_A(\phi \otimes \phi) = 0, \qquad \phi\mu_A(\phi \otimes g) = 0, \qquad \phi\mu_A(g \otimes \phi) = 0.$$

**Definition 2.** *[8]* Let $A$ and $A'$ be two DG-algebras and $c : \{A, A', f, g, \phi\}$ be a contraction. The homotopy operator $\phi$ is said to be *an algebra homotopy* if

$$\phi\mu_A = \mu_A(1_A \otimes \phi + \phi \otimes gf).$$

**Definition 3.** *[19]* Let $A$ and $A'$ be two DGA-algebras and $r : \{A, A', f, g, \phi\}$ a contraction. We say that $r$ is

- *a semi-full algebra contraction* if $f$ is a quasi algebra projection, $g$ is a morphism of DGA-algebras and $\phi$ is a quasi algebra homotopy.
- *an almost-full algebra contraction* if $f$ and $g$ are morphisms of DGA-algebras and $\phi$ is a quasi algebra homotopy.
- *a full algebra contraction* if $f$ and $g$ are morphisms of DGA-algebras and $\phi$ is an algebra homotopy.

Obviously, full and almost-full algebra contractions are, in particular, semi-full algebra contractions. It is not difficult to prove that both sets of semi-full and almost-full algebra contractions are closed by composition and tensor product of contractions.

If $A$ is a commutative DGA-algebra, the contraction (3) is an example of an almost-full algebra contraction.

**Definition 4.** *[7]* Let $A$ and $A'$ be two DG-algebras and $c : \{A, A', f, g, \phi\}$ a contraction. An *algebra perturbation datum* $\delta$ *of* $c$ is a perturbation datum of this contraction which is also a derivation.

The following result tells us that the set of semi-full algebra contractions is closed by homological perturbation. This theorem is used in the proof of some theorems of this paper.

**Theorem 3 (SF-APL).** *([19])*
*Taking as data a semi-full algebra contraction $r$ and an algebra perturbation datum $\delta$ of $r$, the perturbed contraction $r_\delta$ is an algebra contraction of the same type, where the product on $A'_\delta$ is the original product $\mu_{A'}$.*

### 3.1  From Resolutions to Reduced Complexes

Throughout this subsection, $A$ will denote a connected DGA-algebra.

The goal of this subsection is to establish a contraction from $\bar{B}(A)$ to $\bar{X}$ ('reduced complexes') by means of the canonical comparison contraction between the contractile relatively free resolutions $B(A)$ and $(X, t, d)$:

$$\mathcal{C} : \{B(A), (X, t, d), f, g, \phi\}.$$

To this end, we will apply the classical 'neglect' functor on the category of all $A$-modules to the category of all $\Lambda$-modules, $\Lambda \otimes_A -$ and $1 \otimes_A -$, on the complexes and morphisms involved in the above contraction, respectively.

The following properties will play an important role in what follows.

1. $X \cong (\mathrm{Ker}\, \epsilon_A \otimes \bar{X}) \oplus (\Lambda \otimes \bar{X})$.

2. $\Lambda \otimes_A X \cong \Lambda \otimes \bar{X}$ and $1 \otimes_A d_X = (\epsilon_A \otimes 1_{\bar{X}})\, d_X$.

3. $1 \otimes_A g = g|_{\bar{X}}$, $\quad 1 \otimes_A \phi = \phi|_{\bar{B}(A)}$.

4. $1 \otimes_A f = (\epsilon_A \otimes 1_{\bar{X}})\, f$.                                    (8)

5. $f(\bar{b}) = 0 \Rightarrow (1 \otimes_A f)(\bar{b}) = 0$.

6. $d_X(\bar{x}) = 0 \Rightarrow (1 \otimes_A d_X)(\bar{x}) = 0$.

Properties 1, 2, and 4 are deduced from the meaning of tensoring by $A$. Since $g$ and $\phi$ are $A$-lineal and satisfy (7), the third is followed. Properties 5 and 6 are consequences of 1 and 2.

By property 3 we have that $1 \otimes_A g$ and $1 \otimes_A \phi$ are DGA-module morphisms. In spite of the fact that $f$ does not satisfy (7), we will prove that $1 \otimes_A f$ is a morphism of DGA-modules as well.

Firstly, note that

$$(1 \otimes_A f)(1 \otimes_A d) = (\epsilon_A \otimes 1)\, f\, (\epsilon_A \otimes 1)\, d = (\epsilon_A \otimes 1)\, f\, d = (\epsilon_A \otimes 1)\, d\, f,$$

here we have used that $f$ is $A$-lineal and a DGA-module morphism as well.
On the other hand,

$$(1 \otimes_A d)(1 \otimes_A f) = (\epsilon_A \otimes 1)\, d\, (\epsilon_A \otimes 1)\, f,$$

Now, taking into account that the differential $d$ of an $A$-module $X$ satisfies

$$d(a \otimes \bar{x}) = d_A(a) \otimes \bar{x} + (-1)^{|a|} a \otimes d(\bar{x})$$

it is clear that if $a \otimes \bar{x} \in \mathrm{Ker} A \otimes \bar{X}$ then $d(a \otimes \bar{x}) \in \mathrm{Ker} A \otimes \bar{X}$. Thereby,

$$(\epsilon_A \otimes 1)\, d\, f = (\epsilon_A \otimes 1)\, d\, (\epsilon_A \otimes 1)\, f.$$

The properties required for this data set

$$\{\bar{B}(A), \bar{X}, 1 \otimes_A f, 1 \otimes_A g, 1 \otimes_A \phi\}$$

in order to be a contraction are inherited from $\mathcal{C}$ in a straightforward manner. Therefore, we can state:

**Theorem 4.** *The data set $1 \otimes_A \mathcal{C}: \{\bar{B}(A), \bar{X}, 1 \otimes_A f, 1 \otimes_A g, 1 \otimes_A \phi\}$ is a contraction.*

## 3.2    From Reduced Complexes to Resolutions

Taking as input a contraction from the reduced bar construction of a connected DGA-algebra $A$, $\bar{B}(A)$, to a free DGA-module $\bar{X}$, we describe [2] a method for obtaining resolutions which split off of the bar construction. This process plays an important role in the main result of this section.

**Proposition 3.** *[2] Let $A$ be a connected DGA-algebra. Given a contraction $c$ from $(\bar{B}(A), d_{\bar{B}(A)})$ to a DGA-module $(\bar{X}, \bar{d})$, in which the homotopy operator increases the simplicial degree by one, there is a comparison contraction from the bar resolution $B(A)$ to the resolution $X$, where the underlying module in $X$ is just the $A \otimes \bar{X}$ and the differential structure is done via perturbation of $c$.*

**Theorem 5.** *Assuming that $(X, t, d)$ is a contractile relatively free resolution which splits off of the bar construction of a connected DGA-algebra $A$, under the canonical comparison contraction, then $\theta \cap$ is a perturbation datum for*

$$C \colon \{A \otimes (\varLambda \otimes_A B(A)),\ A \otimes (\varLambda \otimes_A X),\ 1 \otimes (1 \otimes_A f),\ 1 \otimes (1 \otimes_A g),\ 1 \otimes (1 \otimes_A \phi)\}$$

*and the perturbed contraction $C_{\theta \cap}$ coincides with the canonical comparison contraction.*

PROOF. Theorem 4 of subsection 3.1 states that from the canonical comparison contraction $\mathcal{C} \colon \{B(A),\ X,\ f,\ g,\ \phi\}$ it is possible to establish a contraction between the reduced complexes

$$\mathcal{C}^{\otimes A} \colon \{\varLambda \otimes_A B(A),\ \varLambda \otimes_A X,\ 1 \otimes_A f,\ 1 \otimes_A g,\ 1 \otimes_A \phi\}.$$

Proposition 3 states that if $A$ is connected then $\theta \cap$ is a perturbation datum of the contraction

$$1 \otimes \mathcal{C}^{\otimes A} \colon \{A \otimes (\varLambda \otimes_A B(A)),\ A \otimes (\varLambda \otimes_A X),\ 1 \otimes (1 \otimes_A f),\ 1 \otimes (1 \otimes_A g),\ 1 \otimes (1 \otimes_A \phi)\}$$

Now, we prove that the perturbed contraction $(1 \otimes \mathcal{C}^{\otimes A})_{\theta \cap}$ coincides with $\mathcal{C}$. To this end, it suffices to show that the formulae

$$
\begin{aligned}
f_{\theta \cap} &= 1 \otimes \bar{f} - (1 \otimes \bar{f})\theta \cap (1 \otimes \bar{\phi}) + (1 \otimes \bar{f})\theta \cap (1 \otimes \bar{\phi})\theta \cap (1 \otimes \bar{\phi}) - \cdots \\
g_{\theta \cap} &= 1 \otimes \bar{g} - (1 \otimes \bar{\phi})\theta \cap (1 \otimes \bar{g}) + (1 \otimes \bar{\phi})\theta \cap (1 \otimes \bar{\phi})\theta \cap (1 \otimes \bar{g}) - \cdots \\
\phi_{\theta \cap} &= 1 \otimes \bar{\phi} - (1 \otimes \bar{\phi})\theta \cap (1 \otimes \bar{\phi}) + (1 \otimes \bar{\phi})\theta \cap (1 \otimes \bar{\phi})\theta \cap (1 \otimes \bar{\phi}) - \cdots \\
d_{\theta \cap} &= \quad D \quad + (1 \otimes \bar{f})\theta \cap (1 \otimes \bar{g}) - (1 \otimes \bar{f})\theta \cap (1 \otimes \bar{\phi})\theta \cap (1 \otimes \bar{g}) + \cdots
\end{aligned}
\tag{9}
$$

coming from the BPL are the morphisms integrating $\mathcal{C}$, where $D$ denotes the usual differential over $A \otimes (\varLambda \otimes_A X)$, and $\bar{h}$ denotes $1 \otimes_A h$.

Let us recall that $1 \otimes_A g = g|_{\bar{X}}$, $1 \otimes_A \phi = \phi|_{\bar{B}(A)}$. Furthermore,

$$
(\theta \cap) s\, (a \otimes [a_1| \dots |a_n]) = 
\begin{cases}
a \otimes [a_1| \dots |a_n] & \text{if } a \otimes [a_1| \dots |a_n] \in \operatorname{Ker}\epsilon_A \otimes \bar{B}(A) \\
\\
0 & \text{elsewhere}
\end{cases}
$$

Taking into account the above identities and the inductive definitions of $f$, $g$, and $\phi$ it follows that

$$(\theta\cap)(1\otimes\bar\phi) = -(\theta\cap)(1\otimes s\phi d+1\otimes sgf)|_B = -(\mu_A\otimes1)|_{A\otimes Ker\,\epsilon_A\otimes\bar B}(1\otimes(\phi d+gf)|_B)$$

and,

$$(\theta\cap)(1\otimes\bar g) = (\theta\cap)(1\otimes sgd|_B) = (\mu_A\otimes1)|_{A\otimes Ker\,\epsilon_A\otimes\bar B}(1\otimes gd|_B).$$

In view of the previous identities and using that $\phi\phi = 0$, $\phi g = 0$ and (7), we have that

$$(1\otimes\bar\phi)\theta\cap(1\otimes\bar\phi) = -(1\otimes\bar\phi)(\mu_A\otimes1)|_{A\otimes Ker\,\epsilon_A\otimes\bar B}(1\otimes(\phi d+gf)|_B)$$

$$= -(\mu_A\otimes1)|_{A\otimes Ker\,\epsilon_A\otimes\bar B}(1\otimes\phi(\phi d+gf)|_B) = 0.$$

Furthermore,

$$(1\otimes\bar\phi)\theta\cap(1\otimes\bar g) = (1\otimes\bar\phi)(\mu_A\otimes1)|_{A\otimes Ker\,\epsilon_A\otimes\bar B}(1\otimes gd|_B)$$

$$= (\mu_A\otimes1)|_{A\otimes Ker\,\epsilon_A\otimes\bar B}(1\otimes\phi gd|_B) = 0.$$

Thus, the formulae (9) may now be rewritten as

$$f_{\theta\cap} = 1\otimes\bar f - (1\otimes\bar f)\theta\cap(1\otimes\bar\phi), \qquad g_{\theta\cap} = 1\otimes\bar g,$$

$$\phi_{\theta\cap} = 1\otimes\bar\phi, \qquad d_{\theta\cap} = D + (1\otimes\bar f)\theta\cap(1\otimes\bar g).$$

Obviously, $g_{\theta\cap}$ and $\phi_{\theta\cap}$ are the $A$-lineal extensions of $\bar g$ and $\bar\phi$, hence $g_{\theta\cap} = g$ and $\phi_{\theta\cap} = \phi$. Working out the second summand of $f_{\theta\cap}$, we have that

$$-(1\otimes\bar f)\theta\cap(1\otimes\bar\phi) = (\mu_A\otimes\bar f)|_{A\otimes Ker\,\epsilon_A\otimes\bar B}(1\otimes(\phi d+gf)|_B)$$

$$= (\mu_A\otimes\bar f)|_{A\otimes Ker\,\epsilon_A\otimes\bar B}(1\otimes\phi d|_B) + (\mu_A\otimes\bar f)|_{A\otimes Ker\,\epsilon_A\otimes\bar B}(1\otimes gf|_B).$$

By property 5 of (8) the first term is zero, since $f\phi = 0$. The second one, acting over an element $\eta_A(1)\otimes[a_1|\ldots|a_n]$ coincides with the summands of $f([a_1|\ldots|a_n])$ which becomes zero when the functor $1\otimes_A -$ is applied over it, since $fgf = f$. Hence, $f_{\theta\cap} = f$.

In a similar way, it is proved that $d_{\theta\cap} = d$.

$\square$

## 3.3    $A_\infty$-Structures and HPT

The notion of an $A_\infty$-(co)algebra was introduced by J. Stasheff [20], which is "roughly speaking" a differential graded (co)algebra in which the (co)associative law holds up to homotopy. Here we deal with the category of $A_\infty$-coalgebras. Given a DG-module $(M,\Delta_1)$ and a sequence of maps $\{\Delta_i \in \mathrm{Hom}^{i-2}(M,M^{\otimes i})\}_{i\geq1}$, $(M,\Delta_i)_{n\geq1}$ is called an $A_\infty$-coalgebra if the relation

$$\sum_{n=1}^{i}\sum_{k=0}^{i-n}(-1)^{n+k+nk}(1^{i-n-k}\otimes\Delta_n\otimes1^k)\Delta_{i-n+1} = 0$$

holds for each $i\geq1$.

The problem of transferring (co)algebra structures up to contraction has been largely considered in the literature. Here we will need the following results:

**Lemma 1.** *[8]     Assuming that $C$ is a coalgebra, $M$ a DGA-module and $r: \{C, M, f, g, \phi\}$ a contraction, and using the tensor trick (see [8, 9, 10]), then $M$ becomes an $A_\infty$-coalgebra.*

Moreover, the maps integrating the $A_\infty$-coalgebra $(M, \Delta_i)_{i \geq 1}$ are shown in [1] to be explicitly

$$\Delta_i = (-1)^{\frac{(i-1)(i-2)}{2}} f^{\otimes i} \circ$$

$$\circ \left[ \sum_{k_2=1}^{2} \sum_{k_3=1}^{k_2+1} \cdots \sum_{k_{i-1}=1}^{k_{i-2}+1} \prod_{j=2}^{i-1} (-1)^{k_j} (1^{\otimes k_j - 1} \otimes \Delta_c \phi \otimes 1^{\otimes j - k_j}) \right] \Delta_c g, \qquad (10)$$

where $\displaystyle\prod_{j=2}^{i-1} h_j$ denotes the composition $h_{i-1} \circ \cdots \circ h_2$.

Given a DGA-algebra $A$ and an $A_\infty$-coalgebra $C$, an $A_\infty$-*twisting cochain* (or $A_\infty$-TTP) $\tau: C \to A$ is a DG-module morphism of degree -1, such that satisfies the following identity

$$d\tau + \sum_{i=1}^{\infty} \mu^{(i)} \tau^{\otimes i} \Delta_i = 0,$$

where $\mu^{(1)} = 1$, $\mu^{(2)} = \mu$, and in general $\mu^{(k)} = \mu(1 \otimes \mu^{(k-1)})$. Analogously to twisting cochain, $\tau: C \to A$ is an $A_\infty$-twisting cochain [18] if and only if $d_\tau = d \otimes 1 + \sum_{i=1}^{\infty} (\mu^{(i)} \otimes 1)(1 \otimes \tau^{\otimes i-1} \otimes 1)(1 \otimes \Delta_i)$ is a differential on $A \otimes C$, which, together with this differential, is denoted by $A \otimes_\tau C$ and is referred to as the $A_\infty$-*twisted tensor product* (along $\tau$).

**Theorem 6.** *[1] Let $t : C \to A$ be a twisting cochain and $c(f, g, \phi) : C \Rightarrow C'$ be a contraction such that $c$ induces on $C'$ an $A_\infty$-coalgebra structure (see Lemma 1). Additionally, assume that $t\phi = 0$ and $(1 \otimes \phi)t\cap$ is pointwise nilpotent. There is a contraction*

$$A \otimes_t C \Rightarrow A \otimes_{\bar{t}} C',$$

*where $\bar{t} = tg$ is an $A_\infty$-twisting cochain and $A \otimes_{\bar{t}} C'$ is an $A_\infty$-twisted tensor product.*

*Remark 3.* The hypotheses of Theorem 6 are satisfied when $C$ is a simply connected DGA-coalgebra. For instance, $\bar{B}(A)$ is a simply connected DGA-coalgebra when $A$ is a connected DGA-algebra.

The following result is a straightforward consequence of Theorem 6, and it is a main result of this paper.

**Algorithm 1. Computing the $A_\infty$-twisting tensor product structure**

INPUT: A contractile relatively free resolution $(X, t, d)$ of $\Lambda$
over a connected DG-algebra $A$ where $d \circ t|_{\bar{x}} = 0$.

Step 1. Form the canonical comparison contraction $(f, g, \phi): B(A) \Rightarrow X$
using the formulas (5), (4) and (6).
Step 2. Form the contraction $(1 \otimes_A f, 1 \otimes_A g, 1 \otimes_A \phi): \bar{B}(A) \Rightarrow \bar{X}$ as in
Theorem 4 .

OUTPUT: The map $\tau = \theta g|_{\bar{x}}$ which is an $A_\infty$-twisting cochain and the
maps $\Delta_i$ (given by formula (10) using the morphisms of the
contraction of Step 2) integrating the $A_\infty$-coalgebra
structure $(\bar{X}, \Delta_i)_{i \geq 1}$.

CORRECTNESS: Let us emphasize that $1 \otimes_A g = g|_{\bar{x}}$ and $1 \otimes_A \phi = \phi|_{\bar{B}(A)}$ since $g$
and $\phi$ are $A$-lineal and $g(\bar{X}) \subseteq \bar{B}(A)$, and $\phi(\bar{B}(A)) \subseteq \bar{B}(A)$. An explicit formula
for $\psi$ is given in [13, 14] which increases the simplicial degree in $\bar{B}(A)$ by one
and $\phi_0 = 0$. Therefore, $\theta \phi|_{\bar{B}(A)} = 0$ since $\theta: \bar{B}(A) \to A$ is the universal twisting
cochain. Now, applying Theorem 6, we have the following $A_\infty$-twisting cochain

$$\tau = \theta g|_{\bar{x}}: \bar{X} \to A$$

The second step is to construct the tensor product contraction

$$A \otimes \bar{B}(A) \Rightarrow A \otimes \bar{X}$$

and to use the Basic Perturbation Lemma with $\theta \cap$ as the perturbation datum
(see Theorem 5). Then, it is straightforward to check that $(1 \otimes \phi|_{\bar{B}(A)}) \theta \cap$ is
pointwise nilpotent. So we obtain the new contraction,

$$A \otimes_\theta \bar{B}(A) \Rightarrow (A \otimes \bar{X}, d_{\theta \cap})$$

Now, using Theorem 6, we have that

$$(A \otimes \bar{X}, d_{\theta \cap}) = A \otimes_\tau \bar{X}$$

where $A \otimes_\tau \bar{X}$ is an $A_\infty$-twisted tensor product.

In the proof of the last identity, we use the special properties of the morphisms
which take part in the canonical comparison contraction.

$\square$

## 4    Algebra Structures in the Comparison of Resolutions

If $A$ is a commutative DGA-algebra, it is well known that it is possible to define
a commutative product $*$ on $\bar{B}(A)$ called shuffle product. Furthermore, $B(A)$
has canonically associated a commutative algebra structure by means of the
morphism $\mu_{B(A)} = (\mu_A \otimes *)(1_A \otimes T \otimes 1_{\bar{B}(A)})$, where $T(\bar{b} \otimes a) = (-1)^{|\bar{b}| |a|} a \otimes \bar{b}$.

Throughout this section, we assume that $A$ is a commutative DGA-algebra, $(X, t, d)$ is a contractile relatively free resolution with $t$ as contracting homotopy which is a quasi algebra homotopy, and there exists $\mathcal{C} \colon \{B(A), X, f, g, \phi\}$ the canonical comparison contraction.

Before giving the main result of this section we need some preliminary results which are easy to prove:

**Lemma 2.** *[19] Let $M$ be a DGA-module and $c \colon \{A, M, f', g', \phi'\}$ be a contraction. If $\phi'\mu_A(g' \otimes g') = 0$, then the morphism $\mu_M = f'\mu_A(g' \otimes g')$ defines a commutative product on $M$. Furthermore, $g'$ is a DGA-algebra morphism with regard to the products $\mu_A$ and $\mu_M$.*

**Lemma 3.** *Let $A'$ be a DG-algebra, and $c \colon \{A, A', f', g', \phi'\}$ be a contraction of DG-modules. Then,*

$$\phi'\mu_A - \mu_A\phi'^{[\otimes 2]} = \phi'\mu_A\phi'^{[\otimes 2]}d^{[2]} - d\phi'\mu_A\phi'^{[\otimes 2]} - g'f'\mu_A\phi'^{[\otimes 2]} \qquad (11)$$

*where $\phi'^{[\otimes 2]}$ and $d^{[2]}$ denote, respectively, $1_A \otimes \phi + \phi \otimes gf$ and $d_A \otimes 1_A + 1_A \otimes d_A$. Assuming that $f'$ is a quasi algebra projection and $\phi'$ is a quasi algebra homotopy,*

$$\mu_A(\phi'(a) \otimes \phi'(b)) = (-1)^{|a|+1}\phi'(\mu_A(\phi'(a) \otimes b)) + \phi'(\mu_A(a \otimes \phi'(b))) \qquad (12)$$

*where $a, b \in A$.*

**Lemma 4.** *The identity $1 = sd\phi d + sdgf$ holds on reduced elements with degree greater than zero.*

**Lemma 5.** *If $\phi_n\mu_{B(A)}(g \otimes g)_n(\bar{x}_1 \otimes \bar{x}_2) = 0$, for any a reduced element $\bar{x}_1 \otimes \bar{x}_2$ of degree $n$ then for any element $x_1 \otimes x_2$ of degree $n$ of $X \otimes X$*

$$\phi_n\mu_{B(A)}(g \otimes g)_n)(x_1 \otimes x_2) = 0.$$

Now, we state the main result of this section:

**Theorem 7.** *The $A$-module $(X, d)$ equipped with the morphism $\mu_X = f * (g \otimes g)$ becomes a commutative DGA-algebra. Furthermore, $\mathcal{C}$ is an almost-full algebra contraction with regard to the products $\mu_{B(A)}$ and $\mu_X$.*

PROOF. The proof will be divided into two parts.

Firstly, we show that $\mu_X$ defines a commutative product on $X$. To this end, we apply Lemma 2, in order to prove by induction that $\phi\mu_{B(A)}(g \otimes g) = 0$ on reduced elements. Then this relation is extended to $X \otimes X$ by Lemma 5. We have $\phi_0 = 0$ by construction and for $n > 0$, on reduced elements,

$$\phi_n\mu_{B(A)}(g \otimes g)_n = -s\phi_{n-1}d_n\mu_{B(A)}(g \otimes g)_n - sg_nf_n\mu_{B(A)}(g \otimes g)_n.$$

This equality comes from the fact that $\mu_{B(A)}|_{\bar{B}(A) \otimes \bar{B}(A)} \subset \bar{B}(A)$ and the property (7) of $g$. By induction hypothesis

$$s\phi_{n-1}d_n\mu_{B(A)}(g \otimes g)_n = s\phi_{n-1}\mu_{B(A)}(g \otimes g)_{n-1}d_n^{[2]} = 0.$$

Now from Lemma 4

$$sg_n f_n \mu_{B(A)} (g \otimes g)_n = sg_n f_n (sd_n \phi_{n-1} d_n + sd_n g_n f_n) \mu_{B(A)} (g \otimes g)_n$$

$$= sg_n f_n (sd_n \phi_{n-1} + sg_{n-1} f_{n-1}) \mu_{B(A)} (g \otimes g)_{n-1} d_n^{[2]}$$

$$= sg_n f_n sd_n \phi_{n-1} \mu_{B(A)} (g \otimes g)_{n-1} d_n^{[2]} + sg_n f_n sg_{n-1} f_{n-1} \mu_{B(A)} (g \otimes g)_{n-1} d_n^{[2]}.$$

By induction and Lemma 5 the first summand is null. Let us notice that in the second summand $sg_{n-1} f_{n-1} \mu_{B(A)} (g \otimes g)_{n-1}$ can be applied on non-reduced elements, so in order to prove that it is null we need a more sophisticated argument.

We show by induction that if $sgf\mu_{B(A)} (g \otimes g)$ is null on any element of degree less than $n$, then it is possible to extend this property to any element in $X \otimes X$ of degree $n$. The first case of the induction is trivial (for $n = 0$). We take a generic element $(a \otimes \bar{x}) \otimes (a' \otimes \bar{x}')$ in degree $n$. Thus,

$$sg_n f_n \mu_{B(A)} (g(a \otimes \bar{x}) \otimes g(a' \otimes \bar{x}'))_n = (-1)^{|a'||\bar{x}|} s(aa' gf \mu_{B(A)} (q(\bar{x}) \otimes g(\bar{r}'))).$$

Now, using again Lemma 4,

$$s(aa' gf \mu_{B(A)} (g(\bar{x}) \otimes g(\bar{x}'))) = s(aa' gf (sd\phi d + sdgf) \mu_{B(A)} (g(\bar{x}) \otimes g((\bar{x})))),$$

let us observe that the summands, $(sd\phi d + sdgf) \mu_{B(A)} (g(\bar{x}) \otimes g(\bar{x}'))$, are zero by induction hypothesis. This fact completes the proof of the first step.

Secondly, we are proving that $\mathcal{C}$ is an almost-full algebra contraction (i.e., $f$ and $g$ are DGA-algebra morphisms and $\phi$ is a quasi algebra morphism with regard to the products $\mu_{B(A)}$ and $\mu_X$). Lemma 2 guarantees that $g$ is a DGA-algebra morphism. In order to prove that $f$ is a DGA-algebra morphism we need to see that $\mu_X (f \otimes f) = f\mu_{B(A)}$. The proof is by induction on reduced elements and then extended $A$-linearly in each degree. Obviously, we have $\mu_X (f_0 \otimes f_0) = f_0 \mu_{B(A)}$ and for $n > 0$,

$$\mu_X (f_{n-i}(\bar{b}) \otimes f_i(\bar{b}')) = \mu_X (t f_{n-i-1} d_{n-i}(\bar{b}) \otimes t f_{i-1} d_i(\bar{b}'))$$

$$= (-1)^{|\bar{b}|} t \mu_X (t f_{n-i-1} d_{n-i}(\bar{b}) \otimes f_{i-1} d_i(\bar{b}')) + t \mu_X (f_{n-i-1} d_{n-i}(\bar{b}) \otimes t f_{i-1} d_i(\bar{b}'))$$

$$= (-1)^{|\bar{b}|} t \mu_X (f_{n-i}(\bar{b}) \otimes f_{i-1} d_i(\bar{b}')) + t \mu_X (f_{n-i-1} d_{n-i}(\bar{b}) \otimes f_i(\bar{b}')).$$

In the second identity above we have taken into account that $t$ is a quasi algebra homotopy with respect to the product $\mu_X$ and we have applied (12).

On the other hand,

$$f_n \mu_{B(A)} (\bar{b} \otimes \bar{b}') = t f_{n-1} d_n \mu_{B(A)} (\bar{b} \otimes \bar{b}')$$

$$= t f_{n-1} \mu_{B(A)} (d_{n-i}(\bar{b}) \otimes \bar{b}') + (-1)^{|\bar{b}|} t f_{n-1} \mu_{B(A)} (\bar{b} \otimes d_i(\bar{b}'))$$

$$= t \mu_X (f_{n-i-1} d_{n-i}(\bar{b}) \otimes f_i(\bar{b}')) + (-1)^{|\bar{b}|} t \mu_X (f_{n-i}(\bar{b}) \otimes f_{i-1} d_i(\bar{b}')),$$

the last identity is obtained from induction hypothesis. So we have actually proved that $f$ is a DGA-algebra morphism.

Finally, we will prove that $\phi$ is a quasi algebra homotopy, i.e., the conditions

$$\phi\mu_{B(A)}(\phi \otimes g) = 0, \quad \phi\mu_{B(A)}(g \otimes \phi) = 0, \quad \phi\mu_{B(A)}(\phi \otimes \phi) = 0 \quad \text{hold.}$$

The proof is by induction. We have $\phi_0 = 0$ by construction, so the above three identities hold. For $n > 0$, on reduced elements, the proof consists in replacing $\phi\mu_{B(A)}$ by

$$-s\phi d\mu_{B(A)} - sgf\mu_{B(A)} = s\phi\mu_{B(A)}(d \otimes 1 + 1 \otimes d) - sg\mu_X(f \otimes f),$$

then, the summands of the form $sg\mu_X(f \otimes f)$ are all null, since $f\phi = 0$. Moreover, the summands which contain $dg \otimes \phi$ or $\phi \otimes dg$ are zero, since $d_n g_n = g_{n-1} d_n$. So it is possible to apply induction hypothesis. To sum up, we must only study the summands of the form:

$$s\phi\mu_{B(A)}(d\phi \otimes g), \qquad s\phi\mu_{B(A)}(g \otimes d\phi), \qquad s\phi\mu_{B(A)}(d\phi \otimes \phi), \qquad s\phi\mu_{B(A)}(\phi \otimes d\phi).$$

Replacing $d\phi$ by $1 - gf - \phi d$ is immediate to see that all summands are null. By $A$-linearity the proof is extended to elements of $B(A)$. This completes the proof.                                                                        □

Finally, we provide the following algorithm for computing (algebra) resolutions which split off of the bar construction of a commutative DGA-algebra $\tilde{A}$, taking as input datum a contractile relatively free resolution of $\Lambda$ over a commutative DGA-algebra $A$, where $A$ and $\tilde{A}$ coincide as graded module.

### Algorithm 2. Computing 'algebra' resolutions which split off

INPUT: A contractile relatively free resolution $(X, t, d)$ of $\Lambda$
       over a commutative DG-algebra $A$ where $d \circ t|_{\bar{X}} = 0$ and $t$ is a
       quasi algebra homotopy.
       A commutative DGA-algebra $\tilde{A}$ which has the same underlying
       graded $\Lambda$-module structure than $A$.

Step 1. Form the canonical comparison contraction $(f, g, \phi) : B(A) \Rightarrow X$
        using the formulas (5), (4) and (6).
Step 2. Construct the bar constructions $(B(\tilde{A}), \partial^+)$ and $(B(A), \partial)$.
        Define the morphism $\delta = \partial^+ - \partial$.
Step 3. Perturb the above contraction using $\delta$, (if $\phi\delta$ is
        nilpotent).
OUTPUT: A semi-full algebra contraction $B(\tilde{A}) \Rightarrow \tilde{X}$. Hence, $\tilde{X}$ is
        an algebra resolution of $\Lambda$ over $\tilde{A}$, where $\mu_{\bar{X}} = f * (g \otimes g)$.

We point out that the contraction of Step 1 is almost-full (see Theorem 7). Furthermore, in the case that $\phi\delta$ is pointwise nilpotent, thus $\delta$ is an algebra perturbation datum of the contraction of Step 1. Hence, using Theorem 3, we conclude with the desired result. The main computational advantage of the algebra structure in $\tilde{X}$ is that it is only necessary to compute the perturbed differential on the generators of $\tilde{X}$ as an algebra, in spite of, on the whole set of generators as a module. We will clarify this aspect in the following example.

# 5   An Example

Now, we give an example in order to illustrate the main results of the paper. We work with the resolution $Q_{(p)}(w, 2n) \otimes E(\sigma(w), 2n+1) \otimes \Gamma(\varphi_p(w), 2np+2)$ (see [5]). Making use of the main results of the paper we reach the same results on the $A_\infty$-structure of this DG-module as Proute in [18]. Furthermore, we prove that this complex is a DGA-algebra. Hence, it is an example of a multiplicative $A_\infty$-twisted tensor product. Moreover, this resolution can be "perturbed" into a resolution of $\mathbf{Z}_{(p)}$ over $\Gamma(w, 2n)$. Notice that the way for obtaining the resolution above is different from that given in [2].

Following Cartan's work in [5], we will use in the sequel the suspension '$\sigma$', $p$-transpotence '$\varphi_p$' and $k$-th divided power '$\gamma_k$' for terming the generators of the DGA-algebras.

Let $p$ be a prime number and $I = (w^p)$ be the ideal generated by $w^p$. Then, $Q_{(p)}(w, 2n) = P(w, 2n)/I$ is the truncated polynomial algebra on one generator $w$ of degree $2n$ with zero differential. We consider here the resolution $X = Q_{(p)}(w, 2n) \otimes E(\sigma(w), 2n+1) \otimes \Gamma(\varphi_p(w), 2np+2)$ where the differential is a derivation and is defined by

$$d(\sigma(w)) = w, \quad d(\gamma_i(\varphi_p(w))) = w^{p-1}\,\sigma(w)\,\gamma_{i-1}(\varphi_p(w)).$$

The following degree one morphism $t \colon X \to X$ linear over $\Lambda$ (but not over $Q_{(p)}(w, 2n)$) defined as

$$t(1) = 0, \quad t(w^k\gamma_i(\varphi_p(w))) = w^{k-1}\sigma(w)\gamma_i(\varphi_p(w))$$

and

$$t(w^k\sigma(w)\gamma_i(\varphi_p(w))) = \begin{cases} \gamma_{i+1}(\varphi_p(w)) & k = p-1 \\ 0 & k \neq p-1 \end{cases}$$

is a contracting homotopy for $X$. This explicit formula for $t$ is crucial to many constructions but it is not widely distributed. Moreover, the data set

$$c_X \colon \{X,\, \Lambda,\, \epsilon,\, \eta,\, t\} \tag{13}$$

is a contraction, where $\epsilon_0 = 1_\Lambda$, $\epsilon_n = 0$, $n > 0$ and $\eta(\lambda) = \lambda \otimes 1 \otimes 1$. Now, by comparison theorem for resolutions and using the formulae (5), (4) and (6) for comparison maps, we have the following $Q_{(p)}(w, 2n)$-lineal morphisms defined on the reduced complexes by

$$f[w^{r_1}|w^{t_1}|\ldots|w^{r_m}|w^{t_m}] = \left\{\prod_{k=1}^{n}\delta_{p, r_k+t_k}\right\}\gamma_m(\varphi_p(w)),$$

$$f[w^l|w^{r_1}|w^{t_1}|\ldots|w^{r_m}|w^{t_m}] = w^{l-1}\left\{\prod_{k=1}^{n}\delta_{p, r_k+t_k}\right\}\sigma(w)\gamma_m(\varphi_p(w)),$$

where the symbols $\delta_{i,j}$ are defined by: $\delta_{i,j} = \begin{cases} 1 & i = j, \\ 0 & i \neq j. \end{cases}$

$$g(\sigma(w)) = [w], \quad g(\gamma_i(\varphi_p(w))) = [w^{p-1}|w|\overset{i\text{-times}}{\cdots}|w^{p-1}|w],$$

$$g(\sigma(w)\gamma_i(\varphi_p(w))) = [w|w^{p-1}|\,\overset{i\text{-times}}{\cdots}\,|w|w^{p-1}|w]$$

and the homotopy operator $\phi$ is defined by $-\psi$, where

$$\psi[\,] = 0, \quad \psi[w] = 0,$$

$$\psi[w^x] = [w^{x-1}|w], \quad 1 < x < p,$$

$$\psi[w^x|w^y] = [w^x|w^{y-1}|w],$$

$$\psi[z|w^x|w^y] = [z|w^x|w^{y-1}|w] + \delta_{p,x+y}[\psi(z)|w^{p-1}|w]$$

for $z \in \bar{B}(Q_{(p)}(w, 2n))$.

It is a straightforward computation to verify that $dt|_{\tilde{x}} = 0$. Then by Theorem 1 we can state that the data set

$$\mathcal{C}_{B-X} \colon \{B(Q_{(p)}(w, 2n)),\, X,\, f,\, g,\, \phi\}$$

is a contraction: *the canonical comparison contraction between* $B(Q_{(p)}(w, 2n))$ and $X$. It is immediate to see that $t$ is a quasi algebra homotopy, then by Theorem 7 we can guarantee that $\mathcal{C}_{B-X}$ is an almost-full contraction.

Now, we can apply Algorithm 1 to the resolution $X$, and define the degree minus one morphism $\tau \colon E(\sigma(w), 2n+1) \otimes \Gamma(\varphi_p(w), 2np+2) \to Q_{(p)}(w, 2n)$ by

$$\tau(\sigma(w)) = \theta g(\sigma(w)) = w \quad \text{and} \quad \tau = 0 \quad \text{otherwise}$$

which is a $A_\infty$-twisting cochain.

Working with coefficients in $\mathbf{Z}_p$, Proute determined in [18] the $A_\infty$-coalgebra structure of $E(\sigma(w), 2n+1) \otimes \Gamma(\varphi_p(w), 2np+2)$ given the following formulae:

$$\Delta_h = 0, \quad h \neq 2, p\,,$$

$$\Delta_2(\sigma(w)^j \gamma_i(\varphi_p(w))) = \sum_{k=0}^{j} \sum_{l=0}^{i} \sigma(w)^k \gamma_l(\varphi_p(w)) \otimes \sigma(w)^{j-k} \gamma_{i-l}(\varphi_p(w)),$$

$$\Delta_p(\sigma(w)^j \gamma_i(\varphi_p(w))) = \sum_{l_1 + \cdots + l_p = i-1} \sigma(w)^{j+1} \gamma_{l_1}(\varphi_p(w)) \otimes \cdots \otimes \sigma(w)^{j+1} \gamma_{l_p}(\varphi_p(w)).$$

In [18, pp.148-149] it is proved that

$$d = (\mu \otimes 1)(1 \otimes \tau \otimes 1)(1 \otimes \Delta_2) + (\mu^{(p)} \otimes 1)(1 \otimes \tau^{\otimes i-1} \otimes 1)(1 \otimes \Delta_p)$$

as the output of the Algorithm 1 states.

Summing up, $X = Q_{(p)}(w, 2n) \otimes_\tau E(\sigma(w), 2n+1) \otimes \Gamma(\varphi_p(w), 2np+2)$ is a multiplicative $A_\infty$-twisted tensor product, i.e., $X$ is a DGA-algebra and an $A_\infty$-TTP simultaneously.

For the remainder of this example we have taken $\mathbf{Z}_{(p)}$ ($\mathbf{Z}$ localized at prime $p$) as the ground ring. In the following, we give the outline of a process for constructing a resolution of $\mathbf{Z}_{(p)}$ over $\Gamma(w, 2n)$. It is obtained by perturbing a resolution of $\mathbf{Z}_{(p)}$ over $\otimes_{i \geq 0} Q_{(p)}(w_i, 2np^i)$.

Firstly, we use an isomorphism of DGA-algebras [19, Prop. 5.24] between $\Gamma(w, 2n)$ and $\tilde{\otimes}_{i \geq 0} Q_{(p)}(w_i, 2np^i)$. As $\mathbf{Z}_{(p)}$-module, this last DGA-algebra is equal to the ordinary tensor product $\otimes_{i \geq 0} Q_{(p)}(w_i, 2np^i)$. Its multiplicative law is given by

$$
w_i^k w_j^h = \begin{cases} w_i^k \otimes w_j^h & \text{if } i \neq j, \\ w_i^{k+h} & \text{if } i = j \text{ and } h + k < p, \\ -pw_i^t w_{i+1} & \text{if } i = j \text{ and } h + k = p + t \end{cases}
$$

From now on, we will identify the generators $w_i$ of the truncated algebras with the elements $\gamma_{p^i}(w)$ of $\Gamma(w, 2n)$; in fact, the image by the isomorphism of $w_i$ coincides with $\gamma_{p^i}(w)$ excluding the coefficient.

Secondly, we give an explicit contracting homotopy $t^{\otimes}$ for the resolution

$$
X^{\otimes} = \otimes_{i \geq 0} Q_{(p)}(w_i, 2np^i) \otimes \left( \otimes_{i \geq 0} E(\sigma(w_i), 2np^i + 1) \otimes \Gamma(\varphi_p(w_i), 2np^i + 2) \right).
$$

To this end, we use that the complex above is just the tensor product complex

$$
\otimes_{i \geq 0} \left( Q_{(p)}(w_i, 2np^i) \otimes E(\sigma(w_i), 2np^i + 1) \otimes \Gamma(\varphi_p(w_i), 2np^i + 2) \right)
$$

and the formula for $t^{\otimes}$ is:

$$
t \otimes 1 \otimes 1 \otimes \ldots + \eta\epsilon \otimes t \otimes 1 \otimes 1 \ldots + \eta\epsilon \otimes \eta\epsilon \otimes t \otimes 1 \otimes 1 \otimes \ldots + \cdots,
$$

thus,

$$
t^{\otimes}(1 \otimes \overset{l\text{-times}}{\cdots} \otimes 1 \otimes x \otimes z) = 1 \otimes \overset{l\text{-times}}{\cdots} \otimes 1 \otimes t(x) \otimes z
$$

where $|x| > 0$, $x \in Q_{(p)}(w_l, 2np^l) \otimes E(\sigma(w_l), 2np^l + 1) \otimes \Gamma(\varphi_p(w_l), 2np^l + 2)$ and $z \in \otimes_{i > l} \left( Q_{(p)}(w_i, 2np^i) \otimes E(\sigma(w_i), 2np^i + 1) \otimes \Gamma(\varphi_p(w_i), 2np^i + 2) \right)$.

Since the contraction (13) is an almost-full algebra contraction, and the class of almost-full contraction is closed by tensor product [19], it follows that $t^{\otimes}$ is a quasi algebra homotopy.

Now, by comparison theorem for resolutions and using the formulae (5), (4) and (6) for comparison maps, it is possible to construct three morphisms denoted by $f^{\otimes}$, $g^{\otimes}$, $\phi^{\otimes}$.

Since $t|_X = 0$, we have that $dt^{\otimes}|_{\overline{X^{\otimes}}} = 0$. Then by Theorem 1 we can state that the data set

$$
\{B(\otimes_{i \geq 0} Q_{(p)}(w_i, 2np^i)), X^{\otimes}, f^{\otimes}, g^{\otimes}, \phi^{\otimes}\} \tag{14}
$$

is a contraction. Besides, it is an almost-full algebra contraction thanks to Theorem 7. Hence, in particular, $g^{\otimes}$ is DGA-algebra morphism, then $g^{\otimes}$ is completely determined by $g$, i.e.,

$$
g^{\otimes}(x_0 \otimes \cdots \otimes x_n) = g(x_0) * g(x_1) * \cdots * g(x_n), \forall n \geq 0;
$$

where $x_i \in E(\sigma(w_i), 2np^i + 1) \otimes \Gamma(\varphi_p(w_i), 2np^i + 2)$ and $*$ denotes the well-known shuffle product in the bar construction.

The projection $f^\otimes$ is given by

$$f^\otimes[w_{i_1}^{k_1} \otimes z_1 | w_{i_2}^{k_2} \otimes z_2 | \dots | w_{i_n}^{k_n} \otimes z_n]$$

$$= \begin{cases} w_{i_1}^{k_1-1} z_1 w_{i_2}^{k_2-1} z_2 \cdots w_{i_n}^{k_n-1} z_n \, \sigma(w_{i_1})\sigma(w_{i_2}) \cdots \sigma(w_{i_n}) & i_1 < i_2 < \cdots < i_n, \\[4pt] \left\{ \prod_{j=0}^{m-1} \delta_{p,k_{2j+1}+k_{2j+2}} \right\} z_1 \cdots z_n \, \gamma_m(\varphi_p(w_{i_1})) & i_1 = i_2 = \cdots = i_{n=2m}, \\[4pt] \left\{ \prod_{j=1}^{m} \delta_{p,k_{2j}+k_{2j+1}} \right\} w_{i_1}^{k_1-1} z_1 \cdots z_n \, \sigma(w_{i_1})\gamma_m(\varphi_p(w_{i_1})) & i_1 = i_2 = \cdots = i_{n=2m+1}, \\[4pt] f^\otimes[a_{i_1}|\dots|a_{i_{l_1}}] \otimes \cdots \otimes f^\otimes[a_{i_{l_h}}|\dots|a_{i_n}] & i_1 = \dots = i_{l_1} < \cdots < i_{l_h} = \dots = i_n \\[4pt] 0 & \text{otherwise,} \end{cases}$$

where $z_j \in \otimes_{i>j} Q_{(p)}(w_i, 2np^i)$ and $a_j = w_{i_j}^{k_j} \otimes z_j$. And the homotopy operator $\phi^\otimes$ is defined by $-\psi^\otimes$,

$$\psi^\otimes[\,] = 0,$$

$$\psi^\otimes[w_{i_1}^{k_1} \otimes z_1] = [w_{i_1}^{k_1-1} z_1 | w_{i_1}],$$

$$\psi^\otimes[w_{i_1}^{k_1} z_1 | w_{i_2}^{k_2} z_2] = \begin{cases} [w_{i_1}^{k_1} z_1 | w_{i_2}^{k_2-1} z_2 | w_{i_2}] + [w_{i_1}^{k_1-1} z_1 w_{i_2}^{k_2-1} z_2 | w_{i_1} | w_{i_2}] - \\ [w_{i_1}^{k_1-1} z_1 w_{i_2}^{k_2-1} z_2 | w_{i_2} | w_{i_1}] & i_1 < i_2, \\[8pt] [w_{i_1}^{k_1} z_1 | w_{i_2}^{k_2-1} z_2 | w_{i_2}] & i_1 \geq i_2. \end{cases}$$

The situation in higher degrees is similar but slightly more complicated and is left to the interested reader.

For the sake of clarity, we will write the DGA-algebras without denoting the degree of the generators.

[19, Prop. 5.24] tells that there is an isomorphism between $\Gamma(w)$ and a tensor product $\tilde{\otimes}_{i\geq 0} Q_{(p)}(\gamma_{p^i}(w))$.

$\tilde{\otimes}_{i\geq 0} Q_{(p)}(\gamma_{p^i}(w)))$ and $\otimes_{i\geq 0} Q_{(p)}(\gamma_{p^i}(w)))$ have the same underlying graded $\mathbf{Z}_{(p)}$-module structure and because of this, $B(\tilde{\otimes}_{i\geq 0} Q_{(p)}(\gamma_{p^i}(w))))$ has the same underlying graded $\mathbf{Z}_{(p)}$-module structure as $B(\otimes_{i\geq 0} Q_{(p)}(\gamma_{p^i}(w))))$. Let $B$ denote this graded $\mathbf{Z}_{(p)}$-module structure for either case. Thus $B$ supports two different differentials, viz., the bar construction differential $\partial^+$ for $\tilde{\otimes}_{i\geq 0} Q_{(p)}(\gamma_{p^i}(w)))$ and the bar construction differential $\partial$ for $\otimes_{i\geq 0} Q_{(p)}(\gamma_{p^i}(w)))$. Let $\delta = \partial^+ - \partial$ be the perturbation of the DG-module $B$. The formula for $\delta$, up to sign, is

$$\delta\left( \otimes_{i\geq 0}(\gamma_{p^i}(w))^{h_{0,i}} [\otimes_{i\geq 0}(\gamma_{p^i}(w))^{h_{1,i}} | \otimes_{i\geq 0}(\gamma_{p^i}(u))^{h_{2,i}} | \dots] \right)$$
$$= \cdots p(\gamma_{p^l}(w))^t \gamma_{p^{l+1}}(w) \cdots [\otimes_{i\geq 0}(\gamma_{p^i}(w))^{h_{2,i}} | \dots]$$
$$+ \otimes_{i\geq 0}(\gamma_{p^i}(w))^{h_{0,i}} [\cdots p(\gamma_{p^j}(w))^t \gamma_{p^{j+1}}(w) \cdots | \dots] + \dots$$

The first summand appears (it is non-zero) if there exists at least one value for $i$ such that $h_{0,i} + h_{1,i} = p + t$ where $0 \leq t \leq p - 2$. The second summand appears if there exists at least one value for $i$ such that $h_{1,i} + h_{2,i} = p + t$ where $0 \leq t \leq p - 2$. And so on.

It is clear that $\delta$ is a derivation and represents the perturbation induced in the differential of $B(\otimes_{i \geq 0} Q_{(p)}(\gamma_{p^i}(w))))$ by the modification produced in the product of the algebra $\otimes_{i \geq 0} Q_{(p)}(\gamma_{p^i}(w)))$. In this situation, there is a formal process (the Basic Perturbation Lemma) which, taking as the input data the contraction (14) and the perturbation $\delta$, when $\delta\phi^{\otimes}$ is nilpotent in each degree, it gives a new contraction

$$B(\tilde{\otimes}_{i \geq 0} Q_{(p)}(\gamma_{p^i}(w)))$$

$$\Downarrow$$

$$\left(\otimes_{i \geq 0} Q_{(p)}(\gamma_{p^i}(w)) \otimes \left(\otimes_{i \geq 0} E(\sigma\gamma_{p^i}(w)) \otimes \Gamma(\varphi_p\gamma_{p^i}(w))\right), d + d_\delta\right) \qquad (15)$$

Our aim here is to verify that $\delta\phi^{\otimes}$ is nilpotent in each degree. To this end, we take an element of the form

$$\otimes_{i=0}^n (\gamma_{p^i}(w))^{h_{0,i}} [\otimes_{i=0}^n (\gamma_{p^i}(w))^{h_{1,i}} | \ldots | \otimes_{i=0}^n (\gamma_{p^i}(u))^{h_{l,i}}] \in B,$$

the number: $\sum_{j=0}^l \sum_{i=0}^n h_{j,i}$ defines a filtration in $B$.

It is easy to see that $\phi^{\otimes}$ does not increase the filtration degree. On the other hand, $\delta$ either lowers the filtration degree or is null, $\forall n$, $\forall l \in \mathbf{N}$. Then, $\delta\phi^{\otimes}$ either lowers filtration or is null, and this means that this composition is nilpotent in each degree.

Taking into account Theorem 3, the contraction above is a semi-full algebra contraction. Notice that the product on the second complex coincides with the product on $X^{\otimes}$. Then, we only need to compute $d_\delta$ on the algebra generators, in order to compute $d_\delta$ on all module generators. Moreover, we shall show that

$$\phi^{\otimes}\delta g^{\otimes} = 0 \qquad (16)$$

Let us observe that $g^{\otimes}$ carries any algebra generator $x$ of the reduced complex into an element $y$ of the form $[\gamma_{p^i}(w)]$ or $[(\gamma_{p^i}(w))^{p-1}|\gamma_{p^i}(w)]$. Now, we study the image of $y$ under $\delta$. It is not difficult to see that this image is zero if $y = [\gamma_{p^i}(w)]$ and $p[\gamma_{p^{i+1}}(w)]$ if $y = [(\gamma_{p^i}(w))^{p-1}|\gamma_{p^i}(w)]$. Since $\phi^{\otimes}[\gamma_{p^{i+1}}(w)] = 0$, we obtain the desired result.

Consequences of (16) are:

$$(g^{\otimes})_\delta = g^{\otimes}, \qquad d_\delta = f^{\otimes}\delta g^{\otimes}.$$

Summing up, (15) is a resolution of $\mathbf{Z}_p$ over $\Gamma(w)$ where

$$d_\delta(\sigma\gamma_{p^i}(w)) \quad = f^{\otimes}\delta g^{\otimes}(\sigma\gamma_{p^i}(w)) = f^{\otimes}\delta[\gamma_{p^i}(w)] = f^{\otimes}(0) = 0,$$

$$d_\delta(\varphi_p\gamma_{p^i}(w)) \quad = f^{\otimes}\delta g^{\otimes}(\varphi_p\gamma_{p^i}(w)) = f^{\otimes}\delta[(\gamma_{p^i}(w))^{p-1}|\gamma_{p^i}(w)]$$

$$= f^{\otimes}(p[\gamma_{p^{i+1}}(w)]) = p\,\sigma\gamma_{p^{i+1}},$$

$$d_\delta(\gamma_k\varphi_p\gamma_{p^i}(w)) = f^{\otimes}\delta g^{\otimes}(\gamma_k\varphi_p\gamma_{p^i}(w)) = p\,\sigma\gamma_{p^{i+1}}(w)\,\gamma_{k-1}\varphi_p\gamma_{p^i}(w).$$

# References

1. Álvarez, V., Armario, J.A., Frau, M.D., Real, P.: Transferring TTP-structures via contraction. Homology, Homotopy Appl. 7 (2) (2005) 41–54
2. Armario, J.A., Real, P., Silva, B.: On $p$–minimal homological models of twisted tensor products of elementary complexes localized over a prime. in: J. McCleary (ed.), Higher Homotopy Structures in Topology and Mathematical Physics (Poughkeepsie, NY, 1996), Contemp. Math.,A.M.S, Providence, RI, 1999, pp. 303–314
3. Barnes, D. W., Lambe, L. A.: A fixed point approach to homological perturbation theory. Proc. Amer. Math. Soc. 112 (3) (1991) 881–892
4. Brown, R.: The twisted Eilenberg-Zilber theorem. Celebrazioni Archimedee del secolo XX, Simposio di topologia (1967) 34–37
5. Cartan, H.: Algèbres d'Eilenberg-Mac Lane. Séminaire H. Cartan 1954/55, (exposé 2 à 11). Ecole Normale Supérieure, Paris, (1956)
6. Eilenberg, S., Mac Lane, S.: On the groups $H(\pi, n)$, I. Annals of Math. 58 (1953) 55-106
7. Gugenheim, V.K.A.M., Lambe, L.A.: Perturbation theory in Differential Homological Algebra I. Illinois J. Math. 33 (4) (1989) 566–582
8. Gugenheim, V.K.A.M., Lambe, L.A., Stasheff, J.D.: Perturbation theory in Differential Homological Algebra II. Illinois J. Math. 35 (3) (1991) 357–373
9. Huebschmann, J., Kadeishvili, T.: Small models for chain algebras. Math. Z. 209 (1991) 245–280
10. Johansson, L., Lambe, L.A.: Transferring algebra structures up to homology equivalence. Math. Scan. 88 (2) (2001) 181–200
11. Johansson, L., Lambe, L.A., Sköldberg, E.: On constucting resolutions over the polynomial algebra Homology, Homotopy Appl., 4 (2) (2002) 315–336
12. Lambe, L.A.: Resolutions via homological perturbation. J. Symbolic Comp. 12 (1991) 71–87
13. Lambe, L.A.: Homological perturbation theory. Hochschild homology and formal groups. In: Proc. Conference on Deformation Theory and Quantization with Applications to Physics, (Amherst, MA, June 1990), American Mathematical Society, Providence, RI, 1992, 183–218
14. Lambe, L.A.: Resolutions which split off of the bar construction. J. Pure Appl. Algebra, 84 (1993) 311–329
15. Lambe, L.A., Stasheff, J.D.: Applications of perturbation theory to iterated fibrations. Manuscripta Math. 58 (1987) 367–376
16. Mac Lane, S.: Homology. Classics in Mathematics Springer-Verlag, Berlin, (1995). Reprint of the 1975 edition.
17. May, J.P.: The cohomology of restricted Lie algebras and of Hopf algebras. J. Algebra, 3 (1966) 123–146
18. Prouté, A.: Algèbres différentielles fortement homotopiquement associatives ($A_\infty$-algèbre). Ph. D. Thesis, Université Paris VII (1984)
19. Real, P.: Homological Perturbation Theory and Associativity. Homology, Homotopy Appli. 2 (2000) 51–88
20. Stasheff, J.D.: Homotopy Associativity of $H$-spaces I, II. Trans. A.M.S, 108 (1963) 275–312

# A Symbolic-Numeric Approach for Solving the Eigenvalue Problem for the One-Dimensional Schrödinger Equation

I.N. Belyaeva[1], N.A. Chekanov[1], A.A. Gusev[2],
V.A. Rostovtsev[2], and S.I. Vinitsky[2]

[1] Belgorod State University, Studentcheskaja St. 14, Belgorod, 308007, Russia
ibelyaeva@bsu.edu.ru, chekanov@bsu.edu.ru
[2] Joint Institute for Nuclear Research, Dubna, Moscow Region 141980, Russia
vinitsky@thsun1.jinr.ru

**Abstract.** A general scheme of a symbolic-numeric approach for solving the eigenvalue problem for the one-dimensional Shrödinger equation is presented. The corresponding algorithm of the developed program EWA using a conventional pseudocode is described too. With the help of this program the energy spectra and the wave functions for some Schrödinger operators such as quartic, sextic, octic anharmonic oscillators including the quartic oscillator with double well are calculated.

## 1  Introduction

For solving a stationary Shrödinger equation a lot of approximate analytical and numerical methods are elaborated and applied because an exact solution in explicit form exists only for some specific hamiltonians [1]. As is known, the more used methods are diagonalization [2,3], quasiclassical approach [4], continuous analogue of Newton's method [5], different versions of perturbation theory [6], normal form method [7,8,9,10,11], finite-element method [12], $1/N$ expansion [13], oscillator representation [14], variational and operational methods [15,16,17], simplectic method [18] and etc.

The method of integration by means of the power series is known to be a simple one but it requires cumbersome work, and difficulties are increasing in the cases when a differential equation has singularities[19]. On the other hand, an application of the modern PC together with packages of symbolic manipulations such as MAPLE, MATHEMATICA, REDUCE enable us to carry out necessary calculations and construct the solution of differential equation in the form of power series up to a desired degree.

In the present work, an analytic numeric approach for solving the time independent Shrödinger equation is proposed. The developed method is based on finding a general solution as a sum of two independent solutions.

By this method the spectra and wave functions for a quartic, sextic, and octic anharmonic oscillators and also for the quartic oscillator with two minima were calculated. Obtained results are in good agreement with the ones available in literature.

V.G. Ganzha, E.W. Mayr, and E.V. Vorozhtsov (Eds.): CASC 2006, LNCS 4194, pp. 23–32, 2006.

## 2   General Scheme of the Method

Let us consider the eigenvalue problem for ordinary differential equation

$$\frac{d^2\psi(x)}{dx^2} + 2(E - V(x))\Psi(x) = 0, \tag{1}$$

where the function $V(x)$ can have a pole not above the second order in a vicinity of the point $x_0$, and independent variable $x$ belong to the real axis or semiaxis.

For fixed value of $E$ differential equation (1) has two linear independent solutions $y_1(x)$ and $y_2(x)$. If the function $V(x)$ is regular at point $x = x_0$ we find $y_1(x)$ and $y_2(x)$ as solutions of Cauchy problem (1) with boundary conditions

$$y_1(x_0) = 1, \quad \left.\frac{dy_1(x)}{dx}\right|_{x=x_0} = 0, \quad y_2(x_0) = 0, \quad \left.\frac{dy_2(x)}{dx}\right|_{x=x_0} = 1 \tag{2}$$

in the form of Taylor series

$$y_1(x) = 1 + \sum_{k=2}^{N} c_k^{(1)}(x - x_0)^k, \quad y_2(x) = (x - x_0) + \sum_{k=2}^{N} c_k^{(2)}(x - x_0)^k. \tag{3}$$

Substituting (3) into (1) we obtain recurrence relations for evaluation of coefficients $c_k^{(1)}$ and $c_k^{(2)}$.

If the function $V(x)$ has a pole of order not higher than the second, i.e.

$$2(E - V(x)) = (x - x_0)^{-2} \sum_{k=0}^{N} f_k(x - x_0)^k,$$

the solutions of (1) will be found in form of generalized power series

$$y(x) = (x - x_0)^{\rho} \sum_{k=0}^{N} c_k(x - x_0)^k. \tag{4}$$

Substituting (4) into (1) we find determining equation

$$\rho(\rho - 1) - f_0 = 0. \tag{5}$$

It is known from the theory of ordinary differential equations [23] that the form of independent solutions depends on the roots $\rho_1$ and $\rho_2$ of Eq. (5).

1. If the difference of roots, $\rho_1 - \rho_2$ is not equal to an integer, then two linear independent solutions can be present in the form

$$y_1(x) = (x - x_0)^{\rho_1} \sum_{k=0}^{N} c_k^{(1)}(x - x_0)^k, \tag{6}$$

$$y_2(x) = (x - x_0)^{\rho_2} \sum_{k=0}^{N} c_k^{(2)}(x - x_0)^k.$$

2. If the difference of roots, $\rho_1 - \rho_2$ is equal to an integer, then two linear independent solutions can be present in the form

$$y_1(x) = (x - x_0)^{\rho_1} \sum_{k=0}^{N} c_k^{(1)} (x - x_0)^k, \tag{7}$$

$$y_2(x) = (x - x_0)^{\rho_2} \sum_{k=0}^{N} c_k^{(2)} (x - x_0)^k + \xi_{-1} y_1(x) \ln(x - x_0),$$

where $\rho_1 > \rho_2$.

In these two cases we also find the recurrence relations for coefficients $c_k^{(1)}$, $c_k^{(2)}$ by substitution of expansion (6) or (7) into differential equation (1).

The general solution of Eq. (1) takes the form

$$\psi(x) = C_1 y_1(x) + C_2 y_2(x). \tag{8}$$

In physical applications, one needs to find bounded solutions, therefore, we truncate an infinite interval to a finite one $x \in (R_{left}, R_{right})$ and supply the following boundary conditions

$$\psi(R_{left}) = 0, \quad \psi(R_{right}) = 0. \tag{9}$$

Then coefficients $C_1$ and $C_2$ satisfy the set of homogeneous algebraic equations

$$C_1 y_1(R_{left}) + C_2 y_2(R_{left}) = 0, \quad C_1 y_1(R_{right}) + C_2 y_2(R_{right}) = 0. \tag{10}$$

A nontrivial solution of system (10) is found from condition of equality to zero of determinant

$$D(E) = \begin{vmatrix} y_1(R_{left}) & y_2(R_{left}) \\ y_1(R_{right}) & y_2(R_{right}) \end{vmatrix} = 0, \tag{11}$$

which is carried out at certain values of energy making an energy spectrum $E = \{E_k\}$ of Schrödinger equation (1). For given $E = E_k$, the coefficients $C_1$ and $C_2$ are calculated from Eq. (10), including an additional normalization condition

$$\int_{R_{left}}^{R_{right}} |\psi(x)|^2 dx = 1 \tag{12}$$

for the wave function (8).

## 3   Description of the Program

Following the description of the method for solving the eigenvalue problem for equation (1), we present below the algorithm EWA. The corresponding program EWA has been implemented in a Maple Package.

**Input:**
$V(x)$ is potential;
$R_{left}$ and $R_{right}$ are bounds of a truncated interval of the independent variable $x$;
$N$ is the number of terms of the power series;
**Output:**
$\{E_k\}$ and $\{\psi_k(x)\}$ are set of energy levels and wave functions of the equation (1);
**The description of the local variables:**
TypeV is the flag, if TypeV=0 then $V(x)$ does not have singularity in $x_0$, if TypeV=1 then $V(x)$ has a pole in $x_0$;
$c_k^{(1)}$ and $c_k^{(2)}$ are coefficients of the two linear independent solutions of the equation (1);
$\rho_1$, $\rho_2$ are the solutions of the determining equation (5);
$\xi_{-1}$ is a coefficient in expansion (7);
$y_{1,2}(R_{left})$, $y_{1,2}(R_{right})$ are values of two linear independent solutions in bounds of interval;
$C_1$ and $C_2$ are auxiliary coefficients;

---

**1:**    $V(x), x_0 \quad \rightarrow \quad TypeV$;
**2:**    if TypeV=0 then
**2.1:**    $y_1(x) \rightarrow 1 + \sum_{k=2}^{N} c_k^{(1)}(x - x_0)^k$;
**2.2:**    $y_2(x) \rightarrow (x - x_0) + \sum_{k=2}^{N} c_k^{(2)}(x - x_0)^k$;
**3:**    else if TypeV=1 then
**3.1:**    $\rho(\rho - 1) - f_0 = 0 \quad \rightarrow \quad \rho_1, \rho_2$;
**3.2:**    $y_1(x) \rightarrow (x - x_0)^{\rho_1} \sum_{k=0}^{N} c_k^{(1)}(x - x_0)^k$;
**3.3:**    $y_2(x) \rightarrow (x - x_0)^{\rho_2} \sum_{k=0}^{N} c_k^{(2)}(x - x_0)^k + \xi_{-1} y_1(x) \ln(x - x_0)$;
         end if
**4:**    $C_1 y_1(R_{left}) + C_2 y_2(R_{left}) = 0$,
         $C_1 y_1(R_{right}) + C_2 y_2(R_{right}) = 0, \quad \rightarrow \quad E_k$;
**5:**    $E_k$ + normalization conditions $\quad \rightarrow \quad C_1, C_2 \quad \rightarrow \quad \psi_k(x)$

---

**Remark:** This program involves the following sequence of steps.
Step 1. Determination of value of flag TypeV.
Step 2. Finding two linear independent solutions if the potential function $V(x)$ does not have a pole.
Step 3. The coefficients $\rho_1$, $\rho_2$, $c_k^{(1)}$, $c_k^{(2)}$ and $\xi_{-1}$ of expansion (3) for regular potential, or (6), (7) depending on the result of the solution for the determining equation (5) for singular potential, are evaluated. At this step, the coefficients $c_k^{(1)}$, $c_k^{(2)}$, and $\xi_{-1}$ depend on $E$ explicitly.
Step 4. Evaluation of the energy spectrum from boundary conditions (10).
Step 5. Evaluation of the eigenfunctions using normalization conditions.

**Table 1.** The comparison of energy levels $E_{EWA}$ with their exact values $E_{exact}$ (14) for $l = 1, 2, 3$ ($R_{left} = 10^{-8}$, $R_{right} = 5.6$, $N = 116$, $\varepsilon = |E_{EWA} - E_{exact}|/E_{exact}$)

| | | $l = 1$ | |
|---|---|---|---|
| $n$ | $E_{EWA}$ | $E_{exact}$ | $\varepsilon, \%$ |
| 0 | 2.00000000044 | 2 | 0.0000000022 |
| 1 | 4.000000017 | 4 | 0.00000043 |
| 2 | 6.0000020 | 6 | 0.000033 |
| 3 | 8.000083 | 8 | 0.0010 |

| | | $l = 2$ | |
|---|---|---|---|
| $n$ | $E_{EWA}$ | $E_{exact}$ | $\varepsilon, \%$ |
| 0 | 3.00000000066 | 3 | 0.000000022 |
| 1 | 5.00000016 | 5 | 0.0000032 |
| 2 | 7.000013 | 7 | 0.00018 |
| 3 | 9.00037 | 9 | 0.0041 |

| | | $l = 3$ | |
|---|---|---|---|
| $n$ | $E_{EWA}$ | $E_{exact}$ | $\varepsilon, \%$ |
| 0 | 4.0000000067 | 4 | 0.00000016 |
| 1 | 6.0000011 | 6 | 0.000018 |
| 2 | 8.000066 | 8 | 0.00082 |
| 3 | 10.0013 | 10 | 0.013 |

## 4 Examples of the EWA Program Runs

To test the EWA program the one-dimensional (1) and radial Shrödinger equations were considered with the potential functions such that solutions are known:

A) infinite rectangular wall $V(x) = \{0, |x| \le R; \infty, |x| > R\}$;
B) harmonic oscillator $V(x) = x^2/2$;
C) the two-dimensional axial symmetric harmonic oscillator $V(r) = 1/2\omega r^2$ (in this case we use $r$ instead of $x$).

In case A), ten lowest energy levels coincide with exact $E_n = \pi^2 n^2/8R^2$, $n = 1, ...10, ...$ to nine significant figure accuracy if $N = 68$ and $R_{left} = -R$, $R_{right} = R$, $R = 1$.

In case B), for $x \in [R_{left}, R_{right}]$ ($-R_{left} = R_{right} = R = 5.9$, $N = 138$) the accuracy $\varepsilon$ of the 10th energy level obtained is less than 0.004% compared with exact value $E_{10} = 10.5$. And absolute difference between calculated and exact wave function $|\psi_{EWA} - \psi_{exact}|$ is less than $10^{-9}$. Note, the accuracy depends on values of $R$ and $N$, and its magnitude may be increased.

In case C), for $\omega = 1$ the differential equation on the semiaxis $r \in [0, \infty)$

$$\frac{d^2 y(r)}{dr^2} + \left(2E + \frac{1 - 4l^2}{4r^2} - r^2\right) y(r) = 0, \tag{13}$$

has the double pole in origin, and its eigenvalues and functions are known [20]

$$E_n = 2n + |l| + 1, \quad n = 0, 1, 2, \ldots, \quad l = 0, \pm 1, \pm 2, \ldots \tag{14}$$

**Table 2.** The comparison of energy levels $2E_{EWA}$ with their values $E_{exact}$ from [3] for different powers $\mu$ ($\alpha = 0.0005$ for $\mu = 4, 6$ and $\alpha = 0.00005$ for $\mu = 8$)

| $\mu = 4$, | $R_{left} = -5.6,\ R_{right} = 5.6,$ | $N = 116$ |
|---|---|---|
| $n\ \ 2E_{EWA}$ | $E_{exact}$ | $\varepsilon$ |
| 0   1.0007486926734 | 1.00074869267319 | 0.000000000019 |
| 1   3.00373974818 | 3.00373974816873 | 0.00000000046 |
| 2   5.009711873 | 5.00971187278811 | 0.0000000080 |
| 3   7.018652599 | 7.01865259205752 | 0.00000010 |
| 4   9.0305496 | 9.03054956607471 | 0.0000011 |
| 5   11.045391 | 11.0453905781793 | 0.0000094 |
| 6   13.06317 | 13.0631635776785 | 0.000065 |
| 7   15.0839 | 15.0838565876260 | 0.00043 |
| 8   17.1078 | 17.1074577926535 | 0.0022 |
| 9   19.134 | 19.1339554918523 | 0.0031 |
| 10  21.164 | 21.1633381057038 | 0.0060 |
| $\mu = 6$, | $R_{left} = -4.7,\ R_{right} = 4.7,$ | $N = 116$ |
| $n\ \ 2E_{EWA}$ | $E_{exact}$ | $\varepsilon$ |
| 0   1.001848816 | 1.0018488155723 | 0.000000045 |
| 1   3.01278097 | 3.0127809606901 | 0.00000056 |
| 2   5.0448002 | 5.0447999257845 | 0.0000060 |
| 3   7.110096 | 7.1100928558609 | 0.000048 |
| 4   9.21860 | 9.2185817487322 | 0.00030 |
| 5   11.3779 | 11.377808617207 | 0.0015 |
| $\mu = 8$, | $R_{left} = -4.6,\ R_{right} = 4.6$, | $N = 108$ |
| $n\ \ 2E_{EWA}$ | $E_{exact}$ | $\varepsilon$ |
| 0   1.00064637 | 1.0006463698740 | 0.00000012 |
| 1   3.00572693 | 3.0057269553512 | 0.00000070 |
| 2   5.0253946 | 5.0253949690878 | 0.0000061 |
| 3   7.07668 | 7.0766689726027 | 0.00016 |
| 4   9.18033 | 9.1802567401069 | 0.00088 |
| 5   11.3565 | 11.356154413293 | 0.0038 |

$$y_{nl}(r) = (-1)^n \sqrt{\frac{2n!}{(n+l)!}} \exp(-\frac{r^2}{2}) \cdot r^{l+\frac{1}{2}} \cdot L_n^{(l)}(r^2), \qquad (15)$$

where $L_n^{(l)}(r^2)$ are the Chebyshev–Laguerre polynomials. For equation (13), the eigenfunctions and the energy spectrum were found with the help of the program EWA. A comparison of the calculated and exact eigenvalues for different values of $l$ is presented in Table 1. As an example, we present the series expansion of two linear independent solutions of equation (13) at point $r = 0$ for $l = 1$ and $N = 8$:

$$y_1(r) = r^{\frac{3}{2}} \left( 1 - \frac{Er^2}{4} + \frac{(E^2 + 2)r^4}{48} - \frac{(E^3 + 8E)r^6}{1152} + \frac{(E^4 + 20E^2 + 24)r^8}{46080} \right),$$

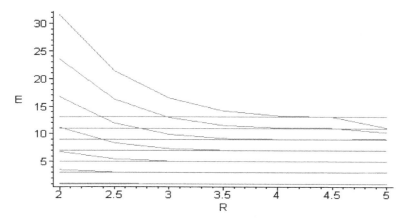

**Fig. 1.** The evaluated energy levels $E_k, k = 1, ..., 7$ of anharmonic oscillator with potential (16) versus the boundary point $R$ for $\mu = 4$, $\alpha = 1$, and $N = 72$. Exact values $E_k^{exact}$ are shown by straight lines.

$$y_2(r) = r^{-\frac{1}{2}} \left( -\frac{1}{2} + \frac{Er^2}{8} - \frac{(1 - E^2)r^4}{16} + \frac{(2E - 11E^3)r^6}{1152} \right) + \frac{E}{2} y_1(r) \ln r.$$

### 4.1 Quartic, Sextic, and Octic Anharmonic Oscillators

The above presented method has been applied for calculation of energy levels and wave functions of anharmonic oscillator with nonlinearity of the fourth, sixth, and eighth degrees so that potential functions $V(x)$ have the following forms

$$V(x) = \frac{x^2}{2} + \alpha x^\mu, \quad \mu = 4, 6, 8, \tag{16}$$

where $\alpha > 0$ is parameter of nonlinearity. Using EWA program for Shrödinger equation (1) with potential functions (16) the energy spectra and wave functions in the form of power series were found. The calculated energy levels are present in Table 2. The behaviour of determinant $D(E)$, and, hence, of roots $\{E_k\}$ of equation (10) strongly depends on the number of terms $N$ in power series (7) and on the value of $R = -R_{left} = R_{right}$. As $N$ increases the values of roots of equation (10) come nearer to true values if as necessary to increase the value of $R$. In Fig. 1, the energy values are shown at fixed $N$ versus the $R$ value.

### 4.2 Anharmonic Oscillator with Two Minima

The potential function of anharmonic oscillator with two minima takes form

$$V(x) = \alpha(x^2 - a^2)^2. \tag{17}$$

Here $\alpha > 0$ is parameter of nonlinearity, and $a$ is parameter that determine the position of two minima of potential function (17). In this case wave functions

**Table 3.** The comparison of energy levels $2E_{EWA}$ of potential function (17) with their values $2E_J$ from [15]

| $a = \sqrt{2},\ \alpha = 1,\ R_{left} = -3.4,\ R_{right} = 3.4,\ N = 180$ | | |
|---|---|---|
| $n\ E_{EWA}$ | $E_J$ | $\varepsilon$ |
| 0  1.80081349 | 1.80081349 | 0 |
| 1  1.89650538 | 1.89650538 | 0 |
| 2  4.37046673 | 4.37046673 | 0 |
| 3  5.57335024 | 5.5733520 | 0.0000007 |
| 4  7.65142527 | - | - |
| 5  9.92036057 | - | - |

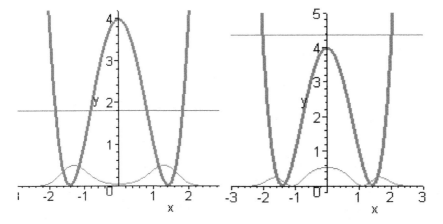

**Fig. 2.** The square of wave function, $|\Psi(x)|^2$ of the ground state (left panel) and for state with $n = 2$ (right panel). The thick lines is the potential function $V(x)$.

and energy spectrum for Schrödinger equation (1) were also calculated by means of the program EWA for different values of parameters $a$ and $\alpha$. As an example, we show the series expansion of two linear independent solutions at point $x = 0$ for $N = 7$:

$$y_1(x) = 1 + (\alpha a^4 - E)x^2 + \frac{1}{6}\left(E^2 - 2E\alpha a^4 + \alpha^2 a^8 - 2\alpha a^2\right)x^4$$

$$+\frac{1}{90}\left(-E^3 + 3E^2\alpha a^4 - 3E\alpha^2 a^8 + 14E\alpha a^2 + \alpha^3 a^{12} - 14\alpha^2 a^6 + 6\alpha\right)x^6,$$

$$y_2(x) = x + \frac{1}{3}\left(\alpha a^4 - E\right)x^3 + \frac{1}{30}\left(E^2 - 2E\alpha a^4 + \alpha^2 a^8 - 6\alpha a^2\right)x^5,$$

$$+\frac{1}{630}\left(-E^3 + 3E^2\alpha a^4 - 3E\alpha^2 a^8 + 26E\alpha a^2 + \alpha^3 a^{12} - 26\alpha^2 a^6 + 20\alpha\right)x^7.$$

The values of lowest energy levels and plots of wave functions are presented in Table 3 and in Fig. 2.

# Acknowledgments

Authors are deeply indebted to Prof. I.V. Puzynin and to the participants of his seminar for a useful discussion. This work is partially supported by BelSU (grant VKG-003-04).

# References

1. Landau, L.D. and Lifshits, E.M.: Quantum Mechanics: Non-Relativistic Theory. Pergamon Press, New York, 1977
2. Wilkinson, J.H., Reinsch, C.: Handbook for Automatic Computation, Vol.2: Linear Algebra. Springer-Verlag, New York, 1971
3. Banerjee, K.: General anharmonic oscillators. In: Proc. Roy. Soc. London, A. **364** (1978) 265–275
4. Maslov, V.P. and Fedoryuk, M.V.: Kvaziklassicheskie pribligeniya dlya uravnenii kvantovoi mekhaniki. Nayka, Moskva, 1976
5. Puzynin, I.V., Amirkhanov, I.V. et al: Continuous Analogue of Newton's Method for the Numerical Investigation of Some Nonlinear Quantum - Field Models.- PEPAN, **30** (1999) 210–265
6. Flugge, S.: Practical Quantum Mechanics. Springer-Verlag, Berlin, Heidelberg, 1971
7. Birkhoff, G.D.: Dynamical Systems. A.M.S. Colloquium Publications. New York, 1927
8. Gustavson, F.G.: On construction of formal integral of a Hamiltonian system near an equilibrium point. Astronom. J. **71** (1966) 670–686
9. Swimm, R.T. and Delos: Semiclassical calculation of vibrational energy levels for nonseparable systems using Birkhoff–Gustavson normal form. J. Chem. Phys. **71** (1979) 1706
10. Ali, M.K.: The quantum normal form and its equivalents. J. Math. Phys. **26** (1985) 25–65
11. Chekanov, N.A.: Kvantovanie normalnoi formy Birkhoff–Gustavson. Jadernaya Fizika **50** (1985) 344–346
12. Abrashkevich, A.G., Abrashkevich, D.G., Kaschiev, M.S., Puzynin, I.V.: FESSDE, a program for the finite-element solution of the coupled-channel Schroedinger equation using high-order accuracy approximations. Comp. Phys. Commun. **85** (1995) 65–74
13. Jaffe, L.G.: Large N limits as classical mechanics, Rev. Mod. Phys. **54** (1982) 407–435
14. Dineykhan, M., Efimov, G.V.: The Schroedinger equation for bound state systems in the oscillator representation. Reports of Math. Phys. **6** (1995) 287–308
15. Jafarpour, M., Afshar, D.: Calculation of energy eigenvalues for the quantum anharmonic oscillator with a polynomial potential. J. Phys. A: Math. Gen. **35** (2002) 87–92
16. Ivanov, I.A.: Sextic and octic anharmonic oscillator: connection between strong-coupling and weak-coupling expansions. J. Phys. A: Math. Gen. **31** (1998) 5697–5704
17. Ivanov, I.A.: Link between the strong-coupling and weak-coupling asymptotic perturbation expansions for the quartic anharmonic oscillator. J. Phys. A: Math. Gen. **31** (1998) 6995–7003

18. Liu, X.S., Su, L.W., Ding, P.Z.: Intern. J. Quantum Chem. (2002) **87** (2002) 1–11
19. Ince, E.L.: Ordinary Differential Equations. Dover Pubns, New York, 1956
20. Abramowitz, M. and Stegun, I.A.: Handbook of Mathematical Functions. Dover, New York, 1968

# Reducing Computational Costs in the Basic Perturbation Lemma*

Ainhoa Berciano[1], María José Jiménez[2], and Pedro Real[2]

[1] Dpto. Matemática Aplicada, Estadística e Investigación Operativa,
Universidad del País Vasco, Barrio Sarriena s/n, 48940 Leioa (Vizcaya), Spain
ainhoa.berciano@ehu.es
[2] Dpto. de Matemática Aplicada I, Universidad de Sevilla,
Avda. Reina Mercedes s/n, 41012 Sevilla, Spain
{majiro, real}@us.es

**Abstract.** Homological Perturbation Theory [11, 13] is a well-known general method for computing homology, but its main algorithm, the *Basic Perturbation Lemma*, presents, in general, high computational costs. In this paper, we propose a general strategy in order to reduce the complexity in some important formulas (those following a specific pattern) obtained by this algorithm. Then, we show two examples of application of this methodology.

## 1 Introduction

Most algorithms in Algebraic Topology and Homological Algebra carry high computational costs. We are concerned with the search of techniques that cut down complexity in processes from these areas. In order to compute homology, two important tools in Homological Algebra are, on the one hand, the notion of *contraction* [6], a special type of homotopy equivalence, between differential graded modules (DG–modules) and, on the other hand, the *Basic Perturbation Lemma* (BPL), which allows one to generate a new contraction by "perturbation" of a previous one (see [2] or [18]). In fact, the BPL, is an algorithm whose input is a contraction between two differential graded modules together with a "perturbation" of the differential structure on the first DG–module; the output is a contraction from the perturbed module onto the second module, whose differential comes out to be modified. The formulas of the latter contraction as well as the modification in the differential structure of the second DG–module often imply high computational costs. Thanks to a tool introduced in this paper, *compatible grading*, we are able to reduce this complexity in time and space. In fact, we establish some proper conditions under which the calculus implied in the computation of several formulas obtained by the BPL, with a specific structure, can be significantly cut down. However, the improvement achieved in computation will depend on the specific morphisms implied, as well as the compatible grading considered.

---

* Partially supported by the PAICYT research project FQM-296 and the UPV–EHU project 00127.310-E-15916.

We describe two important applications, showing the advantages from the computational point of view in each case. The first one is related to the algebraic structure of the homology of a truncated polynomial algebra. Another example of application of the general strategy presented here is the so-called *inversion theory*, which was born in [17] and was later used in [4] and [14] in order to simplify the computation of the perturbed differential of a 1-homological model for a commutative DGA-algebra. We must say that we only recall here these results, under the new viewpoint.

This paper is divided as follows. First, we give some preliminaries and notations. Secondly, we establish the theory of compatible gradings, which represents the main original contribution of this paper. Finally, two different applications of this novel strategy for reducing the computational costs of the BPL are shown in the last two sections.

## 2    Preliminaries and Notations

Here we will collect some basic definitions and results in the context of Homological Algebra, as well as the notations that we will adopt. See [5], [16] or [19] for further information.

Take a commutative ground ring with unit, $\Lambda$. We will work with differential graded modules, *DG–modules*, which are graded modules endowed with a morphism $d$ of graded modules of degree $-1$ such that $d\,d = 0$. A graded module $M$ is *connected* whenever $M_0 = \Lambda$ in which case, the graded module $\bar{M}$ is defined as $\bar{M}_n = M_n$ for $n > 1$ and $\bar{M}_0 = 0$.

We strictly adhere to the *Koszul conventions* with regard to signs, meaning that if $f : M \to M'$ and $g : N \to N'$ are both DG–module morphisms, then

$$(f \otimes g)(x \otimes y) = (-1)^{|g||x|} f(x) \otimes g(y).$$

Given a DG–module $(M, d_M)$, the *suspension* of $M$ is the DG–module $(sM, d_{sM})$, where $(sM)_n = M_{n-1}$ and $d_{sM} = -d_M$. Analogously, the *desuspension* of $M$ is given by $(s^{-1}M)_n = M_{n+1}$ with differential $-d_M$. We will denote by $\uparrow$ and $\downarrow$ the suspension and desuspension morphisms which shift the degree by $+1$ and $-1$, respectively.

Given a DG-module $(M, d)$, the *tensor module* of $M$, $T(M)$, is constructed as follows:

$$T(M) = \bigoplus_{n \geq 0} M^{\otimes n}.$$

The *(tensor) grading* of a homogeneous element of $T(M)$, $a_1 \otimes \cdots \otimes a_n$ is given by $\sum_{i=1}^{n} |a_i|$. The differential structure in $T(M)$ is provided by the *tensor differential*,

$$d_t = \sum_{i=0}^{n-1} 1^{\otimes i} \otimes d_M \otimes 1^{\otimes n-i-1}.$$

Any morphism of DG–modules $f : M \to N$ induces another one $T(f) : T(M) \to T(N)$, being $T(f)|_{M^{\otimes n}} = f^{\otimes n}$.

If a DG–module $A$ is endowed with an associative product with unit, it is called a *DG–algebra*, $(A, d_A, \mu_A)$. Sometimes we will use the notation $*_A$ for the product on $A$. A *DG–coalgebra* is a DG–module provided with a compatible coproduct and counit.

There are several examples of connected commutative DG–algebras with null differential:

- The *polynomial DG–algebra* $P(v, 2n)$, generated by $v$ of degree $2n$, where $n$ is a positive integer. The product is the usual one of monomials i.e., $v^i v^j = v^{i+j}$.
- The *truncated polynomial DG–algebra* $Q_p(v, 2n)$, which is the quotient algebra $P(v, 2n)/(v^p)$.
- The *exterior DG–algebra* $E(u, 2n + 1)$, $n \geq 0$, with algebra generator $u$ of degree $2n + 1$ and trivial product $u^2 = 0$.
- The *divided polynomial DG–algebra* $\Gamma(w, 2n)$, $n \geq 1$, generated by $\gamma_1(w) = w$ ($\gamma_0(w) = 1$) with product the one given by $\gamma_k(w)\gamma_h(w) = \frac{(k+h)!}{k!\,h!}\gamma_{k+h}(w)$;

Given a connected DG–algebra $A$, one can construct the *reduced bar construction* of $A$, $\bar{B}(A)$, whose underlying module is $T(s\bar{A})$. A typical element of $\bar{B}(A)$, is denoted by $\bar{a} = [a_1|\cdots|a_n] \in (s\bar{A})^{\otimes n}$. The total differential $d_B$ is given by $d_B = d_t + d_s$, being $d_t$ the natural one on the tensor module and $d_s$ the *simplicial differential*, that depends on the product on $A$. This DG–module is endowed with a structure of DG–coalgebra by the natural coproduct $\Delta_{\bar{B}}$ defined on the tensor module:

$$\Delta_{\bar{B}}([a_1|\cdots|a_r]) = \sum_{i=0}^{r} [a_1|\cdots|a_i] \otimes [a_{i+1}|\cdots|a_r].$$

If $A$ is a commutative DG–algebra, $\bar{B}(A)$ is endowed with an additional structure of algebra by the *shuffle product*, $\star$.

Let $p$ and $q$ be two nonnegative integers, a $(p, q)$–*shuffle* is defined as a permutation $\pi$ of the set $\{0, 1, \ldots, p + q - 1\}$ such that $\pi(i) < \pi(j)$ whenever $0 \leq i < j \leq p - 1$ or $p \leq i < j \leq p + q - 1$.

Notice that there are $\binom{p+q}{p}$ different $(p, q)$–shuffles.

So, the *shuffle product* $\star$ on $\bar{B}(A)$, is defined (up to sign) by:

$$[a_1|\cdots|a_p] \star [b_1|\cdots|b_q] = \sum_{\pi \in \{(p,q)-shuffles\}} \pm [c_{\pi(0)}|\cdots|c_{\pi(p-1)}|c_{\pi(p)}|\cdots|c_{\pi(p+q-1)}];$$

where $(c_0, \ldots, c_{p-1}, c_p, \ldots, c_{p+q-1}) = (a_1, \ldots, a_p, b_1, \ldots b_q)$.

Given a simply connected DG–coalgebra $C$, the *reduced cobar construction*, $\bar{\Omega}(C)$, is a DG–algebra whose underlying module is $T(s^{-1}\bar{C})$. A typical element of $\bar{\Omega}(C)$ will be written $\bar{c} = \langle c_1|\cdots|c_n \rangle$. The total differential $d_{\bar{\Omega}}$ is given by the sum of the tensor differential and the *cosimplicial differential*:

$$d_{cos} = \sum_{i=0}^{n-1} 1^{\otimes i} \otimes \downarrow^{\otimes 2} \Delta_C \uparrow \otimes 1^{\otimes n-i-1}. \tag{1}$$

The product $\mu_\Omega$ is the natural one on the underlying module, which works by juxtaposition.

Now we briefly recall the main concepts from Homological Perturbation Theory that we will use in this paper.

A *contraction* $c : \{N, M, f, g, \phi\}$ [6, 13], also denoted by $(f, g, \phi) : N \overset{c}{\Rightarrow} M$, from a DG–module $(N, d_N)$ to another one $(M, d_M)$ is a especial type of homotopy equivalence given by the morphisms $f$, $g$ and $\phi$; $f : N_* \to M_*$ and $g : M_* \to N_*$ are two morphisms of DG–modules and $\phi : N_* \to N_{*+1}$ is a homotopy operator. This way, apart from the conditions

$$(c1)\ fg = 1_M, \quad (c2)\ \phi d_N + d_N \phi + gf = 1_N,$$

the following ones must be satisfied

$$(c3)\ f\phi = 0, \quad (c4)\ \phi g = 0, \quad (c5)\ \phi\phi = 0.$$

Given two contractions of DG–modules

$$c_i : \{N_i,\ M_i,\ f_i,\ g_i,\ \phi_i\} \quad i = 1, 2,$$

one can construct the following ones [10, 11]:

1. The *suspension contraction* of $c_1$, $s(c_1)$, which consists in taking the suspension DG–modules and the induced morphisms:

$$s(c_1) : \{s(N_1),\ s(M_1),\ f_1,\ g_1,\ -\phi_1\}.$$

2. The *tensor module contraction*, $T(c_1)$, obtained by taking the tensor modules of $M_1$ and $N_1$ and the induced morphisms,

$$T(c_1) : \{T(N_1), T(M_1), T(f_1), T(g_1), T(\phi_1)\};$$

where

$$T(\phi_1)|_{N_1^{\otimes n}} = \phi_1^{[\otimes n]} = \sum_{i=0}^{n-1} 1^{\otimes i} \otimes \phi_1 \otimes (g_1 f_1)^{\otimes n-i-1}.$$

3. The *tensor product contraction*:

$$c_1 \otimes c_2 : \{N_1 \otimes N_2,\ M_1 \otimes M_2, f_1 \otimes f_2,\ g_1 \otimes g_2,\ \phi_1 \otimes g_2 f_2 + 1_N \otimes \phi_2\}.$$

4. In the case that $N_2 = M_1$, the *composition contraction*, given by:

$$c_2 \circ c_1 : \{N_1,\ M_2,\ f_2 f_1,\ g_1 g_2,\ \phi_1 + g_1 \phi_2 f_1\}.$$

Let $N$ be a graded module and let $f : N \to N$ be a morphism of graded modules. The morphism $f$ is called to be *pointwise nilpotent* if for each $x \in N$, $x \neq 0$, there exists a positive integer $n$ such that $f^n(x) = 0$. A *perturbation of a DG-module* $N$ consists in a morphism of graded modules $\delta : N \to N$ of degree $-1$, such that $(d_N + \delta)^2 = 0$. A *perturbation datum* of the contraction

$c : \{N, M, f, g, \phi\}$ is a perturbation $\delta$ of the DG-module $N$ such that the composition $\phi\delta$ is pointwise nilpotent.

The key in the Homological Perturbation Theory is the **Basic Perturbation Lemma** (briefly, BPL) [18, 2, 8, 15], which is an algorithm whose input is a contraction of DG–modules $c : \{N, M, f, g, \phi\}$ and a perturbation datum $\delta$ of $c$ and whose output is a new contraction $c_\delta$. The only requirement is the pointwise nilpotency of the composition $\phi\delta$, that guarantees that the sums involved on the formulas are finite for each $x \in N$.

where $f_\delta, g_\delta, \phi_\delta, d_\delta$ are given by the formulas

$$d_\delta = f \, \delta \, \Sigma_c^\delta \, g; \qquad f_\delta = f \, (1 - \delta \, \Sigma_c^\delta \, \phi); \qquad g_\delta = \Sigma_c^\delta \, g; \qquad \phi_\delta = \Sigma_c^\delta \, \phi;$$

and $\Sigma_c^\delta = \sum_{i \geq 0} (-1)^i \, (\phi\delta)^i$ .

## 3   Compatible Gradings

Let us consider a contraction of DG–modules $c : \{N, M, f, g, \phi\}$ and a perturbation datum for this contraction, $\delta$. The BPL allows one to construct, under certain conditions, a new contraction $c_\delta : \{(N, d_N + \delta), (M, d_M + d_\delta), f_\delta, g_\delta, \phi_\delta\}$. The first motivation of this paper was the search of a general way of determining "classes" of elements for which the final result in the calculation of $d_\delta$ would be zero. However, the results obtained will mean a possibility of reducing costs in any formula containing the composition $f \, (\delta \, \phi)^i$. In fact, the pattern

$$f \, \delta \, \phi \, \delta \, \phi \cdots \delta \, \phi \, \psi,$$

repeatedly appears inside several formulas obtained by BPL (included $d_\delta$) where $\psi$ can be different compositions of morphisms. Obviously, there is a possibility of reducing the complexity of the morphisms with the pattern above by looking for the elements such that the application of $(\delta \, \phi)^i$ to them belongs to the kernel of $f$. This simple idea is the key of the original theory developed in this section.

**Definition 1.** Let $c : \{N, M, f, g, \phi\}$ be a contraction of DG–modules. Let $S$ be a submodule of $N$ and $\mathcal{G} = \{S_i\}_{i \geq 0}$ be a grading on $S$ over the nonnegative integers, that is, $S = \bigoplus_{i \geq 0} S_i$. This grading is called *c–compatible* if it satisfies the following conditions:

$-\ f(S_i) = 0$ for $i \geq 1$;

$-\ \phi(S_i) \subset \displaystyle\bigoplus_{j \geq i+1} S_j.$

The degree of an element with respect to this new grading will be called $\mathcal{G}$–$degree$.

**Definition 2.** Let $c : \{N, M, f, g, \phi\}$ be a contraction of DG–modules, $S \subset N$ a submodule, $\mathcal{G}$ a $c$–compatible grading on $S$ and $\delta$ a perturbation datum for the contraction $c$. The grading $\mathcal{G}$ on $S$ is called to be $(c, \delta)$–$compatible$ if

$$\delta(S_i) \subset \bigoplus_{j \geq i-1} S_j.$$

Taking into account these definitions, it is immediate to state the following proposition.

**Proposition 1.** Let $c : \{N, M, f, g, \phi\}$ be a contraction of DG–modules, $S \subset N$ a submodule, $\delta$ a perturbation datum for $c$ and $\mathcal{G} = \{S_k\}_{k \geq 0}$ a $(c, \delta)$–compatible grading on $S$. Then,

$$f\,(\delta\,\phi)^i(\bigoplus_{k \geq 1} S_k) = 0 \quad \text{for all } i \geq 0.$$

*Proof.* It is obvious, because of the previous definitions, that the composition $\delta\,\phi$ keeps the $\mathcal{G}$–degree the same or increases it, that is,

$$(\delta\,\phi)^i(S_j) \subset \bigoplus_{k \geq j} S_k$$

for any $i$. This way, if $j \geq 1$, $f(\delta\,\phi)^i(S_j) = 0$.

The desired consequence of this result is to find a reduction in complexity of formulas containing the sequence of morphisms $f\,(\delta\,\phi)^i$ when they are applied to elements of the submodule $S$. For instance, the calculus of the perturbed differential $d_\delta$, as well as the projection $f_\delta$ obtained from the application of the BPL to the initial contraction could be reduced: on the one hand, $(\delta\,\phi)$ will only have to be applied to elements of $\mathcal{G}$–degree zero and, on the other hand, there will only have to consider summands of $\phi$ giving place to elements with $\mathcal{G}$–degree 1, at the same time that $\delta$ must provide elements of $\mathcal{G}$–degree 0.

We now wonder what conditions are needed in order to establish compatible gradings for tensor product and composition of contractions. In the case of the tensor product, the compatible gradings have a nice behavior.

**Theorem 1.** Let $c : \{N, M, f, g, \phi\}$ and $c' : \{N', M', f', g', \phi'\}$ be two contractions, let $\mathcal{G} = \{S_k\}_{k \geq 0}$ and $\mathcal{G}' = \{S'_k\}_{k \geq 0}$ be $c$-compatible and $c'$-compatible gradings of respective submodules $S$ of $N$ and $S'$ of $N'$. Then the natural grading on the tensor product $\mathcal{G} \otimes \mathcal{G}' = \{(S \otimes S')_k\}_{k \geq 0}$, with $(S \otimes S')_k = \displaystyle\bigoplus_{i+j=k} (S_i \otimes S'_j)$,

is $c \otimes c'$-compatible.

*Proof.*   – Obviously, $(f \otimes f')(S_i \otimes S'_j) = 0$ if $i + j \geq 1$;

– $(\phi \otimes g' \, f' + 1 \otimes \phi')(S_i \otimes S'_j) \subset \bigoplus_{k \geq i+j+1} (S \otimes S')_k$, since $\phi(S_i) \subset \bigoplus_{k \geq i+1} S_k$ and

$f'(S'_j) = 0$ if $j \geq 1$ and, on the other hand, $\phi'(S'_j) \subset \bigoplus_{k \geq j+1} S'_k$.

We can extend this result, in a natural way, to the tensor module.

**Corollary 1.** *Let $c : \{N, M, f, g, \phi\}$ be a contraction and $\mathcal{G} = \{S_k\}_{k \geq 0}$ a c-compatible grading. Take the tensor module contraction $T(c)$, then the natural grading on the tensor module $T(S) \subset T(N)$, $T(\mathcal{G}) = \{(T(S))_k\}_{k \geq 0}$, with*
$$(T(S))_k = \bigoplus_{i_1 + \cdots + i_n = k} (S_{i_1} \otimes \cdots \otimes S_{i_n}) \text{ is } T(c)\text{-compatible.}$$

The natural grading on the tensor product of two DG–modules is also compatible with the perturbation naturally induced by both perturbations of the initial contractions.

**Proposition 2.** *Under the conditions in the theorem 1, let $\delta$ and $\delta'$ be perturbation data of $c$ and $c'$, respectively, such that the gradings $\mathcal{G}$ and $\mathcal{G}'$ on $S$ and $S'$ are $(c, \delta)$ and $(c', \delta')$-compatible, respectively. Then, the grading $\mathcal{G} \otimes \mathcal{G}'$ on $S \otimes S'$ is $(c \otimes c', \delta \otimes 1 + 1 \otimes \delta')$-compatible.*

**Corollary 2.** *Let $c : \{N, M, f, g, \phi\}$ be a contraction and $\mathcal{G} = \{S_k\}_{k \geq 0}$ a c-compatible grading on $S \subset N$; let $\delta$ be a perturbation datum of $c$ such that the grading $\mathcal{G}$ is $(c, \delta)$-compatible. Then, the grading $T(\mathcal{G})$ on $T(S)$ is $(T(c), \delta_t)$-compatible, where $\delta_t|_{N^{\otimes n}} = \sum_{i=0}^{n-1} 1^{\otimes i} \otimes \delta \otimes 1^{\otimes n-i-1}$.*

Concerning the composition of contractions, we are able to state some conditions under which it is possible to establish a compatible grading for a contraction $c' \circ c$, starting from a contraction $c'$ with a compatible grading.

**Theorem 2.** *Let $c : \{N, M, f, g, \phi\}$ and $c' : \{M, N', f', g', \phi'\}$ be two contractions, let $\mathcal{G}' = \{M_k\}_{k \geq 0}$ be a $c'$-compatible grading on a submodule of $M$. Take a grading $\mathcal{G} = \{N_k\}_{k \geq 0}$ on a submodule of $N$ such that:*

– $f(\bigoplus_{k \geq 0} N_k) \subset \bigoplus_{k \geq 0} M_k$ *and $f$ is a morphism of DG-modules of degree 0 with respect to both gradings;*

– $g(M_k) \subset \bigoplus_{i \geq k} N_i$;

– $\phi(N_k) \subset \bigoplus_{i \geq k+1} N_i$.

*Then the grading $\mathcal{G}$ is $(c' \circ c)$-compatible.*

*Proof.* Recall that

$$(c' \circ c) : \{N, N', f' \, f, g \, g', \phi + g \, \phi' \, f\}.$$

- Of course, $f' f(N_k) \subset f'(M_k) = 0$ if $k \geq 1$, since $\mathcal{G}'$ is $c'$–compatible;

- Take now $\phi + g\,\phi'\,f$. We must check that $g\,\phi'\,f(N_k) \subset \bigoplus_{i>k+1} N_i$, but this is obvious due to the conditions imposed on $f$ and $g$ and the fact that $\mathcal{G}'$ is $c'$–compatible.

Notice that, given a contraction, one may construct lots of compatible gradings, so an important task is to determine which is a good one for our purposes, depending on the formula we want to reduce. The improvements achieved in each case will depend on the compatible grading considered.

# 4  Application 1: The $A_\infty$-Structure of the Homology of $Q_p(u, 2n)$

Here we present an example of using compatible gradings to reduce complexity in the computation of some morphisms obtained by perturbation theory. We study the $A_\infty$-coalgebra structure of the homology $H_*(Q_p(u, 2n); \mathbf{Z})$ of a truncated polynomial algebra in the ring $\mathbf{Z}$.

An $A_\infty$-coalgebra is a DG-module $(M, \Delta_1)$ endowed with a family of morphisms of graded modules

$$\Delta_i : M \to M^{\otimes i}$$

of degree $i - 2$ such that, for $n \geq 1$,

$$\sum_{n=1}^{i} \sum_{k=0}^{i-n} (-1)^{n+k+nk} (1^{\otimes i-n-k} \otimes \Delta_n \otimes 1^{\otimes k}) \Delta_{i-n+1} = 0. \tag{2}$$

To study the structure of the homology we need to consider the contraction

$$c_{BQ} : \{\overline{B}(Q_p(u, 2n)), E(v, 2n+1) \otimes \Gamma(w, 2np+2), f_{BQ}, g_{BQ}, \phi_{BQ}\}$$

From now on, $E(v, 2n+1) \otimes \Gamma(w, 2np+2)$ will be denoted by $E \otimes \Gamma$ as well as $\bar{B}(Q_p(u, 2n))$ by $\bar{B}(Q_p)$, for short.

Then, an $A_\infty$–coalgebra structure is induced on $E \otimes \Gamma$, via the tensor trick (see [9, 12, 11]) and the BPL (using as a perturbation datum the cosimplicial differential $d_{cos}$). In fact, it is possible to construct a new contraction

$$\overline{\Omega}\bar{B}(Q_p) \Rightarrow \widetilde{\Omega}(E \otimes \Gamma),$$

where $\widetilde{\Omega}(E \otimes \Gamma) = (Ts^{-1}(E \otimes \Gamma), \tilde{d})$ and the morphism $\tilde{d}$, obtained by the perturbation process, induces an $A_\infty$–coalgebra structure $(\Delta_1, \Delta_2, \Delta_3, \ldots)$ on $(E \otimes \Gamma)$. Our aim is to use a compatible grading that allows a complexity reduction for the morphisms $\Delta_i$.

We denote an element of $\bar{B}(Q_p)$ of the form $[u^{r_1}|\ldots|u^{r_m}]$ by $[r_1|\ldots|r_m]$, where $0 \leq r_i < p$.

The explicit morphisms of the contraction $c_{BQ}$ are described below:

- $f_{BQ} : \bar{B}(Q_p) \to E \otimes \Gamma$

$$f_{BQ}[r_1|t_1|\ldots|r_m|t_m] = \{\textstyle\prod_{k=1}^{m} \delta_{p,r_k+t_k}\}\, \gamma_m(w),$$

$$f_{BQ}[l|r_1|t_1|\ldots|r_m|t_m] = \delta_{1,l}\,\{\textstyle\prod_{k=1}^{m} \delta_{p,r_k+t_k}\}\, v \otimes \gamma_m(w),$$

where the symbols $\delta_{i,j}$ are defined by: $\delta_{i,j} = \begin{cases} 0 & \text{if } i \neq j \\ 1 & \text{if } i = j \end{cases}$

- $g_{BQ} : E \otimes \Gamma \to \bar{B}(Q_p))$ is defined over the algebra generators as follows:

$$g_{BQ}(v) = [1], \quad g_{BQ}(\gamma_k(w)) = [1|p-1| \overset{k \text{ times}}{\cdots} |1|p-1].$$

- $\phi_{BQ} : \bar{B}(Q_p) \to \bar{B}(Q_p)$ is defined by:

$$\phi_{BQ}[\,] = 0;$$

$$\phi_{BQ}[x] = -[1|r-1] \quad 1 < x < p;$$

$$\phi_{BQ}[x|y] = -[1|x-1|y];$$

$$\phi_{BQ}([x|y]|z) = -[1|x-1|y]|z - \delta_{p,x+y}\,[1|p-1]|\,\phi(z)$$

where $z \in \bar{B}(Q_p)$ and $|$ denotes the juxtaposing product.

Let us consider the submodule $S$ of $\bar{B}(Q_p(u,2n))$ generated by the elements $[a_1|a_2|\cdots|a_r]$ such that either $a_{2i+1} = 1$ for all $i$ or $a_{2i} = 1$ for all $i$.

We now construct a grading, $\mathcal{G} = \{S_k\}_{k\geq 0}$, over $S$ in the search of a $c_{BQ}$–compatible grading. Take the submodule $S_k \subset S$ described bellow:

- $S_0$ is generated by the elements of the form $[p-1|1|p-1|1\cdots]$, of any simplicial dimension; $[a_1|1|a_2|1\cdots]$, with $a_i = p-1$ except for one of them, $a_j$, which satisfies $1 \leq a_j \leq p-2$; $[1|p-1|1|p-1|1\cdots]$, of any simplicial dimension.
- $S_k$, for $k \geq 1$, is generated by the elements of

  type I: $[1|a_1|\cdots|1|a_r|1]$ or $[1|a_1|\cdots|1|a_r]$, where at least one component $a_i \neq 1$ and there are $k$ components (at even positions) satisfying $1 \leq a_i \leq p-2$;

  type II: $[a_1|1|a_2|\cdots|1]$ or $[a_1|1|a_2|\cdots|a_r]$ where at least one component $a_i \neq 1$ and there are $k+1$ components (at odd positions) satisfying $1 \leq a_i \leq p-2$;

  type III: $[1|1|\cdots|1]$ with simplicial dimension $2k$ or $2k+1$.

Then, it is easy to prove that $\mathcal{G}$ is $c_{BQ}$–**compatible**:

- Obviously, $f_{BQ}(S_k) = 0$ for any $k \geq 1$;

$-\ \phi(S_i) \subset \bigoplus_{j \geq i+1} S_j$. More specifically, it is easy to prove, by induction, that
$\phi(S_i) \subset S_{i+1}$. Moreover, that $\phi(\text{elements of type II}) = (\text{elements of type I})$ and that $\phi(\text{ elements of type I}) = 0 = \phi(\text{ elements of type III})$.

The grading $\mathcal{G}$ on $S$ induces in a natural way, a grading over $Ts^{-1}(S)$, $T(\mathcal{G})$, as indicated in corollary 1, that is $Ts^{-1}c_{BQ}$–**compatible**.

Moreover, the cosimplicial differential, $d_{cos}$, which depends on the coproduct on $\bar{B}(Q_p)$ (see formula 1), is a perturbation datum for the latter contraction and satisfies

$$d_{cos}(T(S)_j) \subset \oplus_{i \geq j-1}T(S)_i.$$

In fact, it is easy to check that $d_{cos}(T(S)_j) \subset T(S)_{j-1} \oplus T(S)_j$.

All these results lead to the following theorem.

**Theorem 3.** *Let $T(\mathcal{G})$ be the grading naturally induced on $Ts^{-1}(S)$ by corollary 1, being $\mathcal{G}$ the grading described above. Then $T(\mathcal{G})$ is $(Ts^{-1}c_{BQ}, d_{cos})$-compatible.*

Now, we are in conditions of analyzing the consequences of having a $(Ts^{-1}c_{BQ}, d_{cos})$-compatible grading in the computation of the $A_\infty$–coalgebra structure on $E \otimes \Gamma$.

The differential obtained by the BPL on $Ts^{-1}(E \otimes \Gamma)$ is

$$\tilde{d} = \sum_{i \geq 0}(-1)^i T(f)d_{cos}(T(-\phi)d_{cos})^i T(g),$$

and the morphisms $\Delta_i : E \otimes \Gamma \rightarrow (E \otimes \Gamma)^{\otimes i}$, providing an $A_\infty$–coalgebra structure on $E \otimes \Gamma$, are given by

$$\Delta_i = (-1)^{[i/2]} \uparrow^{\otimes i} f^{\otimes i} (d_{cos} T(-\phi))^{i-2} d_{cos} g \downarrow \qquad (3)$$

Notice that $\text{Im } g_{BQ} \subset S_0$ and that according to proposition 1,

$$T(f)\,(d_{cos} T(-\phi))^i (\bigoplus_{k \geq 1}(T(S))_k) = 0 \quad \text{ for all } i \geq 0,$$

so we can conclude that, since $T(-\phi)(T(S)_k) \subset T(S)_{k+1}$, we only must consider the summands of $d_{cos}$ such that

$$d_{cos}(T(S)_k) \subset T(S)_{k-1}.$$

This means that any time we apply $d_{cos}$ to an element of cosimplicial dimension $m$, one can consider only **one summand in formula 1 instead of $m$**. That summand is the one that applies $\Delta_{\bar{B}}$ on the factor on which $\phi$ has just been applied. Moreover, since the output of $\phi$ is always a sum of elements of type I, we only need to consider the half of the summands of $\Delta_{\bar{B}}$ (those that decrease the degree by one). So, we could express the "reduced" formula of $\Delta_i$ by

$$\bar{\Delta}_i = (-1)^{[i/2]} \uparrow^{\otimes i} f^{\otimes i} (\bar{d}_{cos} T(-\phi))^{i-2} d_{cos} g \downarrow \qquad (4)$$

where
$$\bar{d}_{cos} = 1^{\otimes k} \otimes \downarrow^{\otimes 2} \bar{\Delta}_B \uparrow \otimes 1^{\otimes n - k - 1}$$

and $k$ depends on the factor on which $\phi$ has been applied in $T(-\phi)$. Then,

$$\bar{\Delta}_B([a_1|\cdots|a_r]) = \sum_{i=1}^{\lfloor r/2 \rfloor} [a_1|\cdots|a_{2i-1}] \otimes [a_{2i}|\cdots|a_r].$$

Finally, we expose some examples in order to illustrate the improvements achieved in complexity. The calculations have been made (up to signs) using MATHEMATICA 4.0, in which we have implemented the formulas of $\Delta_i$ and $\bar{\Delta}_i$, in a Pentium IV 1,6GHz 512MB RAM.

Firstly, we will express the time used in the computation of $\Delta_3(\gamma_k(w))$ for different values of $k$ in the case $p = 3$.

| Time used (Seconds) | $k = 5$ | $k = 10$ | $k = 20$ | $k = 30$ |
|---|---|---|---|---|
| $\Delta_3$ | 0.15 | 1.072 | 50.282 | $882,959$ |
| $\bar{\Delta}_3$ | 0.07 | 0.391 | 5.859 | 77.392 |

The following table shows the time used in computing $\Delta_i(\gamma_5(w))$ versus $\bar{\Delta}_i(\gamma_5(w))$ for different values of $i$ and $p$.

| Time used (seconds) | $i = 3 = p$ | $i = 4 = p$ | $i = 5 = p$ | $i = 6 = p$ |
|---|---|---|---|---|
| $\Delta_i$ | 0.15 | 1.081 | 16.373 | 226.456 |
| $\bar{\Delta}_i$ | 0.07 | 0.311 | 2.043 | 16.213 |

In the last table we expose the number of summands at different stages of the computation of $\Delta_6(\gamma_5(w))$ as well as $\bar{\Delta}_6(\gamma_5(w))$ in the case $p = 6$.

| Number of summands | $i = 1$ | $i = 2$ | $i = 3$ | $i = 4$ |
|---|---|---|---|---|
| $(d_{cos} T(-\phi))^i d_{cos} g(\langle \gamma_5(w) \rangle)$ | 135 | 945 | 4410 | 15876 |
| $(\bar{d}_{cos} T(-\phi))^i d_{cos} g(\langle \gamma_5(w) \rangle)$ | 55 | 315 | 1274 | 4116 |

# 5  Application 2: On the 1-Homological Model of a Commutative Connected DG–Algebra

In [4] and [14] a strategy was developed, called "inversion theory", with the goal of improving the computation of some formulas (obtained by the BPL) involved in the 1–homological model of a commutative DGA–algebra (see [1]). In this section, we will see that this theory fits perfectly in the framework developed before, so we will briefly review the concepts given in the inversion theory under this new viewpoint to realize that, in fact, it can be considered as an application of the general methodology of this paper.

Every commutative DGA-algebra $A$ is quasi-isomorphic (there is a homomorphism that induces an isomorphism in homology) to a twisted tensor product (TTP) of exterior and polynomial algebras, that is, a tensor product $\otimes_{i \in I} A_i$, each $A_i$ being an exterior or a polynomial algebra, endowed with an additional differential structure $\rho$ (see [3]).

Here we will restrict to the case of commutative connected DG–algebras, in order to deal with simpler formulas.

Take, then, a TTP of exterior and polynomial algebras $A = (\otimes_{i \in I} A_i, \rho)$. In [1] a process is described consisting in composition and tensor product of contractions followed by the application of the BPL in order to obtain a new contraction called a 1-*homological model* for $A$:

$$c_\delta : \{(\bar{B}(\otimes_{i \in I} A_i), \delta), \ (\otimes_{i \in I} hBA_i, d_\delta), \ f_\delta, \ g_\delta, \ \phi_\delta\}$$

where $\delta$ is a perturbation on $\bar{B}(\otimes_{i \in I} A_i)$ induced by $\rho$ and $hBA_i$ is an exterior or a divided polynomial algebra, depending on whether $A_i$ was a polynomial or an exterior algebra. The search of a $(c, \delta)$–compatible grading on $\bar{B}(\otimes_{i \in I} A_i)$ is motivated by the high computational cost of $d_\delta$, which is of exponential nature, since shuffles are involved in the formulas.

The construction of $c_\delta$ makes use of three basic contractions:

– The contraction given in [7]:

$$c_{BP} : \{\bar{B}(P(v, 2n)), E(s(v), 2n + 1), f_{BP}, g_{BP}, \phi_{BP}\}.$$

If an element $[v^{r_1} | \cdots | v^{r_m}]$ is denoted by $[r_1 | \cdots | r_m]$ for short, then,

$$f_{BP}([r]) = \begin{cases} 0 & \text{if } r \neq 1 \\ s(v) & \text{if } r = 1 \end{cases} ; \quad f_{BP}([r_1 | \cdots | r_m]) = 0$$

$$g_{BP}(sv) = [1]; \quad \phi_{BP}([r_1 | \cdots | r_m]) = [1 | r_1 - 1 | \cdots | r_m].$$

– The isomorphism of DG-algebras (also described in [7]):

$$c_{BE} : \{\bar{B}(E(u, 2n + 1)), \Gamma(s(u), 2n + 2), f_{BE}, g_{BE}, 0\},$$

where

$$f_{BE}([u | \overset{m \text{ times}}{\cdots} | u]) = \gamma_m(s(u)); \quad g_{BE}(\gamma_m(s(u))) = [u | \overset{m \text{ times}}{\cdots} | u].$$

– Let $A$ and $A'$ be two commutative connected $DG$–algebras. There is a contraction $c_{B\otimes} : \{\bar{B}(A \otimes A'), \bar{B}(A) \otimes \bar{B}(A'), f_{B\otimes}, g_{B\otimes}, \phi_{B\otimes}\}$ (see [7]), whose formulas (for the connected case) are recalled here:

- $f_{B\otimes}$ is null except for the case

$$f_{B\otimes}[a_1 \otimes 1| \cdots |a_i \otimes 1|1 \otimes a'_{i+1}| \cdots |1 \otimes a'_n] = [a_1| \cdots |a_i] \otimes [a'_{i+1}| \cdots |a'_n].$$

- $g_{B\otimes}([a_1| \cdots |a_n] \otimes [a'_1| \cdots |a'_m]) = [a_1| \cdots |a_n] \star [a'_1| \cdots |a'_m].$

- $\phi_{B\otimes}([a_1 \otimes a'_1| \cdots |a_{n-q} \otimes a'_{n-q}|a'_{n-q+1}| \cdots |a'_n])$

$$= \sum_{p=0}^{n-q-1} \sum_{\pi} \pm [a_1 \otimes a'_1| \cdots |a_{\bar{n}-1} \otimes a'_{\bar{n}-1}| \tag{5}$$
$$(a'_{\bar{n}} *_{A'} \cdots *_{A'} a'_{n-q})|c_{\pi(0)}| \cdots |c_{\pi(p+q)}],$$

where $\pi$ runs over the $\{(p+1, q)$-shuffles$\}$, $\bar{n} = n - p - q$ and

$$(c_0, \ldots, c_{p+q}) = (a_{\bar{n}}, \ldots, a_{n-q}, a'_{n-q+1}, \ldots a'_n).$$

Notice that $g_{B\otimes}$ and $\phi_{B\otimes}$ works in exponential time due to the shuffles involved.

So, for a commutative connected DG–algebra $(A \otimes A', \rho)$, being $A$ and $A'$ an exterior or polynomial algebra, one can construct, by composition and tensor product of the previous contractions, the contraction $c = (c_{BA} \otimes c_{BA'}) \circ c_{B\otimes}$,

$$(f, g, \phi) : \bar{B}(A \otimes A') \Rightarrow \bar{B}(A) \otimes \bar{B}(A') \Rightarrow hBA \otimes hBA'$$

where $hBA$ as well as $hBA'$ are either a polynomial or a divided polynomial algebra depending on whether $A$ and $A'$ were, respectively, an exterior or a polynomial algebra. The morphisms are given by

$$f = (f_{BA} \otimes f_{BA'})f_{B\otimes}$$
$$g = g_{B\otimes}(g_{BA} \otimes g_{BA'})$$
$$\phi = \phi_{B\otimes} + g_{B\otimes}(\phi_{BA} \otimes g_{BA'}f_{BA'} + 1_{BA} \otimes \phi_{BA'})f_{B\otimes}$$

Take the contraction $c_{BP}$. Consider the grading $\mathcal{G}_{BP} = \{G_k^{BP}\}_{k\geq 0}$ by which $G_k^{BP}$ is the submodule generated by the elements of the form $[r_1| \cdots |r_{k+1}]$ (that is, those of simplicial degree $k + 1$). It is easy to check that $\mathcal{G}_{BP}$ **is a** $c_{BP}$–**compatible grading**.

Consider, now, the contraction $c_{BE}$. The grading $\mathcal{G}_{BE} = \{G_k^{BE}\}_{k\geq 0}$ with $G_0^{BE} = \bar{B}(E(u, 2n+1))$ and $G_k^{BE} = 0$ for $k \geq 0$ is, trivially, a $c_{BE}$–**compatible grading**.

Then, $\mathcal{G}_{BA} \otimes \mathcal{G}_{BA'}$ **is a** $c_{BA} \otimes c_{BA'}$–**compatible grading**, by theorem 1.

Then, we must look for a grading on $\bar{B}(A \otimes A')$ (or a submodule of it) such that the conditions stated in theorem 2 are satisfied, in order to have a $c$–compatible grading.

Take the notation $\mathcal{G}_{BA} \otimes \mathcal{G}_{BA'} = \{G_k^{\otimes B}\}_{k\geq 0}$, meanwhile the desired grading on $\bar{B}(A \otimes A')$ will be denoted by $\mathcal{G}_{B\otimes} = \{G_k^{B\otimes}\}_{k\geq 0}$.

Let us consider $G_k^{B\otimes}$ as the submodule of $\bar{B}(A \otimes A')$ generated by the elements with $k$ *inversions* as defined in [14]: Let $\bar{a} = [a_1 \otimes a'_1|a_2 \otimes a'_2| \cdots |a_n \otimes a'_n]$ be an homogeneous element of $\bar{B}(A \otimes A')$. The component $a_i \otimes a'_i$ is said to be an inversion if one of the the following cases occurs:

- $a_i = 1$ and there exists an index $j > i$ with $a_j \in \bar{A}$ (we say that $a_i$ is a $\otimes$–inversion);
- $A'$ is a polynomial algebra and $a_{i-1} = 1 = a_i = \cdots = a_n$ (we say that $a_i$ is a $p$–inversion).
- $A$ is a polynomial algebra, $a_i \in \bar{A}$ and there exists an index $j > i$ such that $a_j \in \bar{A}$ (we say that $a_i$ is a $p1$–inversion).

This grading verifies the conditions in theorem 2:

- $f_{B\otimes}$ is a morphism of degree 0 with respect to the gradings $\mathcal{G}_{B\otimes}$ and $\mathcal{G}_{BA} \otimes \mathcal{G}_{BA'}$: This is obvious, since $f_{B\otimes}(\bar{a}) \neq 0$, only for the elements of the form $\bar{a} = [a_1|\cdots|a_j|a_1'|\cdots|a_k']$, with $a_i \in \bar{A}$ and $a_i' \in \bar{A}'$, which has the same degree in $\mathcal{G}_{B\otimes}$ than $[a_1|\cdots|a_j] \otimes [a_1'|\cdots|a_k']$ in $\mathcal{G}_{BA} \otimes \mathcal{G}_{BA'}$.
- $g_{B\otimes}(G_k^{\otimes B}) \subset \bigoplus_{i \geq k} G_i^{B\otimes}$: this is easy to see since $g_{B\otimes}$ makes the shuffle product.
- $\phi(G_k^{B\otimes}) \subset \bigoplus_{i \geq k+1} G_i^{B\otimes}$:

Recall that $\phi = \phi_{B\otimes} + g_{B\otimes}(\phi_{BA} \otimes g_{BA'} f_{BA'} + 1_{BA} \otimes \phi_{BA'})f_{B\otimes}$ and that either $\phi_{B\otimes}$ or $\phi_{BP}$ always produce at least one more inversion [4, 14].

So, $\mathcal{G}_{B\otimes}$ **is a** $c$–**compatible grading** on $B(A \otimes A')$.

Now, take a perturbation $\rho$ of the tensor product $A \otimes A'$, then $\rho$ produces a perturbation

$$\delta = -\sum_{i=0}^{n-1} 1^{\otimes i} \otimes \uparrow \rho \downarrow \otimes 1^{\otimes n-i-1}$$

on $\bar{B}(A \otimes A')$. Then we can state the following result.

**Theorem 4.** *The grading $\mathcal{G}_{B\otimes}$ on $B(A \otimes A')$ described above is $(c, \delta)$–compatible.*

*Proof.* We must check that $\delta(G_k^{B\otimes}) \subset \bigoplus_{i \geq k-1} G_i^{B\otimes}$, for any $k \geq 1$. The key of the proof lays in the fact that each summand of $\delta$ acts only on one component $a_i \otimes a_i'$ of the element $[a_1 \otimes a_1'|\cdots|a_n \otimes a_n']$. So, at most, only one inversion can disappear. On the contrary, if $\rho$ is applied to a component which is not implied in any inversion, the summand obtained could have an amount of inversions greater or equal to the original one.

We then have a $(c, \delta)$–compatible grading, what means that

$$f(\delta \phi)^i \left( \bigoplus_{k \geq 1} G_k^{B\otimes} \right) = 0 \quad \text{for all } i \geq 0.$$

This way, we can conclude that, in order to compute $d_\delta$, whose formula is

$$d_\delta = \sum_{i \geq 0} (-1)^i f \delta (\phi \delta)^i g,$$

we can ignore those summands of $\phi$ that increase more than by one the degree of the element. This conclusion leads to the following result already proved by inversion theory.

**Corollary 3.** *[4, 14] The formula for $\phi$, involved in the perturbed differential, $d_\delta$, of the 1–homological model of a commutative connected DG–algebra $(A \otimes A', \rho)$ can be reduced to the following one:*

$$\phi = \bar{\phi}_{B\otimes} + \bar{g}_{B\otimes}(\phi_{BA} \otimes g_{BA'}f_{BA'} + 1 \otimes \phi_{BA'})f_{B\otimes},$$

*where*

$$- \bar{\phi}_{B\otimes}([a_1 \otimes a_1'| \cdots |a_n \otimes a_n'])$$

$$= \sum_{0 \leq p \leq n-q-1 \leq n-1} \pm [a_1 \otimes a_1'| \cdots |a_{\bar{n}-1} \otimes a_{\bar{n}-1}'|(a_{\bar{n}}' *_{A'} \cdots *_{A'} a_{n-q}')$$

$$|a_{\bar{n}}| \cdots |a_{n-q}|a_{n-q+1}'| \cdots |a_n'].$$

$$- \bar{g}_{B\otimes}([a_1| \cdots |a_n] \otimes [a_1'| \cdots |a_m']) = [a_1| \cdots |a_n|a_1'| \cdots |a_m'].$$

This way, the formula of $\phi_{B\otimes}$ comes out to be of quadratic instead of exponential order. On the other hand, the formula for $g_{B\otimes}$ is reduced to 1 summand in contrast with the original $\begin{pmatrix} m+n \\ n \end{pmatrix}$.

These results can be extended, recursively, to the general case of a commutative connected DG–algebra factored as a twisted tensor product $(\otimes_{i=1}^n A_i, \rho)$, being $A_i$ an exterior or polynomial algebra, $i = 1, 2, \ldots, n$ and $\rho$ a differential, such that $\rho(A_k) \subset \otimes_{i=1}^k A_i$.

# References

1. Álvarez, V., Armario, A., Real, P., Silva, B.: HPT and computability of Hochschild and cyclic homologies of commutative DGA–algebras. In: Conference on Secondary Calculus and Cohomological Physics, Moscow. EMIS Electronic Proceedings, http://www.emis.de/proceedings, 1997
2. Brown, R.: The twisted Eilenberg-Zilber theorem, Celebrazioni Archimedae del Secolo XX, Simposio di Topologia (1967), 34–37
3. Burghelea, D., Vigué Poirrier, M.: Cyclic homology of commutative algebras I, Lecture Notes in Mathematics, Algebraic Topology Rational Homotopy **1318**, 51–72 (1986)
4. C.H.A.T.A. group. Computing "small" 1–homological models for Commutative Differential Graded Algebras. In: Proc. Third Workshop on Computer Algebra in Scientific Computing, Springer-Verlag, Berlin, Heidelberg (2000) 87–100
5. Cartan, H., Eilenberg, S.: Homological Algebra, Princeton Univ. Press, 1956
6. Eilenberg, S., MacLane, S.: *On the groups $H(\pi, n)$, I*, Annals of Math. **58** (1953), 55–106
7. Eilenberg, S., MacLane, S.: On the groups $H(\pi, n)$, II, Annals of Math. **60** (1954), 49–139
8. Gugenheim, V.K.A.M.: On the chain complex of a fibration, Illinois J. Math. **3** (1972), 398–414
9. Gugenheim, V.K.A.M.: *On Chen's iterated integrals*, Illinois J. Math. (1977), 703–715

10. Gugenheim, V.K.A.M., Lambe, L.: *Perturbation theory in Differential Homological Algebra, I*, Illinois J. Math. **33** (1989), 56–582
11. Gugenheim, V.K.A.M., Lambe, L., Stasheff, J.: Perturbation theory in differential homological algebra, II, Illinois J. Math. **35**(3) (1991), 357–373
12. Gugenheim, V. K. A. M., Stasheff, J.: On perturbations and $A_\infty$–structures, Bull. Soc. Math. Belg. **38** (1986), 237–246
13. Huebschmann, J., Kadeishvili, T.: Small models for chain algebras, Math. Zeit. **207** (1991), 245–280
14. Jiménez, M.J., Real, P.: "Coalgebra" structures on 1–homological models for commutative differential graded algebras. In: Proc. Fourth Workshop on Computer Algebra in Scientific Computing, Springer-Verlag, Berlin, Heidelberg (2001) 347–361
15. Lambe, L.A., Stasheff, J.: Applications of perturbation theory to iterated fibrations, Manuscripta Math. **58** (1987), 363–376
16. MacLane, S.: Homology, Classics in Mathematics, Springer-Verlag, Berlin, 1995. Reprint of the 1975 edition
17. Real, P.: Homological perturbation theory and associativity, Homology, Homotopy and Applications **2**(5) (2000), 51–88
18. Shih, W.: Homologie des espaces fibrés, Inst. Hautes Etudes Sci. **13** (1962), 93–176
19. Weibel, C.A.: An introduction to Homological Algebra, Cambridge studies in advanced mathematics **38**, Cambridge University Press, 1994

# Solving Algorithmic Problems on Orders and Lattices by Relation Algebra and RelView

Rudolf Berghammer

Institut für Informatik, Universität Kiel
Olshausenstraße 40, D-24098 Kiel

**Abstract.** Relation algebra is well suited for dealing with many problems on ordered sets. Introducing lattices via order relations, this suggests to apply it and tools for its mechanization for lattice-theoretical problems, too. We combine relation algebra and the specific purpose Computer Algebra system RelView to solve some algorithmic problems.

## 1  Introduction

An ordered set $(X, \sqsubseteq)$ consists of a non-empty set $X$ and a partial order relation $\sqsubseteq$ on $X$. It is a *lattice* if every pair of elements $x, y \in X$ has a greatest lower bound $x \sqcap y$ and a least upper bound $x \sqcup y$. Lattices play an important role in many areas of computer science. These include, e.g., knowledge representation, data mining, information retrieval, cryptography and cryptanalysis, static program analysis, logic programming, algorithmics, and models of computation.

Relation algebra [16, 14] generalizes lattices since it additionally uses complements, compositions with identities, and transpositions. Its use in computer science is mainly due to the fact that many datatypes can be modeled via relations, many problems on them can be naturally specified by relation-algebraic expressions and formulae, and, therefore, many solutions reduce to relation-algebraic computations. Finite relations can be implemented very efficiently. At Kiel University we have developed a Computer Algebra system for the manipulation and visualization of relations and for relational programming, called RelView [1, 5]. It is written in C, uses binary decision diagrams for representing relations (see [11, 4, 12]), and makes full use of the X-windows GUI. The main purpose of RelView is the evaluation of relation-algebraic expressions, which are constructed from the relations of its workspace using pre-defined operations and tests, user-defined relational functions, and user-defined relational programs. Relational functions are defined as is customary in mathematics, where the right-hand sides are relation-algebraic expressions over the relations of the system's workspace and the formal parameters. A relational program is much like a function procedure in Modula 2, except that it only uses relations as datatype.

As demonstrated e.g., in [14], relation algebra is well suited for dealing with many problems concerning order relations in a component-free manner. Taking ordered sets as a starting point for introducing lattices (instead of algebras having two binary operations $\sqcap$ and $\sqcup$), lattices are nothing else than specific

V.G. Ganzha, E.W. Mayr, and E.V. Vorozhtsov (Eds.): CASC 2006, LNCS 4194, pp. 49–63, 2006.
© Springer-Verlag Berlin Heidelberg 2006

partial order relations. This suggests to apply the formal apparatus of relation algebra and tools for its mechanization for lattice-theoretical problems, too. First examples for this approach are [6, 5], where relation algebra and the RELVIEW system are combined for computing and visualizing cut completions and concept lattices. The material presented in this paper is a continuation of [6, 5]. Having a Computer Algebra system for computations on orders and lattices is potentially very useful and in the following we also want to demonstrate that the RELVIEW system is especially suited to this task. The remainder of the paper is organized as follows. First we collect some technical preliminaries of relation algebra and the relation-algebraic treatment of orders, extremal elements, and Hasse-diagrams in Sections 2 and 3. Then we concentrate in the main part (Sections 4 until 7) on some applications in lattice theory. In doing so, we also want to illustrate the advantages of RELVIEW when using it for visualization purposes. Section 8 contains some concluding remarks.

## 2    Relational Preliminaries

We write $R : X \leftrightarrow Y$ if $R$ is a relation with domain $X$ and range $Y$, i.e., a subset of $X \times Y$. If the sets $X$ and $Y$ of $R$'s *type* $X \leftrightarrow Y$ are finite and of size $m$ and $n$, respectively, we may consider $R$ as a Boolean $m \times n$ matrix. Since this interpretation is well suited for many purposes and also used by REL-VIEW, in the following we often use matrix terminology and notation. Especially, we speak about rows and columns and write $R_{x,y}$ instead of $\langle x, y \rangle \in R$. We assume the reader to be familiar with the basic operations on relations, viz. $R^{\mathsf{T}}$ (*transposition*), $\overline{R}$ (*complement*), $R \cup S$ (*union*), $R \cap S$ (*intersection*), and $RS$ (*composition*), the predicate $R \subseteq S$ (*inclusion*), and the special relations $\mathsf{O}$ (*empty relation*), $\mathsf{L}$ (*universal relation*), and $\mathsf{I}$ (*identity relation*).

By $syq(R, S) := \overline{R^{\mathsf{T}}\overline{S}} \cap \overline{\overline{R}^{\mathsf{T}}S} : Y \leftrightarrow Z$ the *symmetric quotient* of two relations $R : X \leftrightarrow Y$ and $S : X \leftrightarrow Z$ is defined. Many properties of this construct can be found in [14]. In this paper we only need that for all $y \in Y$ and $z \in Z$

$$syq(R, S)_{y,z} \iff \forall\, x : R_{x,y} \leftrightarrow S_{x,z}.$$

Given a product $X \times Y$, there are two projection functions which decompose a pair $u = \langle u_1, u_2 \rangle$ into its first component[1] $u_1$ and its second component $u_2$. We consider instead of these functions the corresponding *projection relations* $\pi : X \times Y \leftrightarrow X$ and $\rho : X \times Y \leftrightarrow Y$ such that for all $u \in X \times Y$, $x \in X$, and $y \in Y$ we have $\pi_{u,x}$ iff $u_1 = x$ and $\rho_{u,y}$ iff $u_2 = y$. Projection relations enable us to specify the well known pairing operation of functional programming relation-algebraically as follows: For relations $R : Z \leftrightarrow X$ and $S : Z \leftrightarrow Y$ we define their *fork* $[R, S] : Z \leftrightarrow X \times Y$ by $[R, S] := R\pi^{\mathsf{T}} \cap S\rho^{\mathsf{T}}$. Component-wisely we then have for all $z \in Z$ and $u \in X \times Y$ that $[R, S]_{z,u}$ iff $R_{z,u_1}$ and $S_{z,u_2}$.

---

[1] We denote the first component of a pair $u$ by $u_1$ and the second component of $u$ by $u_2$. In such a case we also write $\sqcap u$ instead of $u_1 \sqcap u_2$ and $\sqcup u$ instead of $u_1 \sqcup u_2$.

There are some relation-algebraic possibilities to model sets. Our first modeling uses *vectors*, which are relations $v$ with $v = v\mathsf{L}$. Since for a vector the range is irrelevant, we consider in the following mostly vectors $v : X \leftrightarrow \mathbf{1}$ with a specific singleton set $\mathbf{1} := \{\bot\}$ as range and omit in such cases the second subscript, i.e., write $v_x$ instead of $v_{x,\bot}$. Such a vector can be considered as a Boolean matrix with exactly one column, i.e., as a Boolean column vector, and *represents* the subset $\{x \in X \mid v_x\}$ of $X$. A non-empty vector $v$ is said to be a *point* if $vv^\mathsf{T} \subseteq \mathsf{I}$, i.e., $v$ is *injective*. This means that it represents a singleton subset of its domain or an element from it if we identify a singleton set with the only element it contains. In the Boolean matrix model, hence, a point $v : X \leftrightarrow \mathbf{1}$ is a Boolean column vector in which exactly one component is true.

As a second way to model sets we will apply the relation-level equivalents of the set-theoretic symbol $\in$, i.e., *membership-relations* $\mathsf{M} : X \leftrightarrow 2^X$. These specific relations are defined by demanding for all $x \in X$ and $Y \in 2^X$ that $\mathsf{M}_{x,Y}$ iff $x \in Y$. A Boolean matrix implementation of $\mathsf{M}$ requires exponential space. However, in [11, 4, 12] an implementation of $\mathsf{M}$ using reduced ordered binary decision diagrams is presented, where the number of vertices is linear in the size of $X$. This implementation is part of RELVIEW.

Finally, we will use injective functions for modeling sets. Given an injective function $\imath : Y \to X$, we may consider $Y$ as a subset of $X$ by identifying it with its image under $\imath$. If $Y$ is actually a subset of $X$ and $\imath$ is given as a relation of type $Y \leftrightarrow X$ such that $\imath_{y,x}$ iff $y = x$ for all $y \in Y$ and $x \in X$, then the vector $\imath^\mathsf{T}\mathsf{L} : X \leftrightarrow \mathbf{1}$ represents $Y$ as a subset of $X$ in the sense above. Clearly, the transition in the other direction is also possible, i.e., the generation of a relation $inj(v) : Y \leftrightarrow X$ from the vector representation $v : X \leftrightarrow \mathbf{1}$ of $Y \subseteq X$ such that for all $y \in Y$ and $x \in X$ we have $inj(v)_{y,x}$ iff $y = x$. A combination of such relations with membership relations allows a *column-wise representation* of sets of subsets. More specifically, if the vector $v : 2^X \leftrightarrow \mathbf{1}$ represents a subset $\mathfrak{S}$ of $2^X$ in the sense above, then for all $x \in X$ and $Y \in \mathfrak{S}$ we get the equivalence of $(\mathsf{M}\,inj(v)^\mathsf{T})_{x,Y}$ and $x \in Y$. This means that the elements of $\mathfrak{S}$ are represented precisely by the columns of the relation $\mathsf{M}\,inj(v)^\mathsf{T} : X \leftrightarrow \mathfrak{S}$.

The choice of a point $point(v)$ contained in a vector $v \neq \mathsf{O}$ is fundamental for relational programming since it corresponds to the choice of an element from a non-empty set. A generalization to arbitrary relations is $atom(R)$, which yields an atom included in $R \neq \mathsf{O}$, i.e., a subrelation of $R$ with exactly one pair. The implementation of these choices in RELVIEW is deterministic and bases upon the fact that all carrier sets are linearly ordered by an internal enumeration.

## 3    Orders

Given a relation $R : X \leftrightarrow X$, it is rather easy to calculate relation-algebraic specifications of reflexivity, antisymmetry, and transitivity from the usual logical formulations of these properties. As result we obtain that $(X, R)$ is an ordered set iff $\mathsf{I} \subseteq R$, $R \cap R^\mathsf{T} \subseteq \mathsf{I}$, and $RR \subseteq R$ hold. Using this relation-algebraic specification, to be an ordered set can be easily tested using the RELVIEW tool.

When dealing with ordered sets, one typically investigates extremal elements. Based upon the representation of sets as vectors, in [14] this is done using relation algebra. The descriptions of [14] lead to the following relation-algebraic specifications; they compute for a partial order relation $R : X \leftrightarrow X$ and a vector $v : X \leftrightarrow \mathbf{1}$ vectors of type $X \leftrightarrow \mathbf{1}$, which represent the set of lower bounds (least element and greatest lower bound, respectively) of the set represented by $v$:

$$lbds(R, v) = \overline{\overline{R}\,v}$$
$$lel(R, v) = v \cap lbds(R, v)$$
$$glb(R, v) = lel(R^{\mathsf{T}}, lbds(R, v))$$

Transposing the relation $R$ immediately yields $upds(R, v) = lbds(R^{\mathsf{T}}, v)$ for the set of upper bounds, $gel(R, v) = lel(R^{\mathsf{T}}, v)$ for the greatest element, and $lub(R, v) = glb(R^{\mathsf{T}}, v)$ for the least upper bound.

Geometrical representations of ordered sets have been used and investigated by mathematicians and computer scientists for centuries, e.g., for visualization, the discovery of new results, and the construction of counter-examples. Usually, one draws the *Hasse-diagram* of a partial order relation $R : X \leftrightarrow X$, which is the geometrical representation of the (unique) least relation $S$ contained in $R$ such that its reflexive and transitive closure $S^* := \bigcup_{n \in \mathbb{N}} S^n$ equals $R$. In [3] it is shown that every discrete (especially every finite) partial order relation has such a relation (called the Hasse-diagram, too), which is computed by the relational function $hasse(R) = R \cap \overline{\mathsf{I}} \cap \overline{(R \cap \overline{\mathsf{I}})(R \cap \overline{\mathsf{I}})}$.

## 4   Lattices

Assume $R : X \leftrightarrow X$ to be a partial order relation. The objective of this section is to obtain relation-algebraic specifications of the two lattice operations $\sqcap$ and $\sqcup$ as relations $Inf(R)$ and $Sup(R)$ of type $X \times X \leftrightarrow X$. We start with the development of $Inf(R)$. Assume $u \in X \times X$ and $x \in X$. Then we have:

$$\begin{aligned} &x \text{ is the greatest lower bound of } u \\ \Leftrightarrow\ & R_{x,u_1} \wedge R_{x,u_2} \wedge \forall\, y : R_{y,u_1} \wedge R_{y,u_2} \rightarrow R_{y,x} \\ \Leftrightarrow\ & R_{x,u_1} \wedge R_{x,u_2} \wedge \neg \exists\, y : R_{y,u_1} \wedge R_{y,u_2} \wedge \overline{R}_{y,x} \\ \Leftrightarrow\ & [R, R]^{\mathsf{T}}_{u,x} \wedge \neg \exists\, y : [R, R]^{\mathsf{T}}_{u,y} \wedge \overline{R}_{y,x} \\ \Leftrightarrow\ & ([R, R]^{\mathsf{T}} \cap \overline{[R, R]^{\mathsf{T}}\, \overline{R}}\,)_{u,x} \end{aligned}$$

If we remove the subscripts $u$ and $x$ from the last expression of this calculation, we get the relation-algebraic specification we are looking for as follows:

$$Inf(R) = [R, R]^{\mathsf{T}} \cap \overline{[R, R]^{\mathsf{T}}\, \overline{R}} \tag{1}$$

The relational function $Sup(R) = Inf(R^{\mathsf{T}})$ for specifying the $\sqcup$-operation as relation of type $X \times X \leftrightarrow X$ is an immediate consequence of (1).

Having obtained relation-algebraic specifications for the two lattice operations $\sqcap$ and $\sqcup$, we also are able to test relation-algebraically an ordered set $(X, R)$ to be a lattice. We have that $(X, R)$ constitutes a lattice iff $\mathsf{L} = Inf(R)\mathsf{L}$ and $\mathsf{L} = Sup(R)\mathsf{L}$, since the right-hand side of this equivalence describes that the two relations $Inf(R)$ and $Sup(R)$ are total.

At this place it should be remarked that all relation-algebraic specifications we have presented so far and we will present in the remainder of the paper can be straightforwardly translated in the programming language of RELVIEW. As an example, the translation of (1) into RELVIEW-code looks as follows:

```
Inf(R) = [R,R]^ & -([R,R]^ * -R)
```

Hence, the system can be used for dealing with ordered sets and lattices. Animation and visualization is also possible since RELVIEW allows besides fully automatic executions also step-wise executions and offers a representation of relations as directed graphs. In addition, sophisticated algorithms for drawing graphs nicely and some possibilities for marking vertices and edges are available.

## 5   Modularity and Distributivity

We now pass from general lattices to specific classes and start with the class of *modular lattices*. These are lattices $(X, R)$ such that for all $x, y, z \in X$ the property $R_{x,z}$ implies the *modular equation* $x \sqcup (y \sqcap z) = (x \sqcup y) \sqcap z$. (Since $R : X \leftrightarrow X$ is the partial order relation of the lattice, $R_{x,z}$ corresponds to the usual notation $x \sqsubseteq z$.) Various equivalent formulations to this definition exist. For developing a relational modularity-test, we use in the following (see e.g., [8]) that a lattice $(X, R)$ is modular iff for all $x, y \in X$ the existence of $z \in X$ with $x \sqcap z = y \sqcap z$ and $x \sqcup z = y \sqcup z$ and $R_{x,y}$ imply $x = y$.

Concentrating on the premise of the implication of the universal quantification, we can calculate as given below. In doing so, we assume $\pi, \rho : X \times X \leftrightarrow X$ to be the projection relations as introduced in Section 2. Furthermore, we abbreviate the relations $Inf(R)$ and $Sup(R)$ for the lattice operations $\sqcap$ and $\sqcup$ by $I$ and $S$ respectively, which yields the equivalence of $\sqcap u = \sqcap v$ and $(II^\mathsf{T})_{u,v}$ and the equivalence of $\sqcup u = \sqcup v$ and $(SS^\mathsf{T})_{u,v}$. Let $x, y \in X$. Then we have:

$$
\begin{aligned}
& (\exists z : x \sqcap z = y \sqcap z \wedge x \sqcup z = y \sqcup z) \wedge R_{x,y} \\
\Leftrightarrow\ & (\exists u : u_1 = x \wedge \exists v : v_1 = y \wedge u_2 = v_2 \wedge \sqcap u = \sqcap v \wedge \sqcup u = \sqcup v) \wedge R_{x,y} \\
\Leftrightarrow\ & (\exists u : \pi_{u,x} \wedge \exists v : \pi_{v,y} \wedge (\rho\rho^\mathsf{T})_{u,v} \wedge (II^\mathsf{T})_{u,v} \wedge (SS^\mathsf{T})_{u,v}) \wedge R_{x,y} \\
\Leftrightarrow\ & (\exists u : \pi^\mathsf{T}_{x,u} \wedge \exists v : (\rho\rho^\mathsf{T} \cap II^\mathsf{T} \cap SS^\mathsf{T})_{u,v} \wedge \pi_{v,y}) \wedge R_{x,y} \\
\Leftrightarrow\ & (\exists u : \pi^\mathsf{T}_{x,u} \wedge ((\rho\rho^\mathsf{T} \cap II^\mathsf{T} \cap SS^\mathsf{T})\pi)_{u,y}) \wedge R_{x,y} \\
\Leftrightarrow\ & (\pi^\mathsf{T}(\rho\rho^\mathsf{T} \cap II^\mathsf{T} \cap SS^\mathsf{T})\pi \cap R)_{x,y}
\end{aligned}
$$

Because of the last expression of this derivation and the above characterization, we get the following relation-algebraic specification of modularity:

$$
(X, R) \text{ is modular } \Leftrightarrow \pi^\mathsf{T}(\rho\rho^\mathsf{T} \cap II^\mathsf{T} \cap SS^\mathsf{T})\pi \cap R \subseteq \mathsf{I} \tag{2}
$$

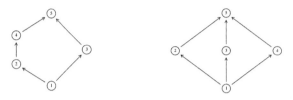

**Fig. 1.** Hasse diagrams of $N_5$ and $M_3$

Distributive lattices form an important subclass of the class of modular lattices. They are important in many applications of computer science and mathematics because they can be easily associated with set families. Originally defined by the distributive law $x \sqcup (y \sqcap z) = (x \sqcup y) \sqcap (x \sqcup z)$ (or its dual version $x \sqcap (y \sqcup z) = (x \sqcap y) \sqcup (x \sqcap z)$) to hold for all $x, y, z \in X$, we apply the following equivalent characterization (see again [8]): A lattice $(X, R)$ is distributive iff for all $x, y \in X$ from the existence of $z \in X$ such that $x \sqcap z = y \sqcap z$ and $x \sqcup z = y \sqcup z$ it follows that $x = y$. Compared with the above characterization of modularity, only the relationship $R_{x,y}$ is missing within the premise. Hence, an immediate consequence of (2) is the following fact:

$$(X, R) \text{ is distributive } \Leftrightarrow \pi^{\mathsf{T}} (\rho\rho^{\mathsf{T}} \cap II^{\mathsf{T}} \cap SS^{\mathsf{T}}) \pi \subseteq \mathsf{I} \tag{3}$$

Other popular characterizations of modular and distributive lattices are given by forbidden substructures. Here two lattices play a decisive role, viz. the "pentagon lattice" $N_5$, whose Hasse-diagram is shown as the left one of the REL-VIEW-pictures of Figure 1, and the "diamond lattice" $M_3$, whose Hasse-diagram is shown as the right RELVIEW-picture of Figure 1. It is well known (cf. again [8]) that a lattice is non-modular iff it contains a sublattice that is isomorphic to $N_5$ and a modular lattice is non-distributive iff it contains a sublattice that is isomorphic to $M_3$. In the remainder of this section we show how relation algebra and RELVIEW can be combined to compute for a non-modular lattice $(X, R)$ a sublattice that is isomorphic to $N_5$ and to visualize its Hasse-diagram in the drawing of $(X, R)$. The technique we use can be applied to a modular and non-distributive lattice, too, to compute and visualize in such a case a sublattice that is isomorphic to $M_3$.

Assume $R : X \leftrightarrow X$ to be the partial order relation of a non-modular lattice $(X, R)$ and $\pi, \rho : X \times X \leftrightarrow X$ to be the projection relations as introduced in Section 2. Furthermore, let $I$ abbreviate $Inf(R)$ and $S$ abbreviate $Sup(R)$. In the first step, we consider the relation

$$U = \overline{R \cup R^{\mathsf{T}}} \tag{4}$$

of type $X \leftrightarrow X$ and develop with its help a relation $C$ of the same type such that

$$C_{x,y} \Leftrightarrow \exists z : x \sqcap z = y \sqcap z \land x \sqcup z = y \sqcup z \land U_{x,z} \land U_{y,z}$$

holds for all $x, y \in X$. Due to definition (4), the relationships $U_{x,z}$ and $U_{y,z}$ of this component-wise specification of $C$ say that the pairs $\langle x, z \rangle, \langle y, z \rangle \in X \times X$

are incomparable with respect to the partial order relation $R$. As a consequence, the intersection $C \cap R \cap \bar{\mathsf{I}}$ contains exactly the pairs $\langle x, y \rangle \in X \times X$ that are elements of a sublattice of $(X, R)$ being isomorphic to $N_5$, such that $x$ corresponds to vertex 2 and $y$ corresponds to vertex 4 of the above RELVIEW-picture of $N_5$. To obtain from the component-wise specification of $C$ a relation-algebraic specification, we only have to modify the above development of the expression $(\pi^{\mathsf{T}}(\rho\rho^{\mathsf{T}} \cap II^{\mathsf{T}} \cap SS^{\mathsf{T}})\pi)_{x,y}$ accordingly. The result is:

$$C = \pi^{\mathsf{T}}(\rho\rho^{\mathsf{T}} \cap II^{\mathsf{T}} \cap SS^{\mathsf{T}} \cap \pi U \rho^{\mathsf{T}} \cap \rho U \pi^{\mathsf{T}})\pi \tag{5}$$

Since the lattice $(X, R)$ is assumed to be non-modular, the relation $C \cap R \cap \bar{\mathsf{I}}$ is non-empty. In the second step, we first select an atom from $C \cap R \cap \bar{\mathsf{I}}$ by means of the operation $atom$ of Section 2 and define afterwards with the atom's help two points $p, q : X \leftrightarrow \mathbf{1}$ in the following way:

$$p = atom(C \cap R \cap \bar{\mathsf{I}})\mathsf{L} \qquad q = atom(C \cap R \cap \bar{\mathsf{I}})^{\mathsf{T}}\mathsf{L} \tag{6}$$

If $p$ represents the element $x \in X$ and $q$ represents the element $y \in X$, then $\langle x, y \rangle$ is a pair from $C \cap R \cap \bar{\mathsf{I}}$. Because of the above described meaning of $x$ and $y$ with respect to the lattice $N_5$, it suffices to compute a third point, say $r : X \leftrightarrow \mathbf{1}$, that represents the element $z$ of a sublattice of $(X, R)$ we are searching for, such that $z$ corresponds to vertex 3 of the RELVIEW-picture of $N_5$. The properties $z$ has to fulfill are $U_{x,z}$, $U_{y,z}$, $x \sqcap z = y \sqcap z$, and $x \sqcup z = y \sqcup z$. This leads to

$$c = Up \cap Uq \cap (A \cap II^{\mathsf{T}}B)^{\mathsf{T}}\mathsf{L} \cap (A \cap SS^{\mathsf{T}}B)^{\mathsf{T}}\mathsf{L} \tag{7}$$

as the vector representation $c : X \leftrightarrow \mathbf{1}$ of all candidates for $z$, where the auxiliary relations $A, B : X \times X \leftrightarrow X$ are defined as $A := \pi p \mathsf{L} \cap \rho$ and $B := \pi q \mathsf{L} \cap \rho$. To verify that $Up : X \leftrightarrow \mathbf{1}$ represents the elements which are incomparable to $x$ and $Uq : X \leftrightarrow \mathbf{1}$ does the same for $y$ is trivial. Here is the justification that $(A \cap II^{\mathsf{T}}B)^{\mathsf{T}}\mathsf{L} : X \leftrightarrow \mathbf{1}$ represents the elements $z \in X$ such that $x \sqcap z = y \sqcap z$:

$$
\begin{aligned}
x \sqcap z = y \sqcap z &\Leftrightarrow \exists u : u_1 = x \wedge u_2 = z \wedge \exists v : v_1 = y \wedge v_2 = z \wedge (II^{\mathsf{T}})_{u,v} \\
&\Leftrightarrow \exists u : (\pi p)_u \wedge \rho_{u,z} \wedge \exists v : (\pi q)_v \wedge \rho_{v,z} \wedge (II^{\mathsf{T}})_{u,v} \\
&\Leftrightarrow \exists u : (\pi p \mathsf{L} \cap \rho)_{u,z} \wedge \exists v : (II^{\mathsf{T}})_{u,v} \wedge (\pi q \mathsf{L} \cap \rho)_{v,z} \\
&\Leftrightarrow \exists u : (A \cap II^{\mathsf{T}}B)^{\mathsf{T}}_{z,u} \wedge \mathsf{L}_u \\
&\Leftrightarrow ((A \cap II^{\mathsf{T}}B)^{\mathsf{T}}\mathsf{L})_z
\end{aligned}
$$

Replacing $I$ by the relation $S$ shows that $(A \cap SS^{\mathsf{T}}B)^{\mathsf{T}}\mathsf{L} : X \leftrightarrow \mathbf{1}$ represents the elements $z \in X$ such that $x \sqcup z = y \sqcup z$. Finally, $r$ is obtained by selecting it as a point from the vector $c$ of (7) using the operation $point$ of Section 2:

$$r = point(c) \tag{8}$$

The third step consists of the application of the two relational functions $glb$ and $lub$ of Section 3, to get a further vector $v : X \leftrightarrow \mathbf{1}$ as follows:

$$v = p \cup q \cup r \cup glb(R, p \cup r) \cup lub(R, q \cup r) \tag{9}$$

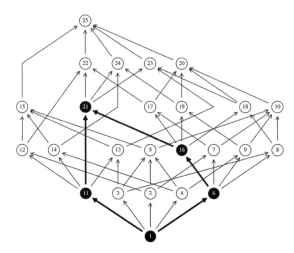

**Fig. 2.** The non-modular product lattice $N_5 \times M_3$

A little reflection shows that the two elements represented by the two points $glb(R, p \cup r) : X \leftrightarrow \mathbf{1}$ and $lub(R, q \cup r) : X \leftrightarrow \mathbf{1}$ correspond exactly to the vertices 1 and 5, respectively, of the lattice $N_5$ in the RELVIEW-picture of Figure 1. Hence, the vector $v : X \leftrightarrow \mathbf{1}$ of (9) represents the carrier set of a "pentagon sublattice" of the lattice $(X, R)$ we are searching for. The relation of this sublattice is immediately obtained (as a subrelation of $R$ to facilitate a RELVIEW-visualization via the marking of edges) as $R \cap vv^{\mathsf{T}} : X \leftrightarrow X$.

It is straightforward to translate (4) to (9) into a RELVIEW-program, that computes for the input relation $R$ the vector $v$ of (9). The RELVIEW-picture of Figure 2 demonstrates how then the system can be used for visualization purposes. It is the graphical representation of the Hasse-diagram of the direct product of the lattices $N_5$ and $M_3$. The product-lattice $N_5 \times M_3$ is non-modular. In the graph this fact is visualized by the Hasse-diagram of a sublattice that is isomorphic to $N_5$, which is emphasized by bold-face edges and black vertices.

## 6 Pseudo Complements

For dealing with intuitionistic propositional logic, (relative) pseudo complements and Heyting algebras (and some variants) have been introduced in the following sense: Given a lattice $(X, R)$ and $x, y \in X$, the greatest element of $\{z \in X \mid R_{z \sqcap x, y}\}$ is called the *relative pseudo complement* of $x$ with respect to $y$ and denoted as $x * y$ (or $x \rightarrow y$). If $(X, R)$ is a distributive lattice, has a least element $\perp$ and a greatest element $\top$, and the relative pseudo complement $x * y$ exists for all pairs $x, y \in X$, then the lattice is called a *Heyting algebra*. In Heyting algebras $x * \perp$ is defined as the *pseudo complement* $x^*$ of $x$.

In the following, we consider a lattice $(X, R)$ with least element $\perp$ and greatest element $\top$. Furthermore, we define two points $b, t : X \leftrightarrow \mathbf{1}$ by $b := lel(R, \mathsf{L})$ and $t := gel(R, \mathsf{L})$, where $\mathsf{L} : X \leftrightarrow \mathbf{1}$. Hence, $b$ represents the least element

**Fig. 3.** (Relative) pseudo complement relation of $N_5$

$\perp$ and $t$ represents the greatest element $\top$ of the lattice. Finally, we assume $\pi, \rho : X \times X \leftrightarrow X$ to be the projection relations on $X \times X$. Our goal is to develop relation-algebraic specifications $RelPcompl(R) : X \times X \leftrightarrow X$ for relative pseudo complements and $Pcompl(R) : X \leftrightarrow X$ for pseudo complements, which allow to compute and visualize these constructions using RELVIEW and which also lead to tests for a lattice to be a Heyting algebra or some of its variants.

We start with the relative pseudo complement and calculate a relation-algebraic specification of the partial function that maps a pair to the relative pseudo complement. Given $u \in X \times X$ and $c \in X$, the component-wise specification of symmetric quotients of Section 2 yields

$$c \text{ relative pseudo complement of } u_1 \text{ wrt. } u_2 \Leftrightarrow syq(S, R)_{u,c},$$

where the auxiliary relation $S : X \leftrightarrow X \times X$ satisfies $S_{z,u}$ iff $R_{z \sqcap u_1, u_2}$ for all $u \in X \times X$ and $z \in X$. It remains to develop a relation-algebraic specification of $S$. To reach this goal, we calculate for all $u \in X \times X$ and $z \in X$ as follows:

$$R_{z \sqcap u_1, u_2} \Leftrightarrow \exists a : a = z \sqcap u_1 \wedge R_{a, u_2}$$
$$\Leftrightarrow \exists v : v_1 = z \wedge v_2 = u_1 \wedge \exists a : a = \sqcap v \wedge R_{a, u_2}$$
$$\Leftrightarrow \exists v : \pi_{v,z} \wedge (\rho \pi^{\mathsf{T}})_{v,u} \wedge (Inf(R) R \rho^{\mathsf{T}})_{v,u}$$
$$\Leftrightarrow (\pi^{\mathsf{T}}(\rho \pi^{\mathsf{T}} \cap Inf(R) R \rho^{\mathsf{T}}))_{z,u}$$

A removal of the subscripts $u$ and $z$ from the last expression of this calculation yields $S = \pi^{\mathsf{T}}(\rho \pi^{\mathsf{T}} \cap Inf(R) R \rho^{\mathsf{T}})$. If we unfold this description of $S$ in $syq(S, R)_{u,c}$ and remove after that the two subscripts $u$ and $c$, we get the following relation-algebraic specification:

$$RelPcompl(R) = syq(\pi^{\mathsf{T}}(\rho \pi^{\mathsf{T}} \cap Inf(R) R \rho^{\mathsf{T}}), R) \qquad (10)$$

Pseudo complements are relative pseudo complements with respect to the least element $\perp$. Relation-algebraically this specialization is described by the expression $[\mathsf{I}, \mathsf{L}b^{\mathsf{T}}] RelPcompl(R)$. To enhance efficiency, in the following we develop a specification of the pseudo complement relation which does not use a fork and the relative pseudo complement relation. We start with the same idea as in the case of relative pseudo complements and obtain for all $c, x \in X$ that

$$c \text{ pseudo complement of } x \Leftrightarrow syq(S, R)_{x,c},$$

where now the auxiliary relation $S : X \leftrightarrow X$ is component-wisely defined by demanding for all $z, x \in X$ that $S_{z,x}$ iff $R_{z \sqcap x, \perp}$. Similar to the case above we can

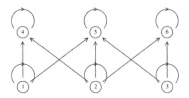

**Fig. 4.** An ordered set with 6 elements

show that the equivalence of $R_{z \sqcap x, \perp}$ and $(R^{\mathsf{T}}(R \cap \overline{b}\, \mathsf{L}))_{z,x}$ holds for all $z, x \in X$, yielding the following relation-algebraic specification:

$$Pcompl(R) = syq(R^{\mathsf{T}}(R \cap \overline{b}\, \mathsf{L}), R) \tag{11}$$

We have translated (10) and (11) into RELVIEW-programs and these applied to the partial order relation of the lattice $N_5$. The results are shown in the pictures of Figure 3. On the left we present the Boolean matrix of the relative pseudo complement relation of $N_5$. For reasons of space it is depicted in its transposed form, i.e., with 5 rows and 25 columns. The picture on the right shows the pseudo complement relation of $N_5$ as Boolean $5 \times 5$ matrix. A black square of such a RELVIEW-matrix means that the elements are in relationship and a white square means that they are not. So, e.g., from the pictures we see that $\langle 4, 2 \rangle$ is the only pair without a relative pseudo complement.

## 7   Completions

Embeddings into complete lattices play an important role in order and lattice theory. In this section we show how such "completions" can be computed and visualized by means of relation algebra and RELVIEW..

Assume $(X, R)$ to be an ordered set. A subset $Y$ of $X$ is called a *cut* of $(X, R)$ if it coincides with the lower bounds of its upper bounds. It is well known (see e.g., [8]) that the set $\mathfrak{C}$ of all cuts together with set inclusion constitutes a complete lattice $(\mathfrak{C}, \subseteq)$, called the (Dedekind-McNeille) *cut completion* of $(X, R)$ since it contains a suborder that is order-isomorphic to $(X, R)$. This suborder is given by the the the set of all *principal cuts* $[x] := \{y \in X \mid R_{y,x}\}$, where $x \in X$. The isomorphism is established via

$$\sigma : X \to \mathfrak{C} \qquad \sigma(x) = [x], \tag{12}$$

as this function is an *order-embedding* of $(X, R)$ into $(\mathfrak{C}, \subseteq)$, i.e., satisfies $R_{x,y}$ iff $\sigma(x) \subseteq \sigma(y)$ for all $x, y \in X$.

Using the operations $syq$ and $inj$ of Section 2 and the membership-relation $\mathsf{M} : X \leftrightarrow 2^X$, in [6] the following three relational functions for computing the cut completion of $(X, R)$ are developed from formal predicate logic specifications:

$$\begin{aligned}
CutList(R) &= \mathsf{M}\, inj((syq(\mathsf{M}, lbds(R, upds(R, \mathsf{M}))) \cap \mathsf{I})\mathsf{L})^{\mathsf{T}} \\
CutLat(R) &= \overline{CutList(R)^{\mathsf{T}}\, \overline{CutList(R)}} \\
Sigma(R) &= syq(R, CutList(R))
\end{aligned} \tag{13}$$

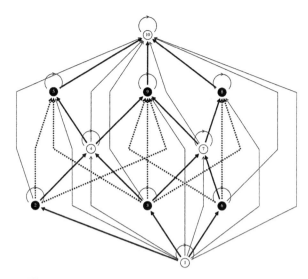

**Fig. 5.** ...the order as part of the cut completion

The columns of $CutList(R) : X \leftrightarrow \mathfrak{C}$ represent the subset $\mathfrak{C}$ of $2^X$ as explained in Section 2, $CutLat(R) : \mathfrak{C} \leftrightarrow \mathfrak{C}$ is the relation-algebraic description of the set inclusion on $\mathfrak{C}$, and $sigma(R) : X \leftrightarrow \mathfrak{C}$ specifies the order-embedding (12) relation-algebraically, i.e., for all $x \in X$ and $Y \in \mathfrak{C}$ we have $sigma(R)_{x,Y}$ iff $\sigma(x) = Y$.

It is straightforward to translate the relational functions of (13) into the programming language of RELVIEW. This allows to compute and visualize cut completions by means of the system. We demonstrate this using a small example. As starting point we consider an ordered set, the graphical representation of which in RELVIEW looks as given in Figure 4. The RELVIEW-picture of Figure 5 shows the cut completion of this ordered set as a directed graph and emphasizes the completion's Hasse-diagram as bold-face edges. Furthermore, it visualizes the embedding of the ordered set of Figure 4 into the cut completion by drawing the elements of the suborder that is order-isomorphic to it as black vertices and the Hasse-diagram of the suborder as dotted bold-face edges.

It is also well known that the cut completion $(\mathfrak{C}, \subseteq)$ of an ordered set $(X, R)$ is in a certain sense the least complete lattice containing a suborder that is order-isomorphic to $(X, R)$. Formally, this is expressed by the following fact (a proof can be found in [15] for example): If $(X, R)$ is order-isomorphic to a suborder of a complete lattice $(V, Q)$ via the order-embedding $\phi : X \to V$, then also $(\mathfrak{C}, \subseteq)$ is order-isomorphic to a suborder of $(V, Q)$ via the order-embedding

$$\psi : \mathfrak{C} \to V \qquad \psi(Y) = \bigsqcup \{\phi(x) \mid x \in Y\} \qquad (14)$$

and, furthermore, $\phi(x) = \psi(\sigma(x))$ holds for all $x \in X$.

As continuation of the work of [6], in the following we show how to calculate a relation-algebraic specification of the order-embedding (14) as a relation of type $\mathfrak{C} \leftrightarrow V$ and how to apply it in the context of another completion. In doing so, we abbreviate the column-wise representation $CutList(R)$ of the set $\mathfrak{C}$ of all cuts

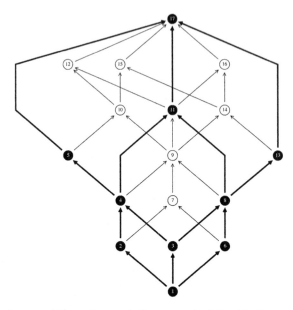

**Fig. 6.** ... and the cut completion as part of the ideal completion

of $(X, R)$ as $C$ and assume the function $\phi : X \to V$ to be given as a relation $\varPhi : X \leftrightarrow V$. The latter means that we assume the equivalence of $\phi(x) = z$ and $\varPhi_{x,z}$ for all $x \in X$ and $z \in V$. Then, for all $Y \in \mathfrak{C}$ and $z \in V$ we obtain:

$$
\begin{aligned}
\psi(Y) = z \;&\Leftrightarrow\; \bigsqcup\{\phi(x) \mid x \in Y\} = z \\
&\Leftrightarrow\; (\forall\, y : y \in Y \to Q_{\phi(y),z}) \land (\forall\, x : (\forall\, y : y \in Y \to Q_{\phi(y),x}) \to Q_{z,x}) \\
&\Leftrightarrow\; (\neg\exists\, y : C_{y,Y} \land \overline{\varPhi Q}_{\,y,z}) \land (\forall\, x : (\neg\exists\, y : C_{y,Y} \land \overline{\varPhi Q}_{\,y,x}) \to Q_{z,x}) \\
&\Leftrightarrow\; \overline{C^\mathsf{T}\,\overline{\varPhi Q}}_{\,Y,z} \land (\forall\, x : \overline{C^\mathsf{T}\,\overline{\varPhi Q}}_{\,Y,x} \to Q_{z,x}) \\
&\Leftrightarrow\; \overline{C^\mathsf{T}\,\overline{\varPhi Q}}_{\,Y,z} \land (\neg\exists\, x : \overline{C^\mathsf{T}\,\overline{\varPhi Q}}_{\,Y,x} \land \overline{Q}^{\mathsf{T}}_{\,x,z}) \\
&\Leftrightarrow\; (\overline{C^\mathsf{T}\,\overline{\varPhi Q}})_{Y,z} \land (\,\overline{\overline{C^\mathsf{T}\,\overline{\varPhi Q}}\;\overline{Q}^{\mathsf{T}}}\,)_{Y,z} \\
&\Leftrightarrow\; (\overline{C^\mathsf{T}\,\overline{\varPhi Q}} \cap \overline{\overline{C^\mathsf{T}\,\overline{\varPhi Q}}\;\overline{Q}^{\mathsf{T}}}\,)_{Y,z}
\end{aligned}
$$

Note that the equivalence of $y \in Y$ and $C_{y,Y}$ (which we have used in the third step) follows from the fact that $C : X \leftrightarrow \mathfrak{C}$ represents $\mathfrak{C}$ column-wisely. If we remove the subscripts $Y$ and $z$ from the last expression of the calculation, we get the following relation-algebraic specification of the order-embedding (14):

$$
psi(R, \varPhi, Q) = \overline{C^\mathsf{T}\,\overline{\varPhi Q}} \cap \overline{\overline{C^\mathsf{T}\,\overline{\varPhi Q}}\;\overline{Q}^{\mathsf{T}}} \tag{15}
$$

To demonstrate an application of (15), we now consider a further completion procedure for embedding an ordered set $(X, R)$ into a complete lattice. As carrier set of the complete lattice we take the set $\mathfrak{I}$ of all *order ideals*, i.e., all subsets $I$ of $X$ such that for all $x \in I$ and $y \in X$ from $R_{y,x}$ it follows $y \in I$. Like cuts, ideals are ordered by set inclusion. The complete lattice $(\mathfrak{I}, \subseteq)$ is called the (order) *ideal completion* of $(X, R)$. Again a suborder that is order-isomorphic to $(X, R)$ is given by the set of principal cuts – in this context called *principal ideals* – and the isomorphism is established via $\phi : X \to \mathfrak{I}$, where $\phi(x) = \{y \in X \mid R_{y,x}\}$.

It is not hard to calculate a vector representation of the subset $\mathfrak{I}$ of $2^X$. Assuming a set $I \in 2^X$, we have the following equivalence:

$$\forall\, x, y : x \in I \wedge R_{y,x} \to y \in I \iff \forall x : x \in I \to \neg \exists y : R_{y,x} \wedge y \notin I$$

$$\iff \forall x : \mathsf{M}_{x,I} \to \overline{\overline{\mathsf{M}}^\mathsf{T} R}_{I,x}$$

$$\iff \neg \exists x : \mathsf{M}^\mathsf{T}_{I,x} \wedge (\overline{\overline{\mathsf{M}}^\mathsf{T} R})_{I,x}$$

$$\iff \overline{(\mathsf{M}^\mathsf{T} \cap \overline{\overline{\mathsf{M}}^\mathsf{T} R})\mathsf{L}}_I$$

Because of the last expression of this calculation, the vector $\overline{(\mathsf{M}^\mathsf{T} \cap \overline{\overline{\mathsf{M}}^\mathsf{T} R})\mathsf{L}}$ : $2^X \leftrightarrow \mathbf{1}$ represents the set $\mathfrak{I}$ of all order ideals of $(X, R)$. The following relation-algebraic specifications of the column-wise representation of $\mathfrak{I}$, the inclusion order on $\mathfrak{I}$, and the order-embedding $\phi$ are immediate consequences:

$$IdealList(R) = \mathsf{M}\, inj(\,\overline{(\mathsf{M}^\mathsf{T} \cap \overline{\overline{\mathsf{M}}^\mathsf{T} R})\mathsf{L}}\,)^\mathsf{T}$$
$$IdealLat(R) = \overline{\overline{IdealList(R)}^\mathsf{T}\, IdealList(R)} \tag{16}$$
$$phi(R) = syq(R, IdealList(R))$$

Translating (16) into RELVIEW-code, we have computed the ideal completion of the ordered set of Figure 4 and have compared it with the cut completion (Figure 5). The marked directed graph of Figure 6 visualizes the containedness of the cut completion in the ideal completion. To obtain this picture, in a first step we have computed the Hasse-diagrams $H_\mathcal{I}$ of the ideal completion of the original 6-element ordered set as well as the Hasse-diagrams $H_\mathcal{C}$ of its cut completion. Then we have adapted the type of $H_\mathcal{C}$ to the type of $H_\mathcal{I}$. In Boolean matrix terminology this means that we have translated the $10 \times 10$ matrix $H_\mathcal{C}$ into a $17 \times 17$ matrix $\tilde{H}_\mathcal{C}$ by adding empty rows and columns for all elements of the ideal completion not being contained in the cut completion. This easily is possible using the relation-algebraic specification (15) of the order embedding $\psi$, since $\tilde{H}_\mathcal{C} = \Psi^\mathsf{T} H_\mathcal{C} \Psi$ with $\Psi$ being the relation corresponding to $\psi$. In the third step we have joined $H_\mathcal{I}$ and $\tilde{H}_\mathcal{C}$. Finally, we have drawn this union as a directed graph and marked in the graph the arcs corresponding to $\tilde{H}_\mathcal{C}$ by boldface arcs and the elements represented by the vector $\Psi^\mathsf{T}\mathsf{L}$ as black vertices.

## 8  Conclusion

In this paper we have used relation algebra and the specific purpose Computer Algebra system RELVIEW for solving order- and lattice-theoretic problems and for visualizing their solutions. We have demonstrated this fruitful combination by means of some small examples. Space restrictions did not allow to represent

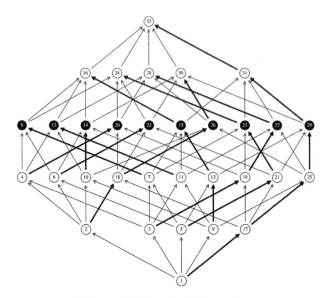

**Fig. 7.** A Dilworth chain partition

larger and more impressive examples for RELVIEW's computational power and
visualization possibilities, like Dilworth chain partitions (based upon a maximum
bipartite matching program), the more efficient computation of cut completions
and ideal completions as sequences of vectors generated by union from a basis
(inspired by [13]), the construction of specific lattices (e.g., subgroup lattices,
lattices of maximum antichains, and lattices obtained by doublings), and the
computation/approximation of free lattices from partial order relations on the
set of generators (guided by [9]). To give a least an impression of the potential of
RELVIEW regarding the visualization of such advanced applications, in Figure 7
the Dilworth chain partition of a 32-element Boolean lattice is shown. The black
vertices of this Hasse-diagram depict a maximum antichain of size $\binom{5}{3} = 10$ and
the boldface arcs represent the 10 chains of the chain partition.

Of course, in spite of the fact that the system implements relations very
efficiently with the help of reduced binary decision diagrams frequently REL-
VIEW-programs cannot compete with special programs tailored for problems of
the kind we just have mentioned – although in the case of NP-hard problems
or problems with a result set of potentially exponential size (like the set of all
cuts, all order ideals, and all extremal chains/antichains) the complexities are
usually the same. Nevertheless, a lot of experiments have shown that precisely
the computation of huge result sets via membership relations is a strength of
RELVIEW. See [11, 4, 6, 5, 2, 7] for some examples.

Nowadays, systematic experiments are accepted as a way for obtaining new
mathematical insights. As a consequence, tools for symbolic manipulation, proto-
typic computation, animation, and visualization become increasingly important
as one proceeds in investigations. We believe that the attraction of RELVIEW in
this area lies in its flexibility, its large application area, its computational power

when dealing with enumerations of huge sets of "interesting objects" (e.g., to verify an example or to construct a counter-example), its manifold animation and visualization possibilities, and the concise form of its programs. Of course, applications cover all relation-based discrete structures, but also objects which at first glance do not seem to be closely connected to relations. See e.g., [7], where RELVIEW is used as a SAT-solver. New properties and types of and problems on such structures and objects are introduced (discovered and investigated, respectively) all the time and RELVIEW proved to be an ideal tool for experimenting with a lot of the new concepts while avoiding unnecessary overhead. RELVIEW-programs are built very quickly and, combining relation algebra with predicate logic and other formal tools (e.g., assertion logics or fixed point calculus), their correctness is guaranteed by the completely formal developments.

# References

1. Behnke R. et al.: RELVIEW — A system for calculation with relations and relational programming. In: Asteslano E. (ed.): Proc. 1st Conf. *Fundamental Approaches to Software Engineering*, LNCS 1382, Springer, 318-321 (1998).
2. Berghammer R., Fronk A.: Exact computation of minimum feedback vertex sets with relational algebra. Fund. Informaticae 70, 301-316 (2006)
3. Berghammer R., Schmidt G.: Discrete ordering relations. Discr. Math. 43, 1-7 (1983).
4. Berghammer R., Leoniuk B., Milanese U.: Implementation of relation algebra using binary decision diagrams. In: de Swart H. (ed.): Proc. 6th Int. Workshop *Relational Methods in Computer Science*, LNCS 2561, Springer, 241-257 (2002).
5. Berghammer R., Neumann F.: RELVIEW– An OBDD-based Computer Algebra system for relations. In: Gansha V.G. et al. (eds.): Proc. 8th Int. Workshop *Computer Algebra in Scientific Computing*, LNCS 3718, Springer, 40-51 (2005)
6. Berghammer R.: Computation of cut completions and concept lattices using relational algebra and RELVIEW. J. Relat. Meth. in Comput. Sci. 1, 50-72 (2004).
7. Berghammer R., Milanese U.: Relational approach to Boolean logic problems. In: MacCaull W. et al. (eds.): Proc. 8th Int. Workshop *Relational Methods in Computer Science*, LNCS 3929, Springer, 48-59 (2006)
8. Birkhoff G.: Lattice theory. American Math. Society Coll. Publ. Volume XXV, American Math. Society, 3rd edition (1967).
9. Freese R., Jezek J., Nation J.B.: Free lattices. Mathematical Surveys and Monographs, Volume 42, American Math. Society (1995).
10. Ganter B., Wille R.: Formal concept analysis: Mathematical foundations. Springer, (1999).
11. Leoniuk B.: ROBDD-based implementation of relational algebra with applications (in German). Ph.D. thesis, Univ. Kiel (2001).
12. Milanese U.: On the implementation of a ROBDD-based tool for the manipulation and visualization of relations (in German). Ph.D. thesis, Univ. Kiel (2003).
13. Nourine L., Raynaud O.: A fast algorithm for building lattices. Inform. Proc. Let. 70, 259-264 (1999)
14. Schmidt G., Ströhlein T.: Relations and graphs. Discrete Mathematics for Computer Scientists, EATCS Monographs on Theoret. Comput. Sci., Springer (1993).
15. Skornjakow L.A.: Elements of lattice theory (in German). Akademie-Verlag (1973).
16. Tarski A.: On the calculus of relations. J. Symbolic Logic 6, 73-89 (1941).

# Intervals, Syzygies, Numerical Gröbner Bases: A Mixed Study

Marco Bodrato and Alberto Zanoni

Dipartimento di Matematica "Leonida Tonelli" – Università di Pisa
Largo B. Pontecorvo 5 – 56127 Pisa, Italy
{bodrato, zanoni}@posso.dm.unipi.it

**Abstract.** In Gröbner bases computation, as in other algorithms in commutative algebra, a general open question is how to guide the calculations coping with numerical coefficients and/or not exact input data. It often happens that, due to error accumulation and/or insufficient working precision, the obtained result is not one expects from a theoretical derivation. The resulting basis may have more or less polynomials, a different number of solution, roots with different multiplicity, another Hilbert function, and so on. Augmenting precision we may overcome algorithmic errors, but one does not know in advance how much this precision should be, and a trial–and–error approach is often the only way to follow. Coping with initial errors is an even more difficult task. In this experimental work we propose the combined use of syzygies and interval arithmetic to decide what to do at each critical point of the algorithm.

**AMS Subject Classification:** 13P10, 65H10, 90C31.

**Keywords and phrases:** Gröbner bases, numerical coefficients, syzygies.

## 1 Introduction

For a general reference to Gröbner bases, we cite [1], [6], [5], [8] and [9]. Numerical stability for Gröbner bases computation has been studied by, among others, Stetter [14], [15] who considers coefficient sizes to influence term ordering. Shirayanagy [13] gives a theoretical basis to floating point computation for systems with exact coefficients, using a sufficiently high working precision. His stabilization technique is based on a clever rewriting rule concerning zero testing. Unfortunately, no upper bound on the initial sufficient precision is known. Migheli [12] uses Hybrids coefficients $H = (n, f)$ – where $n$ is a number modulo a prime and $f$ a floating point value – to measure the stabilization with respect to small variations. Zanoni [20], [21] mimics Migheli's approach using double–floats coefficients $F2 = (f_1, f_2)$, where two different precisions are used at the same time in the computation to control the behaviour of Buchberger algorithm. Traverso [16] presents an idea which is later partially developed in [3].

V.G. Ganzha, E.W. Mayr, and E.V. Vorozhtsov (Eds.): CASC 2006, LNCS 4194, pp. 64–76, 2006.

Being many concepts of algebra intrinsically discrete (dimensions of vector spaces, matrices rank, root multiplicity, etc.), when working with real or complex coefficients, usually the critical points arise when we have to decide if a coefficient $c$ is zero or not. It's what we called the *zero test* (ZT). We are facing a *bifurcation*, and in the general case the algorithm under consideration can follow two completely different paths according to $c = 0$ or $c \neq 0$.

The weak point in easily definable ZTs is lack of flexibility. If we define a too strict one, we could keep too many things that should instead be thrown away. On the contrary, a ZT discarding too many things may equally lead to a not correct behaviour. For example, one may obtain $\langle 1 \rangle$ as final basis.

Moreover, there may be cases in which the *same* ZT is sometimes good, sometimes bad. Infact, it is usually based on some parameters (coefficient sizes, number of correct digits, initial precision, and so on), and to the best of our knowledge nowadays there is no automatic procedure to detect which are the best values (when they exist) for a system to be correctly treated. Some heuristics may be used, but, again, trial–and–error is still the only general method to analyze all the cases which are treatable with a ZT with fixed behaviour.

Let's now consider our case, the Buchberger algorithm. What one looks for is an *adaptive* test, defining a well–determined more general procedure to decide *case by case* the result of the ZT. In other words, we'd like that *the system itself* imposes the conditions that should be satisfied to fulfil, if possible, the ZT. This particularly in the case when initial coefficients are not exact, but known to belong to an interval of possible values, usually determined by available precision.

A zero value is a very particular case, indicating that the system has some hidden relations among its coefficients, which may be not clear from the beginning, because of the limited initial precision. The first problem is to distinguish between possible zeroes, moving initial coefficients in the interval, and values which seems to be zero just because of computational errors. Our philosophy would then be that of trying to (find and) impose these relations during the way, such that at each point of the algorithm in which a coefficient can be zero, it is forced to zero. In a certain sense we are looking for the most "degenerate" polynomial system, having coefficients compatible with initial precision.

With this in mind, in [16] the use of syzygies is proposed. Giving relations expressing the "history" (trace) of all the performed computations, they seem to be a good tool to analyze the current situation when a ZT has to be applied. A prototype implementation for the first experiments is being developed using the C++ PoSSoLib library, result of the FRISCO [10] project.

## 2   Syzygies

Let $\mathbb{K}$ be a field and $\mathcal{F} = \{f_1, \ldots, f_s\} \subset \mathbb{K}[X] = \mathbb{K}[x_1, \ldots, x_n]$ a list of polynomials representing the initial system. In this paper we consider the field of real numbers, $\mathbb{K} = \mathbb{R}$. Syzygies express polynomial relations among the $f_i$.

**Definition 1.** *A* syzygy *for* $\mathcal{F}$ *is a* $s$–tuple $\mathcal{H} = (h_1, \ldots, h_s) \subset \mathbb{K}[X]^s$ *such that*

$$\mathcal{H} \cdot \mathcal{F} = \sum_{i=1}^{s} h_i(X) \cdot f_i(X) = 0$$

Let $\mathcal{G} = \{g_1, \ldots, g_t\} \subset \mathbb{K}[X]$ be a system obtained from $\mathcal{F}$ at a certain point of a Buchberger algorithm application. The idea is to keep track of the steps to derive $\mathcal{G}$ from $\mathcal{F}$, in a similar way as in the extended Euclid algorithm for Bezout's identity. In other words, we look for $k_{ij}(X) \in \mathbb{K}[X]$ with

$$g_j(X) = \sum_{i=1}^{s} k_{ij}(X) \cdot f_i(X) \qquad j = 1, \ldots, t \qquad (1)$$

We can obtain syzygies and $k_{ij}$ by using a variant of Buchberger algorithm itself (see [7]). Look at $f_1, \ldots, f_s \in \mathbb{K}[X]$ as vectors $(f_1, 1, 0, \ldots, 0), \ldots, (f_s, 0, \ldots, 0, 1)$ $\in \mathbb{K}[X]^{s+1}$, considered as a $\mathbb{K}[X]$–module, with a term ordering in which comparisons are on pairs $(t, i)$ – where $t$ is a term and $i$ is a position index – such that any term in initial position is greater than whatever term in any other position.

## 3    Multi-coefficients and Numerical Buchberger Algorithm

The fundamental idea for the `mCoeff` type is to take benefit from the combined use of the floating point and interval arithmetic. A `mCoeff` $m = (m_S, m_L, m_i, m_s)$ is an "enriched" representation of a real number. It has two almost equal floats with different precisions, called short $(m_S)$ and long $(m_L)$ part, respectively, and an interval $m_I = [m_i, m_s]$ containing both $m_S$ and $m_L$.

**Definition 2.** *Let* $x \in \mathbb{R} \setminus \{0\}$. *The natural number* $n = \text{size}(x)$ *such that*

$$x = a \cdot 2^n \qquad \text{with} \qquad \frac{1}{2} \leqslant |a| < 1.$$

*is called the* size *of* $x$.

In our implementation, the initial interval $m_I = [m_i, m_s]$ containing $v$ is computed by default as follows: if $s(v)$ denotes the sign of $v$ we have

$v > 0 : [v(1 - 2^{-\omega}), v(1 + 2^{-\omega})]$
$v < 0 : [v(1 + 2^{-\omega}), v(1 - 2^{-\omega})]$  or  $v \neq 0 : [v(1 - s(v) \cdot 2^{-\omega}), v(1 + s(v) \cdot 2^{-\omega})]$
$v = 0 : [0, 0]$                                        $v = 0 : [0, 0]$

In a future version, we will be possible to set interval width for each coefficient independently. In some cases a trivial interval $[v, v]$ is necessary: e.g. when we know in advance that a (initial) coefficient is exact, or when dividing a coefficient by itself. Even if in general it is difficult to detect such cases, there is one in which this is evident: when making a polynomial monic (leading coefficient becomes *exactly* 1).

To cope with easily detectable zeroes, we propose the below test. We plan to avoid using user-defined parameters or to change the definition when we'll have evidence of the effectiveness of the approach proposed in the following sections.

**ZERO TEST** : A mCoeff $m$ is considered to be zero when (see [20], [21])

- its short or long part is exactly zero, or
- it is the result of an algebraic sum in which the size drops too much, or
- size($m_S$) − size($m_L$) grows too much (indicating that all the meaningful digits disappeared, and only garbage remained).

The "too much" quantities are controlled by user-defined integer parameters.

**Definition 3.** *An interval $I = [a, b] \subset \mathbb{R}$ is* dangerous *when $0 \in I$. A mCoeff $m$ is dangerous when $m_I$ is dangerous. A polynomial $p$ with mCoeffs involved in the Buchberger algorithm is dangerous when its leading coefficient is dangerous and $p$ is no more head–reducible with respect to the current basis.*

Let $\mathcal{F}$ be given, together with the finite precision determining the width of the initial intervals $I_i$ for its coefficients. Any system $\mathcal{F}'$ obtained from $\mathcal{F}$ slightly moving the coefficients inside the corresponding $I_i$ is considered as an equally valid representation of the problem to be solved, indistinguishable from $\mathcal{F}$ (we say it is *near* $\mathcal{F}$). This is the freedom we have in looking for the most interesting representative, that is the most *unstable* one, having presumably more interesting properties (such as positive root multiplicity, etc.) than all the near ones. The main point in the Numerical Buchberger Algorithm is the ZT in (IV). Details about NBA are explained in section 6.

| **Numerical Buchberger Algorithm (NBA)** |

I   Construct the $\overline{\mathcal{F}}$ system with mCoefficients, and start Buchberger algorithm.

II   If there is a remaining S-polynomial, compute $r$, its complete reduction with respect to the current basis, otherwise go to step V.

III   If $r = 0$ or its head coefficient $c$ is not dangerous, update the data structures as usual and go to step II, otherwise to IV.

IV   Decide if $c$ can really be or is surely different from 0. Update data structures and in the first case modify $\mathcal{F}$ and go to I, otherwise continue from II.

V   Extract the final polynomials $g_i$ from the obtained basis, and output them.

## 4 The Zero Test

Let $\alpha, \beta, \gamma, \cdots \in \mathbb{N}^n$ be multiindexes, $\mathbb{T} = \{X^\delta \mid |\delta| = 0, 1, ...\}$ the term basis, and $lt(r) = X^\rho$ the leading term of $r \in \mathbb{K}[X]$. We indicate with $\mathbf{1}$ the term with multiexponent $(0, \ldots, 0)$. We consider relations (1) concerning $r$

$$r(X) = \sum_{\gamma} r_{\gamma} X^{\gamma} = \sum_{i=1}^{s} k_i(X) \cdot f_i(X)$$

Let $K_i^{\alpha}$ be the not zero coefficient of $k_i$ in the monomial containing the term $X^{\alpha}$, and $F_i^{\beta}$ similarly for $f_i$. Abusing notation, we also introduce new variables $F = \{F_i^{\beta} \mid \beta \in B_i, i = 1, \ldots, s\}$ and $K = \{K_i^{\alpha} \mid \alpha \in A_i, i = 1, \ldots, s\}$ corresponding to these coefficients. Thanks to the mCoeff approach, interval limits for $F, K$ unknowns are also available. If we expand the right hand side of this relation and equalize coefficients, we obtain the following system

$$S_{\rho,c} = \left\{ 0 = \sum_{\substack{i=1 \\ \alpha+\beta=\gamma}}^{s} K_i^{\alpha} F_i^{\beta} \quad \gamma > \rho \; ; \quad r_{\gamma} = \sum_{\substack{i=1 \\ \alpha+\beta=\gamma}}^{s} K_i^{\alpha} F_i^{\beta} \quad \gamma \leqslant \rho \right\} \quad (2)$$

The zero test asks if:

$\boxed{\mathcal{P}_{\rho}}$ : is there a set of values for $K_i^{\alpha}$ and $F_i^{\beta}$ inside their definition intervals satisfying the "$= 0$" equations of $S_{\rho,c}$ and such that $lc(r) = r_{\rho} = c = 0$ ?

This means to detect an initial system near $\mathcal{F}$ and a set of syzygy values letting the computation trace up to now still be valid, but such that (iterating the process) we can force to zero as many coefficients as possible. We use the following notation for easiness of reference in the following:

1. $\underline{F}_i^{\beta}$, $\overline{F}_i^{\beta}$ : whose entries are the initial intervals limits for $F_i^{\beta}$, for all $i, \beta$.
2. $\underline{K}_i^{\alpha}$, $\overline{K}_i^{\alpha}$ : similarly for $K_i^{\alpha}$, for all possible $i$ and $\alpha$.

By definition, 0 is never contained in the initial intervals for $\mathcal{F}$. This prevents the trivial null solution to be admissible. One could be tempted to write/solve

$$\mathcal{P}_{\rho} : \begin{cases} \min \; c = \left| \sum_{\substack{i=1 \\ \alpha+\beta=\rho}}^{s} K_i^{\alpha} F_i^{\beta} \right| & (O) \\[2em] 0 = \sum_{\substack{i=1 \\ \alpha+\beta=\gamma}}^{s} K_i^{\alpha} F_i^{\beta} \quad \gamma > \rho & (V_1) \\[2em] \left. \begin{array}{l} \underline{F}_i^{\beta} \leqslant F_i^{\beta} \leqslant \overline{F}_i^{\beta} \\ \underline{K}_i^{\alpha} \leqslant K_i^{\alpha} \leqslant \overline{K}_i^{\alpha} \end{array} \right\} \forall \, i, \alpha, \beta & (B) \end{cases} \quad (3)$$

From now on we will call $r_{\rho}$ the *objective function* (o.f.) and the restrictions $(B)$ for $F$, $K$ the $F$- and $K$-box, respectively $(B = B_F \cup B_K)$.

# 5   System Solving

We have a quadratic optimization problem with quadratic restrictions. The general analysis can proceed following two main directions to nullify the o.f.:

**Symbolic (S) :**  Extract the coefficient relations (polynomials in $F$) that were used in the algorithm. That is, make explicit w.r.t. $F$ the conditions to fulfil for the trace to remain valid up to the current point and, in addition, the new relation (in $F$ again) which represents the o.f., to be forced to zero.

**Numeric (N) :**  Find numerically only some particular values for the $K_i^\alpha, F_i^\beta$ satisfying above relations.

Obtaining exact, symbolic relations in $F$ is the best way to understand what's going on and set appropriately the initial values for $\mathcal{F}$. Tuning the $F$ such that these relations are satisfied *exactly*, we force the critical dangerous $c$ coefficients to be zero, skipping critical points. Note that each time that new $F$–relations are determined, they must be still verified in all of the following computations. The wish is that after some solving of $\mathcal{P}_\rho$ critical point systems, we have sufficient new relations in $F$ variables forming a zero dimensional system, whose solution(s) correspond(s) to one (or many) distinguished system $\mathcal{F}'$ near $\mathcal{F}$ with more interesting properties.

We consider the system composed by $(V_1)$ equations as living in $\mathbb{K}[F][K]$ instead than $\mathbb{K}[K, F]$, that is, with $F$ variables considered as parameters. The system becomes then a sparse parametric linear one.

$$\mathcal{M}_{\mathcal{F}} \cdot K = \begin{pmatrix} \cdot & \cdot & \cdot \\ \cdot & F_{ij} & \cdot \\ \cdot & \cdot & \cdot \end{pmatrix} \begin{pmatrix} \cdot \\ K_{ij} \\ \cdot \end{pmatrix} = 0$$

$\mathcal{M}_{\mathcal{F}}$ entries are monic $F$-monomials. Regarding symbolic manipulations, we can consider the system as temporarily living in $\mathbb{Z}[F][K]$, and pass to $\mathbb{K}$ only if necessary. There is no predefined term ordering for $F$ and $K$: this freedom will be fruitful.

The symbolic approach ($K$ variables ordering, system reduction, etc.) was described in [3]: we will refer here mainly to the numeric approach, apart from some initial symbolic management we report below, which can anyway be done.

## 5.1   Preprocessing

Let $lt(f_i) = X^{\delta_i}$: looking for a simplification of the system shape, we render the initial polynomials $f_i$ monic. This gives the advantage, all leading coefficients being exactly equal to 1, that $F_i^{\delta_i}$ variables are now fixed, and there is then no reason at all even to introduce them. We therefore have *once and for all* reduced in a trivial way the research space dimension: $F_i^{\delta_i} = F_{i0} (= 1)$ variables will never appear in $V_1$.

**Definition 4.** *A polynomial $p$ (an equation $p = 0$) is* mute *if $p$ is a linear binomial with its two coefficients equal to 1.*

Mute equations are the simplest ones that can appear in the system (we are working with monic $f_i$ polynomials !). Due to their nothing more than "renaming" function, we can substitute (we use only one index for simplicity) as follows

$$K_i + K_j = 0 \quad \Longrightarrow \quad K_i = -K_j$$

deleting some $K_i$ once and for all from the system. This helps in reducing the number of variables to be treated, saving thus space and elaboration time for data structures management representing polynomials on a computer.

It helps from the numerical point of view, too. Frequently the intervals of the two variables $K_i$ and $K_j$ are widely different, e.g one is dangerous and the other is not. The simple equality relation allows us to consider the *intersection* of the intervals, and store the new obtained ones for the forthcoming computations.

Generalizing the mute equations interval analysis, for equations like $(F_1)K_1 + (F_2)K_2 = 0$, using the two derived expressions

$$K_1 = -\frac{F_2}{F_1}K_2 \quad ; \quad K_2 = -\frac{F_1}{F_2}K_1$$

we obtain easy alternative formulae to recompute intervals for two $K$ variables, and possibly refine them by intersection.

**Definition 5.** *A variable $v$ is* single *for a linear system $S$ ($S$-single, or simply* single *if $S$ is clear from the context) if it appears only once in $S$ (in polynomial $p_v$).*

If $K_1$ is $V_1$-single, we may consider a block term ordering (possibly changing the one we're currently using) such that it is the greatest variable. Then $p_{K_1}$ will have it as leading term, and it will not be used at all, for there is nothing to reduce by it. We can then consider it as virtually discarded from $V_1$, and look now at $V_1' = V_1 \setminus \{p_{K_1}\}$. It may happen that some other $K_2$ variable appeared two times in the system, one in $p_{K_1}$ and one elsewhere. Having deleted $p_{K_1}$, $K_2$ is now single for $V_1'$, and we can discard $p_{K_2}$, too. Proceeding this way – possibly continuing changing ordering – until possible, we reduce the size of the really meaningful part of the system (less equations and variables), simulating non-effective reductions at practically no cost.

## 5.2   Looking for a Minimum

Working numerically, we must look in the $F$-box for an optimal point $\pi$ for which the o.f. computed absolute value $c_\pi$ is zero or minimal.

Being ours a (numerical) point-wise approach, we must necessarily enter a cycle of o.f. evaluations to analyze behaviour. Various zero searching criteria – grid analysis, descent gradient, etc. – may be followed, but there may be problems for all of them passing from the continue to the discrete environment (local minima, saddles, etc.) In any case, for all of them we need a procedure to compute the o.f. pointwise for various points in the $F$-box.

Having performed all the computations with MCoeffs, we obtained the desired intervals for every coefficient, but also a distinguished value inside it (approximated by the $m_S$, $m_L$ corresponding value). We indicate these values for $K$ variables with $\sigma = (\sigma_{ij})$. In order to work mainly with $F$ variables (which $K$ ones really depend on), the body of the cycle may be composed by the following black–box procedure: given an admissible point $\pi = (\pi_{ij})$ *in the F-box*, the corresponding $c_\pi$ value is returned.

- Specialize $\mathcal{M_F}$ entries with $\pi$ ($F_{ij} = \pi_{ij}$), obtaining $\mathcal{M}_\pi$.
- Solve the system $\mathcal{M}_\pi \cdot K = 0$, that is, find $\mathcal{K} = \ker(\mathcal{M}_\pi)$ in the form $K_v = N(\pi) \cdot K_p$, where $K_p$ is the set of remaining "free parameters".
- Eliminate $K_v$ variables in the o.f. by means of the found expressions in terms of $K_p$ ones.
- Now the o.f. depends only on $K_p$ variables: instantiate them with corresponding $\sigma_{ij}$ values.

We note that, given $\mathcal{F}$, we always use the same $\sigma$ values for $K_p$, for every $\pi$. When there is only one free parameter $K_p = \{\overline{K}\}$ we have

$$c_\pi = D(\pi) \cdot \overline{K}$$

where $D$ is a rational function, indicating explicitly that and how $c$ depends on initial coefficients, remembering the always present degree of freedom deriving from the possibility to multiply a polynomial for a not zero scalar, signed by the surviving $\overline{K}$ variable. In this case it is more evident that what really counts is essentially working on $D(\pi)$.

If we find an admissible point $\pi_1$ such that $s(c_{\pi_1}) \neq s(c_{\pi_0})$ – where $\pi_0$ is the one we have obtained in our particular computation – we may easily recover with the desired precision a root of the o.f., because the admissible region ($F$-box) is *convex*. We consider the segment connecting the two points, reducing thus to the univariate case. Because of the zero theorem for continue functions, there exists $\pi_D = \pi_0 + t \cdot (\pi_1 - \pi_0)$ solving the problem, with $t \in (0,1)$, and we can approximate it e.g. by successive bisections.

## 6 The Procedure

In the general setting of the NBA, we detail here the things to be done. First of all, we attach a label (a natural number) to every reduced S-polynomial in the course of the computation: we indicate with $S_i$ the $i^{\text{th}}$ reduced S-polynomial.

We introduce two data structures (lists): $\mathcal{A}$ (the *agenda*), and $\mathcal{O}$ (the *restrictions*), in which we sign all the information we obtained up to now.

▷ $\mathcal{A}$ contains triples: $\mathcal{A} = \{a_j\} = \{(i_j, c_j, t_j)\}$ where $i_j \in \mathbb{N}$ are labels ($j < \ell \Rightarrow i_j < i_\ell$) and $c_j, t_j$ individuate the "actual" $lc(S_{i_j}), lt(S_{i_j})$, respectively (see below). We say $a_j$ has label $i_j$.

▷ $\mathcal{O}$ contains triples: $\mathcal{O} = \{(o_j, \mathcal{V}_j, \sigma_j)\}$, where $o_j = \sum\limits_{i,\alpha+\beta=\rho_j} K_i^\alpha F_i^\beta$ are o.f. expressions, $\mathcal{V}_j$ are the corresponding ($V_1$) equations and $\sigma_j$ the found values for the $K$ variables.

What does "actual" mean ? The idea is the following one. We may have obtained from precedent computations that, for a specific critical point, the head of the reduced S-polynomial $r$ could effectively be set to zero. Because of numerical errors, however, the cancellations that should have taken place were not exact, and we obtained again the leading dangerous coefficient. But we know that it *must* be zero, because the actual $F$ values were set such that it should. The same may happen for other monomials beyond the head. The "actual" head is the first monomial $m$, starting from the leading one, such that the answer to the corresponding $\mathcal{P}_\rho$ problem was not "$c_j = 0$" (that is, either it's $c_j \neq 0$ or $\mathcal{P}_\rho$ was still not solved). In both cases we record in the corresponding entry of the agenda the coefficient and term of the actual head. If all $r$ coefficients can be set to zero at the same time, we use the default monomial $m_0 = 0 \cdot 1 \implies (c_j, t_j) = (0, 1)$, where $\mathbf{1}$ is the constant term.

Let's see how $\mathcal{A}$, $\mathcal{O}$ are updated. At the beginning, they are both empty. When we find a dangerous polynomial, let $i$ be the current label and $r = S_i$. Look in $\mathcal{A}$:

1. If no $a \in \mathcal{A}$ has label $i$ (in particular, if $\mathcal{A} = \emptyset$), set $r' = r, c' = lc(r)$, go to **6**.
2. If $\mathcal{A} \ni a = (i, 0, 1)$, set $r = 0$ and continue the algorithm with no updating.
3. Otherwise we must have $\mathcal{A} \ni a = (i, c, t)$ with $c \neq 0$, then remove from $r$ all the monomials $m_i = c_i \cdot t_i$ with $t_i > t$, obtaining $r'$.
4. If $c$ is not dangerous (it's an already discussed critical point), substitute with $c$ the coefficient related to the leading term $t$ of $r'$ and continue with the algorithm.
5. ($c$ is dangerous, that is, we must still decide) Let $c' = lc(r')$. If $c'$ is not dangerous, update $a$ with $(i, c', t)$ and continue with the algorithm.
6. If $c'$ is dangerous, set up and solve a $\mathcal{P}_\rho$–like problem (see details below). Let $\pi \in F$-box be the minimum point of the o.f. $c_\pi = $ o.f.$(\pi)$.
6a. If $c_\pi = 0$ (the coefficient can be set to zero), do
   ⋆ update: set (or add to $\mathcal{A}$, if not present) $a = (i, \tilde{c}, \tilde{t})$, where $(\tilde{c}, \tilde{t})$ are respectively the coefficient and term in the successive monomial of $r'$, or $(i, 0, \mathbf{1})$ if there are none left. Add $(o, \mathcal{V}_i, \sigma)$ to $\mathcal{O}$, where $o$ is the o.f. equation and $\mathcal{V}_i, {}^i K$ its related data, as explained above.
   ⋆ set initial values $(m_S, m_L)$ for $\mathcal{F}$ coefficients as $\pi$ values, obtaining $\mathcal{F}'$.
   ⋆ clear all data structures except $\mathcal{A}$, $\mathcal{O}$: restart all the computation from $\mathcal{F}'$.
6b. If $c_\pi \neq 0$ (the coefficient is surely different from zero), we must refine $c'$ interval $I' = [c'_i, c'_s]$:
   ⋆ if $c_\pi > 0$ set $c'_i = c_\pi$, otherwise set $c'_s = c_\pi$. Let $c''$ be the result, with $I'' = I_{c''}$
   ⋆ update (or add to $\mathcal{A}$) $a = (i, c'', t)$ , substitute with $c''$ the coefficient related to the leading term $t$ of $r'$ and continue with the algorithm.

We now specify the details for the $\mathcal{P}_\rho$-like problem in point **6**. If $\mathcal{O} = \emptyset$, we have the $\mathcal{P}_\rho$ problem introduced in section 4. If $\mathcal{O} \neq \emptyset$ we must take into account all the precedent added "= 0" conditions. Unfortunately, simply adding the corresponding precedent o.f. equations in $(F, K)$ to the set of restrictions $V_1$ is generally not possible, because, given a point $\pi$ in the $F$-box, we are not sure if it satisfies these added conditions. It would only if it lies on the variety associated to the set of polynomials expressing the dangerous coefficients in terms of the initial ones (this is the power of symbolic approach, that we would explicitly obtain these relations!). With the numerical approach it is practically impossible to consider exclusively points on this variety inside the $F$-box.

We propose a possible workaround: we consider a $\mathcal{P}_\rho$ problem with a modified o.f., that considers all the precedent obtained conditions ($\mathcal{O}$ entries). Let $\omega = {}^{\#}\mathcal{O}$: a modified o.f. may be

$$O = \sum_{i=1}^{\omega} o_i^2 + |o| \tag{4}$$

We squared the $o_i$ to let $\mathcal{O}$ being differentiable (apart from the absolute value, difficulty solvable with a simple case distinction). Similar functions with analogous behaviour may be equivalently used. In this way, if a point $\pi$ is found such that the sum is zero, then it satisfies *all* the restrictions, and therefore it is admissible. If the sum is not zero, then if the last term $|o_\pi|$ has a positive value $v$, then the coefficient under study is different from zero, and we can use $v$ as new interval (first or second) extreme to exclude 0 from it. If $|o_\pi|$ is zero another $\pi$ must be tested.

We again underline that after the first $\mathcal{P}_\rho$-like problem (really, a $\mathcal{P}_\rho$ problem), we cannot guarantee that $V_1$ equations have a solution inside the $(F, K)$ box, because of new restrictions overlapping. This case must still be deeply investigated by the authors, but we underline that there is the possibility that some o.f. cannot be computed, and therefore the above procedure does not apply. In this case some other kind of ZT must be defined by the user.

## 7   Examples

We present here some preliminary partial results concerning easy system examples, to show what may happen in practical cases. For simplicity, we indicate only the floating point values of the coefficients. Intervals are determined as explained in section 3.

For each example, we use e.g. the [ M(11,96,256,10,3,0,27) DRL ] notation to indicate the use of PL_MCoeffs with 7 parameters – of which the first one (a prime number) is not used now – and DRL for the degree reverse lexicographic term ordering (L for lexicographic).

$\boxed{1}$ [ M(32003,128,256,10,3,0,3) DRL]    $E_1 = \begin{cases} z - 1000 & = 0 \\ x^2 y + zx + x = 0 \\ xy^2 + zy & = 0 \end{cases}$

In this simple example we have that the S-poly $s = S(x^2y + zx + x, xy^2 + zy) = xy$. So where is the problem ? The algorithm added and subtracted $xy(z - 1000)$, and therefore the head coefficient *seems* to be zero. The resulting system is:

$$O = (F_{1,2})K_{1,0} \quad ; \quad V_1 = \begin{cases} K_{1,0} + K_{2,0} = 0 \\ (F_{1,1})K_{1,0} + (F_{2,1})K_{2,0} = 0 \end{cases}$$

From this we can see that the objective function can not be zero (indeed, the interval for $K_{1,0}$ does not contain zero). Simplifying it, we obtain the relation $F_{1,1} - F_{2,1} = 0$ (which was quite easy to detect looking at the initial system). In this case, we have a critical point which is really a "false alarm", being due only to useless computations. Doing the same calculation using syzygies eliminates the uncertainty.

$$\boxed{2} \; [\text{ M(32003,128,256,10,3,0,3) DRL }] \quad E_2 = \begin{cases} z - 10 = 0 \\ z^5 + 20x^2y + 21xy = 0 \\ z^5 + 21xy^2 + 20y^2 = 0 \end{cases}$$

Buchberger algorithm simplifies $z^5$ to $10z^4$ and so on till we obtain $10^5$. This process gives us all the $K_{0,j}$ variables, which are completely independent from computation of critical heads. The first S-polynomial is

$$s = S(20x^2y + 21xy + 10^5, 21xy^2 + 20y^2 + 10^5) = (20 \cdot 20 - 21 \cdot 21)xy^2 + \ldots$$

Since these values are really intervals, it is possible to have that the obtained head coefficient equals zero. This is not the relation we see as an objective function, because the algorithm does continue, simplifying $xy^2$. That's why we have to study a function of the form $F \cdot K$, where $K$ is dangerous.

More precisely, we have that for $6^{th}$ critical pair we find a dangerous situation:

$$O = (F_{2,2})K_{2,1}$$

$$V_1 = \begin{cases} K_{0,0} + K_{2,0} & = 0 \; (1) \quad (F_{0,1})K_{0,2} + K_{0,5} = 0 \; (7) \\ K_{0,1} + K_{1,0} & = 0 \; (2) \quad (F_{0,1})K_{0,3} + K_{0,6} = 0 \; (8) \\ (F_{0,1})K_{0,0} + K_{0,2} & = 0 \; (3) \quad (F_{0,1})K_{0,4} + K_{0,7} = 0 \; (9) \\ (F_{0,1})K_{0,1} + K_{0,3} & = 0 \; (4) \quad (F_{0,1})K_{0,5} + K_{0,8} = 0 \; (10) \\ K_{0,4} + K_{2,1} & = 0 \; (5) \quad (F_{0,1})K_{0,6} + K_{0,9} = 0 \; (11) \\ (F_{1,1})K_{1,0} + (F_{2,1})K_{2,0} & = 0 \; (6) \quad (F_{0,1})K_{0,7} + K_{0,10} = 0 \; (12) \\ (F_{1,2})K_{1,0} + (F_{2,2})K_{2,0} + (F_{2,1})K_{2,1} & = 0 \; (13) \end{cases}$$

We show here in some detail the effect of the preprocessing: to do this we numbered $V_1$ equations for clarity. Equations (1), (2) and (5) are mute. $K_{0,8}$ is single: after having removed (10), $K_{0,5}$ becomes single, and (7) is removed, too. Continuing in this way we see that the variables (and equations in which they appear) we can avoid to consider are

$$\{ K_{0,8}, \; K_{0,5}, \; K_{0,2}, \; K_{0,9}, \; K_{0,6}, \; K_{0,3}, \; K_{0,10}, \; K_{0,7} \}$$

Performing all the simplifications, the new "initial" system is surprisingly simple

$$V_1 = \begin{cases} (F_{1,1})K_{1,0} + (F_{2,1})K_{2,0} = 0 \\ (F_{1,2})K_{1,0} + (F_{2,2})K_{2,0} + (F_{2,1})K_{2,1} = 0 \end{cases}$$

from the computation, we have that $K_{1,0}$ and $K_{2,0}$ are not dangerous, while $K_{2,1}$ is. We then change ordering, with $K_{2,1}$ as greatest variable. We finally have

$$\overline{\mathcal{M}_{\mathcal{F}}} \cdot K = \begin{pmatrix} F_{1,1}F_{2,1} & 0 & \left| F_{1,1}F_{2,2} - F_{1,2}F_{2,1} \right. \\ 0 & F_{1,1} & F_{2,1} \end{pmatrix} \begin{pmatrix} K_{2,1} \\ K_{1,0} \\ K_{2,0} \end{pmatrix} = \begin{pmatrix} 0 \\ 0 \end{pmatrix}$$

and the o.f. $O$ becomes $O = \dfrac{N(F)}{D(F)} K_{2,0} = \dfrac{F_{1,1}F_{2,2} - F_{1,2}F_{2,1}}{F_{1,1}F_{2,1}} K_{2,0}$.

As expected, $N(F)$ represents a determinant. We translated the uncertainty problem from $K$ space ($K_{2,1}$ is dangerous) to a convex 4-dimensional $F$ subspace.

$\boxed{3}$ [ M(32003,64,128,10,3,0,22) L ] Windsteiger's system: the exact version is

$$E_5 = \begin{cases} -4 + 3\left(\dfrac{172966043}{174178537}x - \dfrac{42176556}{358072327}y\right)^2 + \left(\dfrac{1}{3} + \dfrac{42176556}{358072327}x + \dfrac{172966043}{174178537}y\right)^2 = 0 \\ -4 + \left(\dfrac{1}{3} - \dfrac{42176556}{358072327}y + \dfrac{172966043}{174178537}x\right)^2 + 4\left(\dfrac{172966043}{174178537}y + \dfrac{42176556}{358072327}x\right)^2 = 0 \end{cases}$$

and the approximated one (which we use) is

$$E_5' = \begin{cases} 10277480y^2 - 4678710xy + 29722520x^2 + 6620260y + 785252x - 38888890 = 0 \\ 39583780y^2 + 7018070xy + 10416220x^2 - 785252y + 6620260x - 38888890 = 0 \end{cases}$$

We report the obtained condition (partially factorized) and after substituting the exact values for $F_{i,j}$: we see that

$$O = (F_{1,5} - F_{0,5})(F_{0,1} - F_{1,1})^2 + (F_{0,4} - F_{1,4})(F_{0,1} - F_{1,1})(F_{0,2} - F_{1,2}) + $$
$$(F_{1,3} - F_{0,3})(F_{0,2} - F_{1,2})^2 \simeq 3.4125131480145708084595190432555 \cdot 10^{-16}$$

A very small value which justifies the point to be critical.

## 8  Conclusions

In this paper we presented a possible approach to zero testing in numerical Gröbner bases computations with not exact initial coefficients. The combined use of the ad–hoc introduced *multi-component* coefficients and of syzygies permitted to obtain equations to be fulfilled by the coefficients of the initial polynomials and syzygies. If, within interval tolerances, the doubtful coefficient can be zero, a new equation/restriction must be taken care of, otherwise intervals can be refined and computations proceed.

Both symbolic and numeric analysis are possible, the former slow but giving more information, the latter computationally faster but giving only punctual results. A deeper numerical analysis behaviour on real–life examples is planned.

The addition of a modular part to multi-component coefficients might also be used to apply the Gröbner trace algorithm proposed in [17]. This could lead to further improvements for guided floating point computations/analysis.

# References

1. Adams, W. W., Loustaunau, P.: *An Introduction to Gröbner bases*, Graduate Studies in Mathematics, Volume 3, AMS (1994)
2. Alefeld, G., Herzberger, J.: *Introduction to Interval Computations*, Academic Press, New York (1983)
3. Bodrato, M., Zanoni, A.: *Numerical Gröbner bases and syzygies: an interval approach*. Proceedings of the $6^{th}$ SYNASC Symposium. T. Jebelean, V. Negru, D. Petcu, D. Zaharie ed. Mirton, Timisoara, Romania, (2004) 77 - 89
4. Bonini, C., Nischke, K.–P., Traverso, C.: *Computing Gröbner bases numerically: some experiments*, Proceedings SIMAI (1998)
5. Buchberger, B.: *Introduction to Gröbner Bases* Gröbner Bases and Applications (B. Buchberger, F. Winkler eds.), London Mathematical Society Lecture Notes Series 251, Cambridge University Press, (1998) 3–31
6. Becker, T., Weispfenning, V.: *Gröbner Bases: A Computational Approach to Commutative Algebra*, Graduate Studies in Mathematics, Volume 141, Springer Verlag, (1993) (second edition, 1998).
7. Caboara, M., Traverso, C.: *Efficient Algorithms for ideal operations*, Proceedings ISSAC 98, ACM Press (1998) 147–152
8. Cox, D., Little, J., O'Shea, D.: *Ideals, Varieties, and Algorithms*, Springer–Verlag, (1991) (second corrected edition, 1998).
9. Cox, D., Little, J., O'Shea, D.: *Using algebraic geometry*, Springer-Verlag (1998)
10. FRISCO: *A Framework for Integrated Symbolic/Numeric Computation*, ESPRIT Project LTR 21024, European Union (1996–1999)
11. FRISCO test suite: `http://www.inria.fr/saga/POL/`
12. Migheli, L.: *Basi di Gröbner e aritmetiche approssimate*, Tesi di Laurea, Università di Pisa (in italian) (1999)
13. Shirayanagi, K.: *Floating Point Gröbner Bases*, Journal of Mathematics and Computers in Simulation 42, (1996) 509–528
14. Stetter, H.J.: *Stabilization of polynomial system solving with Gröbner bases*, Proceedings ISSAC (1997) 117–124
15. Stetter, H. J.: *Numerical Polynomial Algebra* SIAM, (2004)
16. Traverso, C.: *Syzygies, and the stabilization of numerical buchberger algorithm* Proceedings LMCS, RISC-Linz (2002), 244–255
17. Traverso, C.: *Gröbner trace algorithms* Proceedings ISSAC 88, LNCS 358, Springer Verlag, (1988) 125–138
18. Traverso, C., Zanoni, A.: *Numerical Stability and Stabilization of Groebner Basis Computation*, Proceedings of the 2002 International Symposium on Symbolic and Algebraic Computation, Université de Lille, France, ACM Press, Teo Mora editor, (2002) 262–269
19. Weispfenning, V.: *Gröbner Bases for Inexact Input Data* Computer Algebra in Scientific Computation - CASC 2003, Passau, TUM, (2003) 403–412
20. Zanoni, A.: *Numerical stability in Gröbner bases computation*, Proceedings of the $8^{th}$ Rhine Workshop on Computer Algebra. H. Kredel, W. K. Seidler, ed. (2002) 207–216
21. Zanoni, A.: *Numerical Gröbner bases*, PhD thesis, Università di Firenze, Italy (2003)

# Application of Computer Algebra
# for Construction of Quasi-periodic Solutions
# for Restricted Circular Planar
# Three Body Problem

V.P. Borunov[1], Yu. A. Ryabov[2], and O.V. Surkov[2]

[1] Computing Center of RAS, Moscow
[2] Moscow Auto and Highway Construction Inst. (Technical University)

**Abstract.** The algorithm is realized (with the help of computer algebra methods) for construction of numeric-analytical quasi-periodic solutions of precise(!) equations of restricted planar circular three-body
problem (Sun–Jupiter-small planet) for an arbitrary sufficiently wide variety of initial data. This algorithm and corresponding exe-code allows
us to obtain solutions in automatic mode (certainly, approximate but
satisfying the motion equations with user-specified high precision) represented by twofold Fourier polynomials. Besides, the development of
so-called perturbation function is not required (essential fact). These
solutions are valid in principle for infinite time interval unlike known
classical solutions of such problem. Such solutions are obtained for the
first time.

## 1 Problem Formulation

The following precise(not averaged) equations of restricted planar circular three-
body problem (Sun–Jupiter–asteroid) are considered that determine the orbital
elements:

$$\frac{dp}{d\theta} = \mu F_1(p, e, G, L), \quad \frac{dG}{d\theta} = 1 - \mu F_3(p, e, G, L),$$

$$\frac{de}{d\theta} = \mu F_2(p, e, G, L), \quad \frac{dL}{d\theta} = 1 - F_4(p, e, G), \tag{1}$$

where $\theta$ is the asteroid longitude measured from some constant direction, $p$ is
the orbital parameter, $e$ is the orbital eccentricity, $G = \theta - g$ is the true anomaly
(the difference between $\theta$ and the longitude of orbit perihelion $g$), $L = \theta - n_j t$
(the difference between $\theta$ is the longitude of Jupiter moving uniformly at the
angular velocity $n_j$), and $\mu$ is the mass of Jupiter. Units of mass and time are
chosen to be defined in the manner that the gravitational constant $k^2 = 1$, the
semiaxis of the circular orbit of Jupiter orbit $a_j = 1$, the Jupiter angular velocity
$n_j = 1$, the sum of the mass of the Sun $m_s$, and the mass of Jupiter $\mu$ is equal

V.G. Ganzha, E.W. Mayr, and E.V. Vorozhtsov (Eds.): CASC 2006, LNCS 4194, pp. 77–88, 2006.
© Springer-Verlag Berlin Heidelberg 2006

to 1. The functions in the right-hand sides of equations (1) may be expressed by following formulae:

$$F_1 = 2pT_1, \qquad F_2 = (2e + 2\cos G - e\sin^2 G)T_1 + T_2 \sin G,$$

$$F_3 = \tfrac{1}{e}(2 + e\cos G)\sin G \cdot T_1 - \tfrac{1}{e}\cos G \cdot T_2,$$

$$F_4 = p^{3/2}(1 + e\cos G)^{-2}, \tag{2}$$

where

$$T_1 = p^2[(1 + e\cos G)^{-3} - A^{-3/2}]\sin L,$$

$$T_2 = 1 - p^2(1 + e\cos G)^{-2}\cos L + p^2[(1 + e\cos G)\cos L - p]A^{-3/2},$$

$$A = p^2 + (1 + e\cos G)^2 - 2p(1 + e\cos G)\cos L. \tag{3}$$

Equations (1) and formulae (2), (3) are derived from Clairau–Laplace equations (see [1]) with the perturbation function

$$R = -\frac{1}{r} - r\cos L + \frac{1}{\Delta}, \tag{4}$$

where

$$\Delta = (1 + r^2 - 2r\cos L)^{1/2}. \tag{4'}$$

These equations determine the orbital elements $p, e, g$ of asteroid and the variable $L$ as the functions of the longitude $\theta$. We do not know the papers of other authors, where right these equations are considered. The time $t$ is considered also as the function of $\theta$ according to the relation

$$t = \theta - L(\theta). \tag{4''}$$

This relation is some analog of Kepler equation well known in celestial mechanics, and it is possible to obtain the explicit expression $\theta(t)$ after obtaining $L(\theta)$ from initial equations (1). So, we can determine after solving eq. (1) the values of $p(\theta), e(\theta), g(\theta)$ for every given $\theta$ and determine the corresponding moment of time $t$ from (4"). Conversely, we can determine (if the expression $\theta(t)$ is obtained) for every moment of time $t$ the orbital elements $p, e, g$ and other characteristics of motion. We consider in this paper, namely, the construction of solutions of eq. (1) since these solutions allow us to analyse the asteroid orbit (within our astronomical scheme) on infinite time interval. It is important that we can determine the osculating mean motion $n(\theta)$ along the orbit. We have for every $\theta = \theta_*$

$$n(\theta_*) = \left(\frac{1 - e(\theta_*^2)}{p(\theta_*)}\right)^{3/2}$$

according to the formulae of unperturbed motion.

The structure of our equation (1) is close to the structure of Hamiltonian systems and contains two positional variables $p(\theta), e(\theta)$ and two angular variables

$G(\theta), L(\theta)$. The functions $F_1$ and $F_2$ are odd functions of the angular variables $G, L$, and the functions $F_3, F_4$ are even, and we have every reason to construct the solution of our initial equation (1) with given initial data $p(0), e(0), G(0), L(0)$ in the form:

$$p = U_0 + \sum_{\|k\|=1}^{N} U_k \cos(k, \psi), \quad G = \psi_1 + \sum_{\|k\|=1}^{N} W_K \sin(k, \psi),$$

$$e = V_0 + \sum_{\|k\|=1}^{N} V_k \cos(k, \psi), \quad L = \psi_2 + \sum_{\|k\|=1}^{N} S_k \sin(k, \psi),$$

(5)

where $N$ is the given maximal (large enough) order of harmonics, $k$ is the integer vector $(k_1, k_2)$, and $\psi_1, \psi_2$ are components of the vector $\psi = (\psi_1, \psi_2)$,

$$\|k\| = |k_1| + |k_2|, \quad (k, \psi) = k_1 \psi_1 + k_2 \psi_2,$$

$$\psi_j = \omega_j \theta + \psi_{j0}, \quad j = 1, 2, \quad \psi_0 = (\psi_{10}, \psi_{20}).$$

(6)

The coefficients $U_0, U_k, \ldots S_k$, the frequencies $\omega_1, \omega_2$, and quantities $\psi_{10}, \psi_{20}$ are our unknowns. We seek the numerical values of these unknowns if the numerical values of mass $\mu$ and initial $p(0) = p_0, e(0) = e_0, G(0), L(0)$ (initial value of $\theta$ is zero) are given.

The algorithm for the construction of such solution (two variants) was proposed in [2] and later in [3,4,5] (in more detailed and specific form). We consider here the realization of simpler variant (so-called simple iterations) corresponding, in essence, to the classical successive approximations and to the series in small parameter $\mu$.

## 2    Basic Equations for Coefficients

Substituting (5) into (1) we obtain the relations

$$-\sum_{\|k\|=1}^{N} (k, \omega) U_k \sin(k, \psi) = \mu F_1(p, e, G, L),$$

$$-\sum_{\|k\|=1}^{N} (k, \omega) V_k \sin(k, \psi) = \mu F_2(p, e, G, L),$$

$$\omega_1 + \sum_{\|k\|=1}^{N} (k, \omega) W_k \cos(k, \psi) = 1 - \mu F_3(p, e, G, L),$$

$$\omega_2 + \sum_{\|k\|=1}^{N} (k, \omega) S_k \cos(k, \psi) = 1 - F_4(p, e, G),$$

(7)

where $\omega_1, \omega_2$ are components of the vector $\omega = (\omega_1, \omega_2)$ and $(k, \omega) = k_1 \omega_1 + k_2 \omega_2$, and arguments $p, e, G, L$ of the functions $F_1, \ldots, F_4$ are represented by polynomials (5). In addition, we have the relations connecting the initial values $p(0), \ldots, L(0)$ and quantities $U_0, V_0, \psi_{10}, \psi_{20}$:

$$p(0) = U_0 + \sum_{\|k\|=1}^{N} U_k \cos(k, \psi_0),$$

$$G(0) = \psi_{10} + \sum_{\|k\|=1}^{N} W_k \sin(k, \psi_0),$$

$$e(0) = V_0 + \sum_{\|k\|=1}^{N} V_k \cos(k, \psi_0),$$

$$L(0) = \psi_{20} + \sum_{\|k\|=1}^{N} S_k \sin(k, \psi_0). \tag{8}$$

The functions $F_1, \ldots, F_4$ can be represented (theoretically) by Fourier polynomials (if neglecting the harmonics of order higher than $N$):

$$F_l(p, e, G, L) = \sum_{\|k\|=1}^{N} F_{lk}(U, V, W, S) \sin(k, \psi), l = 1, 2,$$

$$F_3(p, e, G, L) = \sum_{\|k\|=0}^{N} F_{3k}(U, V, W, S) \cos(k, \psi),$$

$$F_4(p, e, G) = \sum_{\|k\|=0}^{N} F_{4k}(U, V, W) \cos(k, \psi), \tag{9}$$

where $F_{lk}$, $l = 1, 2, 3, 4$, $\|k\| \leq N$ are the corresponding Fourier coefficients depending on all $U_k, V_k, 0 \leq \|k\| \leq N$, $W_k, S_k, 1 \leq \|k\| \leq N$. By $U, \ldots, S$ we denote the vectors with components depending on all $U_k, \ldots, S_k$ respectively. Note, we will not need the explicit expressions for $F_{lk}$ and the development of so-called perturbation function of the three-body problem.

Formulas (7) and (9) imply the algebraic equations with respect to $U_k, V_k, W_k$, $S_k$, $1 \leq \|k\| \leq N$, $\omega_1, \omega_2$:

$$-(k, \omega)U_k = \mu F_{1k}(U, \ldots, S), \quad -(k, \omega)V_k = \mu F_{2k}(U, \ldots, S),$$

$$\omega_1 = 1 - \mu F_{30}(U, \ldots, S), \quad (k, \omega)W_k = -\mu F_{3k}(U, \ldots, S),$$

$$\omega_2 = 1 - F_{40}(U, V, W), \quad (k, \omega)S_k = -F_{4k}(U, V, W). \tag{10}$$

These algebraic equations together with relations (8) are basic. We will seek their solution by means of sequential approximations.

## 3   Zero Approximation

This approximation corresponds to the known formulae of unperturbed motion ($\mu = 0$), where $p(\theta) \equiv p_0, e(\theta) \equiv e_0, G(\theta) = \theta + G(0)$. Using the superscript (0) we obtain

$$U_0^{(0)} = p(0) = p_0, \quad V_0^{(0)} = e(0) = e_0,$$

$$G^{(0)} = \psi_1^{(0)} = \theta + \psi_{10}^{(0)}, \quad \psi_{10}^{(0)} = G(0), \quad \omega_1^{(0)} = 1,$$

$$U_k^{(0)} = V_k^{(0)} = W_k^{(0)} = 0, \quad 1 \leq \|k\| \leq N, \tag{11}$$

The variable $L^{(0)}(\theta)$ satisfies the equation

$$\frac{dL^{(0)}}{d\theta} = F_4(p_0, e_0, \psi_1^{(0)}) = 1 - \frac{p_0^{3/2}}{(1 + e_0 \cos \psi_1^{(0)})^2}. \tag{12}$$

The Fourier development of right-hand side in $\psi_1^{(0)}$ is well known [6]. Its coefficients are equal to $F_{4k}(U^{(0)}, V^{(0)}, W^{(0)})$. We obtain for $L^{(0)}$ the Fourier polynomial

$$L^{(0)} = \psi_2^{(0)} + \sum_{n=1}^{N} S_n^{(0)} \sin(n, \psi_1^{(0)}), \tag{13}$$

where

$$\psi_2^{(0)} = \omega_2^{(0)} \theta + \psi_{20}^{(0)}, \qquad \omega_2^{(0)} = 1 - \left(\frac{p_0}{1 - e_0^2}\right)^{3/2} \tag{14}$$

and $S_n^{(0)}$ are numerical coefficients, and the constant $\psi_{20}^{(0)}$ is to be defined according to (8):

$$\psi_{20}^{(0)} = L(0) - \sum_{n=1}^{N} S_n^{(0)} \sin(n\psi_{10}^{(0)}). \tag{15}$$

## 4   Higher-Order Approximations

The sequential approximations of orders $j + 1, j = 0, 1, 2, \ldots$

$$U_k^{(j+1)}, \ldots, \psi_{10}^{(j+1)}, \psi_{20}^{(j+1)}$$

are defined according to (9) and (7) by following formulae:

$$\begin{aligned}
U_k^{(j+1)} &= -\frac{\mu}{(k, \omega^{(j)})} F_{1k}(U^{(j)}, \ldots, S^{(j)}), \\
V^{(j+1)} &= -\frac{\mu}{(k, \omega^{(j)})} F_{1k}(U^{(j)}, \ldots, S^{(j)}), \\
W_k^{(j+1)} &= -\frac{\mu}{(k, \omega^{(j)})} F_{3k}(U^{(j)}, \ldots, S^{(j)}), \\
S_k^{(j+1)} &= -\frac{1}{(k, \omega^{(j)})} F_{4k}(U^{(j)}, V^{(j)}, W^{(j)})
\end{aligned} \tag{16}$$

$$1 \leq \|k\| \leq N$$

$$\begin{aligned}
U_0^{(j+1)} &= p(0) - \sum_{\|k\|=1}^{N} U_k^{(j+1)} \cos(k, \psi_0^{(j)}), \\
V_0^{(j+1)} &= e(0) - \sum_{\|k\|=1}^{N} V_k^{(j+1)} \cos(k, \psi_{(0)}^{j}),
\end{aligned} \tag{17}$$

$$\omega_1^{(j+1)} = 1 - \mu F_{30}(U^{(j)}, \ldots, S^{(j)},$$
$$\omega_2^{(j+1)} = 1 - F_{40}(U^{(j)}, V(j), W^{(j)}),$$
$$\psi_{10}^{(j+1)} = G(0) - \sum_{\|k\|=1}^{N} W_k^{(j+1)} \sin(k, \psi_0^{(j)}), \tag{18}$$
$$\psi_{20}^{(j+1)} = L(0) - \sum_{\|k\|=1}^{N} S_k^{(j+1)} \sin(k, \psi_0^{(j)}).$$

The quantities $F_{lk}(U^{(j)}, \ldots, S^{(j)}), l = 1, \ldots, 4$ are the corresponding coefficients of two-fold Fourier polynomials for functions $F_1, \ldots, F_4$ obtained after substituting the approximation $p^{(j)}, e^{(j)}, G^{(j)}, L^{(j)}$ of order $j$, i.e., the polynomials of form (5) with numerical coefficients $U_k^{(j)}, \ldots, S_k^{(j)}$, into these functions $F_1, \ldots, F_4$. The construction of such Fourier polynomials is to be realized by means of algebraic operations (the basic and multiply repeated operation is multiplication of Fourier polynomials).

Let us consider, for example, the function

$$F_1(p^{(j)}, \ldots, L^{(j)}) =$$
$$2(p^{(j)})^3 \left[ \left(1 + e^{(j)} \cos G^{(j)}\right)^{-3} - A_j^{-3/2} \right] \sin L^{(j)}, \tag{19}$$

where

$$A_j = (p^{(j)})^2 + (1 + e^{(j)} \cos G^{(j)})^2 - 2p^{(j)}(1 + e^{(j)} \cos G^{(j)}) \cos L^{(j)}. \tag{20}$$

It is necessary, first of all, to represent by Fourier polynomials in $\psi_1, \psi_2$ (neglecting the superscript) the functions $\cos L^{(j)}, \sin L^{(j)}$ and also $\cos G^{(j)}, \sin G^{(j)}$. We use the following procedure:

We have $L^{(j)} = \psi_2 + X$, where $X$ is the Fourier polynomial in $\psi_1, \psi_2$, and we can represent $\cos L^{(j)}$ as

$$\cos L^{(j)} = \cos \psi_2 \cos X - \sin \psi_2 \sin X,$$

and in analogous form $\sin L^{(j)}$. The construction of Fourier polynomials for $\sin X$ and $\cos X$ is realized with the help of Taylor formulae in powers of $X$. Multiplying these polynomials by $\sin \psi_2, \cos \psi_2$ and neglecting harmonics of order higher than $N$, we obtain the required polynomials.

The most important operation is the construction of the Fourier polynomials for $A_j^{-3/2}$. At first, we construct the Fourier polynomial for $A_j$, thereafter (19) by corresponding multiplications of polynomials for $p^{(j)}, e^{(j)}, \cos G^{(j)}, \cos L^{(j)}$. The second (principal (!)) step is the construction of the Fourier polynomial for $A_j^{-1/2}$ by means of the Newton iterations. If $Z = A_j^{-1/2}$, then we have the equation $Z^{-2} - A_j = 0$ and the iterative formula

$$Z_{n+1} = (3/2)Z_n - (1/2)Z_n^3, n = 0, 1, 2, \ldots$$

The initial $Z_0 = 1$ for $j = 0$ and $Z_0 = A_{j-1}$ for $j = 1, 2, \ldots$. We further obtain $A_j^{-3/2} = Z^3$.

(Note, this algorithm is revealed as effective in the case $p(0) < 0.6$. We have met difficulties if $p(0) > 0.6$ and intend to improve the algorithm.)

The Fourier polynomial for $(1 + e^{(j)} \cos G^{(j)})^{-3}$ is constructed also by means of the Newton algorithm for $Z = (1 + e^{(j)} \cos G^{(j)})^{-1}$.
We obtain the required polynomial

$$F_1(p^{(j)}, ..., L^{(j)}) = \sum_{\|k\|=1}^{N} F_{1k}(U^{(j)}, ..., S^{(j)}) \sin(k, \psi)$$

and required coefficients $F_{1k}(U^{(j)}, ..., S^{(j)})$ after realization of all multiplication operations in accordance with (19).

In a similar way, one can construct Fourier polynomials for functions $F_l(p^{(j)}, ..., L^{(j)}), l = 2, 3, 4$ and obtain the numerical values of $F_{lk}(U^{(j)}, ...), 0 \leq \|k\| \leq N$. Further, formulae (15)–(17) allow us to determine all $U_k^{(j+1)}, ..., S_k^{(j+1)}$ as well as

$$\omega_1^{(j+1)}, \omega_2^{(j+1)}, \psi_{10}^{(j\,|\,1)}, \psi_{20}^{(j+1)}.$$

Thus, we obtain the approximation $p^{(j+1)}, ..., L^{(j+1)}$ in form (5) with numerical coefficients.

## 5   Results

The algorithm considered above was programmed with use of the language $C^{++}$. The two-fold Fourier polynomials are stored in computer memory as triangular matrices. Indices of matrix entries are the numbers $k_1, k_2$ in expression $k_1\psi_1 + k_2\psi_2$, and the numerical values of matrix entries are equal to the sought coefficients of polynomials (5). The main operation is the multiplying of Fourier polynomials. The harmonics of order higher than $N$ were discarded automatically. The precision testing of constructed Fourier polynomials for $\cos(L^{(j)})$, $\sin(L^{(j)})$, $\cos(G^{(j)}, \sin(G^{(j)}$ was as follows. The procedure preceding the calculations of $L^{(j)}, \cos(L^{(j)}, \sin(L^{(j)}$, etc. was used for given $\theta = \theta_*$ and for the comparison of obtained values with the values of constructed polynomials for $\cos(L^{(j)})$, $\sin(L^{(j)})$, etc. and $\theta = \theta_*$. The same procedure was used for the precision testing of all constructed polynomials in cases of functions $F_l(p^{(j)}, ..., L^{(j)}), l = 1, .., 4$.

The convergence of approximations was tested by calculations of the residuals (discrepancies) $\delta_U^{(j)}, ..., \delta_S^{(j)}$, which are defined by the formulae

$$\delta_U^{(j)} = \sum_{\|k\|=0}^{N} |U_k^{(j+1)} - U_k^{(j)}|, ..., \delta_S^{(j)} = \sum_{\|k\|=1}^{N} |S_k^{(j+1)} - S_k^{(j)}|.$$

These residuals (the sums of differences between corresponding coefficients of polynomials for neighbouring approximations) estimate sufficiently well the differences between left- and right-hand sides of initial equations (1) obtained after substituting the approximation $p^{(j+1)}, ..., L^{(j+1)}$.

The calculation process runs after we enter the initial data $p(0), ..., L(0)$, the $\mu$, the maximal order $N$ of harmonics and the maximum $\varepsilon$ of admissible residuals. The calculations of the zero approximation and of further approximations proceed in automatic mode till the residuals exceed $\varepsilon$. As soon as all $\delta_U, ..., \delta_S$ become less $\varepsilon$, the process is stopped. The quantities $U_0, V_0, \omega_1, \omega_2, \psi_{10}, \psi_{20}$ and all polynomials for $p(\theta), ... L(\theta)$ are printed, where all coefficients are retained, which exceed modulo $\varepsilon$. Besides, the program allows us to calculate $p(\theta), ..., L(\theta)$ for any value of $\theta$.

The model calculations were carried out on PC Pentium 3, 1600 MHz with initial data corresponding to some real and some hypothetic asteroids. We were interested, first of all, in the practical convergence of algorithm, i.e., the possibility of constructing sufficiently precise solution in the cases of initial data $p(0), e(0)$ and mass $\mu$ corresponding to known data for minor planets and Jupiter.

We have mentioned above that two variants of algorithm were developed: 1) by the method of simple iterations and 2) by the method of iterations possessing the quadratic convergence.

According to known theoretical results, the classical method of sequential approximations or series in $\mu$ and consequently the method of simple iterations lead to the divergent process from strict mathematical point of view. So, it is not possible to obtain exact solution of equations (1) in form (5) with $N = \infty$ using the simple iterations. At the same time, according to the results of KAM - theory, the solution of equations (1) in form (5) with $N = \infty$ exists in every case if $\mu$ is sufficiently small and initial $p(0), e(0)$ do not belong to some region of acute commensurability between initial mean angular motions of Jupiter and of asteroid. But this solution is to be obtained, namely, with the help of iterations possessing the quadratic convergence.

Besides, the known specific theories of motion for many celestial bodies were constructed, namely, with the help of classical methods, and these theories contain the approximate solutions representing the real motions with high precision. So, we could hope that our algorithm of simple iterations possesses the practical convergence for $\mu = 0.0014735$ (astronomical data for Jupiter) and for $p(0), e(0)$ far from resonance values. We have nevertheless considered as the first examples the motions (correctly speaking, the model motions within planar circular scheme) of minor planets Flora (N8) and Hestia (N 46) and also of some hypothetical asteroids with $p(0), e(0)$ near the "bad" values.

We have used the following data: for Flora $p(0) = 0.41285, e(0) = 0.157$ and for Hestia $p(0) = 0.470942, e(0) = 0.1723$. These data correspond approximately to known astronomical data. The mean motion of Jupiter (at epoch 2000.0.) is equal [7] $n_j = 299.''1283$ (sec. of arc per astronomical ephemerid day). The mean motion of Flora at initial $\theta = 0$ calculated with initial $p(0), e(0)$ by formula

$$n_j \left( \frac{(1 - e(0)^2)}{p(0)} \right)^{3/2},$$

is equal to $n_0 = 1086.''300$, and this value differs from $4n_j = 1196.''513$ nearly by 10%. The calculated mean motion of Hestia at $\theta = 0$ is equal to $n_0 = 884.''655$

and this value differs from $3n_j = 897.''385$ nearly by 1.4%. So, we have the weak commensurability 1:4 in the case of Flora and sufficiently acute commensurability 1:3 in the case of Hestia.

The calculations were carried out for above mentioned initial $p(0), e(0)$, initial $G(0) = 0.2, L(0) = 0, N = 32$, and $\varepsilon = 10^{-14}$. Such precision was reached after 12 iterations (required computer time was about 1–2 min.) in the case of Flora and after 39 iterations (required time equalled 4 min.) in the case of Hestia. The polynomial for each variable $(p(\theta), e(\theta), G(\theta)$ and $L(\theta)$ ) contains more than 500 harmonics with amplitudes exceeding $10^{-15}$.

Analogous calculations with $\varepsilon = 10^{-14}, N = 32$ were carried out for the series of hypothetical asteroids with initial $p(0), e(0)$ corresponding to a more acute resonance of type 1:3 and to values sufficiently far from resonance values. For example, we have considered the model asteroids with following data:

I. $p(0) = 0.468, e(0) = 0.1723, n_0 = 893.''010$
(132 iterations, required time $\approx 10$ min. The deviation from resonance value $\Delta n_0 \approx -0.48\%$,,
II. $p(0) = 0.464, e(0) = 0.1723, n_0 = 904.''58$,
(125 iterations, time $\approx 11$ min., $\Delta n_0 \approx +0.8\%$, )
III. $p(0) = 0.430, e(0) = 0.1723, n_0 = 1013.''963$
(13 iterations, time $\approx 1.5$ min., $\Delta n_0 \approx +11.5\%$ from res. value 1:3 and $\Delta n_0 \approx -18\%$ from res. value 1:4).

Note, the amplitudes of resonance harmonics exceed essentially the amplitudes of neighbouring non-resonance. For example, the coefficient affecting $\cos(2x-3y)$ in the polynomial for $p(\theta)$ in the case of Hestia is of order $10^{-3}$ and coefficients of neighbouring harmonics are of order $10^{-6}$. The coefficient affecting $\sin(2x - 3y)$ in the polynomial for $L(\theta)$ is of order $10^{-2}$, and neighbouring coefficients are of order $10^{-6}$.

So, the proposed algorithm of simple iterations and corresponding computer program allows us to construct (in automatic regime and without using any development of the perturbation function) sufficiently precise solutions of initial equations (1) in form (5) for arbitrary initial data in every case in region $p(0) \leq 0.55$ and different $e$ from $e = 0.05$ to some $e_*$ for $p(0), e(0)$ being far from resonance values and being enough near such values. For example, $e_* \approx 0.25$ if $p(0) = 0.55, e_* \approx 0.3$ if $p(0) = 0.5, e_* \approx 0.35$ if $p(0) = 0.4, e_* \approx 0.55$ if $p(0) = 0.2$. It is sufficient to use computers of mean capacity. We can give a set of initial data corresponding to many real and hypothetical asteroids and to obtain a set of their calculated orbits. These solutions describe the quasi- periodic motions of asteroids on (theoretically) infinite interval of time.

The polynomials for $p(\theta), e(\theta)$ allow us to analyse the evolution of asteroid orbit on infinite time interval (certainly, within our astronomical scheme). Really, according to (5) ,the quantities $U_0$ and $V_0$ in (5) are the "secular" mean values of $p(\theta), e(\theta)$ to which comparatively small bi-frequency oscillations are added. So, the quantities $n_j[(1 - V_0^2)/U_0]^{3/2}$ may be considered as the "secular" mean motions $\tilde{n}$ of asteroids. For example, we have in the cases of asteroids Flora and Hestia

$\tilde{n} = 1085.''933$ and $\tilde{n} = 882.''780$ correspondingly. So, the secular $\tilde{n}$ are smaller and farther from the resonance values than initial mean motions $n$ in both case.

Similar calculations in cases of the hypothetical asteroids of Hestia type mentioned above show that only a very narrow interval (stripe) of initial motions from 894" to 903" is, so to say, inaccessible for our algorithm. Hence, the motions of numerous asteroids with mean motions $897.''4 \pm 10''$ could exist, and the effect of moving away of such asteroids from this interval is absent (theoretically). Such effect is to be theoretically explained by the analysis of more complicated astronomical scheme: the space elliptical three-body problem, the four-body problem, etc.

Note, although our polynomials (5) are valid theoretically on infinite time interval but if desired precision of orbit elements $p, e, G, L$ is given beforehand, then our polynomials allow us to calculate these elements only on a finite time interval because the expressions for $G(\theta), L(\theta)$ contain the linear members $\omega_1\theta, \omega_2\theta$ and obtained frequencies $\omega_1, \omega_2$ are not precise. For example, we have obtained these frequencies by our calculation in the case of asteroid Hestia with precision $10^{-14}$. Hence, if $\theta = 10^{-8}$, then the errors in $L$ and $G$ calculated from (5) for given $\theta$ after $10^8/2\pi$ revolution of Hestia would be $10^{-6}$ and more.

In conclusion, we would like to mention some further research within our methods.

I. Improvement of algorithm and of computer program described in this paper (the extension for the cases $p(0) > 0.6, 0 < e(0) < 1$, the construction of sequential approximations in form of series in $\mu$ only, using more real astronomical schemes (elliptical planar, space circular, space elliptical three-body scheme, four-body scheme, etc.).

II. The analysis of asteroid motions near the resonances 1:2, 2:3 etc.

III. Certainly, it might be as well to realize the algorithm possessing the quadratic convergence described in [5].

We give below also (for the sake of illustration) a small part of our results for asteroid Hestia that were printed according to our $C^{++}$ program.

Initial data: $p(0) = 0.470942, e(0) = 0.1723, G(0) = 0.2, L(0) = 0$.

Notations: $x = \omega_1\theta + \psi_{10}, y = \omega_2\theta + \psi_{20}$

Results

The basic parameters
$\omega_1 = 9.9969871021443868\mathrm{e}{-001}$
$\omega_2 = 6.6000213545385944\mathrm{e}{-001}$
$\psi_{10} = 2.2368221799861840\mathrm{e}{-001}$
$\psi_{20} = -3.7429070992677602\mathrm{e}{-002}$

Discrepancies
Discrepancy U $= 1.3325126798223544\mathrm{e}{-016}$
Discrepancy V $= 4.2010902592515010\mathrm{e}{-016}$
Discrepancy W $= 1.2361243536082627\mathrm{e}{-015}$
Discrepancy S $= 4.9388172709049770\mathrm{e}{-015}$

Polynomial for U

$+4.7264659004119086e - 001 \cos(0x + 0y)$

$-8.8740208086624291e - 006 \cos(1x + 0y)$

$+5.9768048998398947e - 005 \; cos(0x + 1y)$

$+4.3356195476691453e - 006 \cos(2x + 0y)$

$-1.2998721469472006e - 005 \cos(1x + 1y)$

$+2.9617640367125670e - 005 \cos(1x - 1y)$

$+2.1866320866149674e - 004 \cos(0x + 2y)$

$-7.9199064908986962e - 007 \cos(3x + 0y)$

$+3.8739325909845057e - 006 \cos(2x + 1y)$

$+2.5033054382310416e - 006 \cos(2x - 1y)$

$-3.1490339464068798e - 005 \cos(1x + 2y)$

$-4.0022276764026937e - 004 \cos(1x - 2y)$

$+9.2220159910696509e - 005 \cos(0x + 3y)$

$+1.3385738723935309e - 007 \cos(4x + 0y)$

$-7.2871065570478712e - 007 \cos(3x + 1y)$

$-4.3600582125007851c \quad 007 \cos(3x - 1y)$

$+4.7926786029828064e - 006 \cos(2x + 2y)$

$-5.2054863489293787e - 005 \cos(2x - 2y)$

$-1.6963905649415880e - 005 \cos(1x + 3y)$

$-1.0579545169801604e - 004 \cos(1x - 3y)$

$+4.0859499630751454e - 005 \cos(0x + 4y)$

$-2.1193516300245409e - 008 \cos(5x + 0y)$

$+1.2239262066424798e - 007 \cos(4x + 1y)$

$+8.2708623006819236e - 008 \cos(4x - 1y)$

$-7.0774891946825232e - 007 \cos(3x + 2y)$

$+4.4924155193403603e - 006 \cos(3x - 2y)$

$+2.8534114822717900e - 006 \cos(2x + 3y)$

$-1.8474434344936317e - 003 \cos(2x - 3y)$

$-8.8563600381889473e - 006 \cos(1x + 4y)$

.............................................................

Calculations for different $\theta$

$\theta = 1.0000000000000000e+000$

p = 4.7074624074386301e-001

e = 1.7301891543308384e-001

G = 1.2092830749470154e+000

L = 7.4448631501811335e-001

$\theta = 2.0000000000000000e+000$

p = 4.7062684159624679e-001

e = 1.7423712414455980e-001

G = 2.2098014462610927e+000

L = 1.4035043601425521e+000

$\theta = 3.0000000000000000e+000$

p = 4.7082845445487997e-001

e = 1.7444966403232179e-001

G = 3.2022861280344577e+000
L = 1.9537733541542606e+000

$\theta$ = 1.0000000000000000e+004
p = 4.7089640453062359e-001
e = 1.7206334702343623e-001
G = 9.9972190971244945e+003
L = 6.6000388748859841e+003

...............................................

# References

1. Subbotin, M.F.: Introduction in Theoretical Astronomy. Nauka, Moscow, 1968 (in Russian)
2. Grebenikov, E., Ryabov, Yu.: Constructive Methods in the Analysis of Nonlinear Systems. Mir, Moscow, 1983
3. Ryabov, Yu.A. Analytic-numerical solutions of the restricted three-body problem. Rom. Astron. J. **6**(1) (1996) 53–59
4. Ryabov, Yu.A.: Analytic-numerical solutions of restricted non-resonance planar three-body problem. In: Dynamics, Ephemerides and Astronometry of the Solar System, S.Ferraz-Mello et al. (eds). IAU. Printed in the Netherlands (1996) 289–292
5. Grebenikov, E.A., Mitropolsky, Yu.A., Ryabov, Yu.A.: Asymptotic Methods in Resonance Analytical Dynamics. Chapman and Hall/CRC Press, Boca Raton, London, New York, Washington, 2004
6. Szebehely, V.: Theory of Orbit (Russian translation from English), Nauka, Moscow, 1982
7. Simon, J.I.,Bretagnon, P. et al.: Astron. Astrophys. **282** (1994) 663–683

# Efficient Preprocessing Methods for Quantifier Elimination

Christopher W. Brown[1,*] and Christian Gross[2,**]

[1] Department of Computer Science, United States Naval Academy, U.S.A.
wcbrown@usna.edu
[2] Institut für Informatik II, Universität Bonn, Römerstr. 164, 53117 Bonn, Germany
grossc@cs.uni-bonn.de

**Abstract.** This paper presents a framework and prototype implementation for preprocessing quantified input formulas that are intended as input for quantifier elimination algorithms. The framework loosely follows the AI search paradigm — exploring the space of formulas derived from the input by applying various rewriting operators in search of a problem formulation that will be good input for the intended Q.E. program. The only operator provided by the prototype implementation presented here is substitution for variables constrained by equations in which they appear linearly, supported by factorization and a simple check for non-vanishing of denominators in substitutions. Yet we present examples of quantified formulas which can be reduced by our preprocessing method to problems solvable by current quantifier elimination packages, whereas the original formulas had been inaccessible to those.

## 1   Introduction

Ever since Collins [10] developed and, with his students, implemented quantifier elimination by cylindrical algebraic decomposition (CAD) quantifier elimination became more than just a theoretical tool. This was reinforced especially by the development and implementation of partial CAD by Hong [9] and later Brown [2,1] in Qepcad. The additional approach of quantifier elimination by virtual term substitution by Weispfennig [18,14] and its implementation in Redlog by Dolzmann and Sturm [11] improved the computational time needed for elimination for formulas satisfying certain degree restrictions. Other quantifier elimination algorithms have been proposed and, to some degree, implemented as well. Unfortunately, quantifier elimination programs often fail to produce results in a reasonable amount of time and space, so it is necessary to develop and implement further algorithms.

Both Redlog and Qepcad are mature implementations of quantifier elimination, yet both are quite sensitive to the way in which users phrase problems. Qepcad, for instance, simply applies CAD-based quantifier elimination, so the variable $x$ in a linear constraint like $x = 3b + a$ is not eliminated by the obvious

---

* This work was supported in part by NSF grant number CCR-0306440.
** Major parts of research done while visiting United States Naval Academy.

V.G. Ganzha, E.W. Mayr, and E.V. Vorozhtsov (Eds.): CASC 2006, LNCS 4194, pp. 89–100, 2006.

(and efficient) substitution, and instead becomes part of the CAD construction process, which is inefficient. Both systems allow assumptions on free variables to be stated explicitly, which can then be used to speed up the computation, often dramatically, but users have to be knowledgeable and sophisticated to use these facilities well. The same holds for the variable ordering, which plays a substantial role in both systems.

The present paper is a first step towards the goal of creating a "preprocessor" for quantifier elimination programs — i.e. a program that takes input from possibly inexpert users and tries to rewrite it in a form that is more suited for the quantifier elimination algorithm it will be sent to. The eventual goal is to have a simple and extensible architecture for a preprocessor that efficiently finds the kinds of rephrasings of problems that experts would make — perhaps even doing better than an expert would.

Perhaps it is not surprising, given that we are trying to write a program that solves the task of an human expert, our general approach is based on the standard AI search paradigm. We construct a tree of equivalent formulas whose root is the input, applying rewriting "operators" to generate children of nodes in the tree. For a given formula and set of operators, the full tree would be huge, so only part of the full tree is actually generated. A *heuristic function* is used to determine which nodes in the tree will be *expanded* — i.e. will have children generated — and expanded *next*. For this paper, the only "operator" we consider is substitution for variables that appear in linear equational constraints. One key property of our substitution operator, however, is that it examines the formula context surrounding the equation to determine whether the leading coefficient of the variable being eliminated can be easily determined to be non-zero, which then reduces the size of the substituted result substantially.

Although sophisticated substitution and simplification techniques are realized in Redlog nevertheless we have found examples like the SEIT model [6] that could not be solved with Redlog but for which our preprocessing method generated equivalent formulas whose quantifiers could then be eliminated by Redlog or Qepcad B [5] rather easily.

This paper describes the approach outlined above in more detail, and presents a prototype implementation of the preprocessor in Maple. Four example problems are examined in depth, all of which are "real" quantifier elimination problem inputs, i.e. not inputs artificially manufactured by the authors. Each was stated by people doing research on quantifier elimination but, none the less, the preprocessor is easily able to rewrite the formula in a way that makes it a better input to quantifier elimination algorithms.

## 2   Generating Equivalent Formulas

We only apply our rules to formulas which are existentially quantified and in DNF. Especially all formulas are disjunctions of quantified conjunctions. Furthermore all equations are assumed to be polynomial. We will refer to formulas which fulfill these assumptions as *normalized*.

**Lemma 1.** *Let $f(\mathbf{x})$ be a normalized formula of the following form*

$$f(\mathbf{x}) := \exists (x_1', ..., x_k')[g(\mathbf{x})] \vee h(\mathbf{x}) \tag{1}$$

*with $k \in \mathbb{N}$, $h(\mathbf{x})$ in normalized form, $\mathbf{x} = (x_1, ...x_n)$, $n \in \mathbb{N}$ and $\{x_1', ..., x_k'\} \subseteq \{x_1, ..., x_n\}$ the variables, and let*

$$g(\mathbf{x}) := g_1(\mathbf{x}) \wedge ... \wedge g_l(\mathbf{x}) \quad l \in \mathbb{N} \tag{2}$$

*where $\exists i, j : g_i(\mathbf{x}) \equiv (\pi(\mathbf{x})x_j = \mu(\mathbf{x}))$ where $\pi(\mathbf{x})$ and $\mu(\mathbf{x})$ are polynomials independent of $x_j$. Then the following holds*

$$f(\mathbf{x}) \equiv f'(\mathbf{x}) := \begin{cases} G_1(\mathbf{x}) \vee G_3(\mathbf{x}) \vee h(\mathbf{x}) & \text{if } x_j \notin \{x_1', ..., x_k'\} \\ G_2(\mathbf{x}) \vee G_3(\mathbf{x}) \vee h(\mathbf{x}) & \text{if } x_j \in \{x_1', ..., x_k'\} \end{cases} \tag{3}$$

*with $G_1, G_2, G_3$ defined with $\gamma(\mathbf{x}) := \frac{\mu(\mathbf{x})}{\pi(\mathbf{x})}$ as follows*

$$G_1(\mathbf{x}) := \exists (x_1', ..., x_k')[$$
$$g_1[x_j := \gamma(\mathbf{x})](\mathbf{x}) \wedge ... \wedge g_{i-1}[x_j := \gamma(\mathbf{x})](\mathbf{x}) \wedge (x_j = \gamma(\mathbf{x})) \wedge$$
$$g_{i+1}[x_j := \gamma(\mathbf{x})](\mathbf{x}) \wedge ... \wedge g_l[x_j := \gamma(\mathbf{x})](\mathbf{x}) \wedge (\pi(\mathbf{x}) \neq 0)$$
$$]$$

$$G_2(\mathbf{x}) := \exists (x_1', ..., x_{j-1}', x_{j+1}', ...x_k')[$$
$$g_1[x_j := \gamma(\mathbf{x})](\mathbf{x}) \wedge ... \wedge g_{i-1}[x_j := \gamma(\mathbf{x})](\mathbf{x}) \wedge$$
$$g_{i+1}[x_j := \gamma(\mathbf{x})](\mathbf{x}) \wedge ... \wedge g_l(\mathbf{x}) \wedge (\pi(\mathbf{x}) \neq 0)$$
$$]$$

$$G_3(\mathbf{x}) := \exists (x_1', ..., x_k')[$$
$$g_1(\mathbf{x}) \wedge ... \wedge g_{i-1}(\mathbf{x}) \wedge (0 = \mu(\mathbf{x})) \wedge$$
$$g_{i+1}(\mathbf{x}) \wedge ... \wedge g_l(\mathbf{x}) \wedge (\pi(\mathbf{x}) = 0)$$
$$]$$

*where $g_k[x_j := \gamma(\mathbf{x})](\mathbf{x})$ means that all occurrences of $x_j$ in $g_k$ are substituted by $\gamma(\mathbf{x})$ before calculation with $\mathbf{x}$.*

*Proof.* The equation $\pi(\mathbf{x})x_j = \mu(\mathbf{x})$ gives us that $g(\mathbf{x})$ can only be true if

$$((x_j = \gamma(\mathbf{x})) \wedge (\pi(\mathbf{x}) \neq 0)) \vee ((\mu(\mathbf{x}) = 0) \wedge (\pi(\mathbf{x}) = 0)) \tag{4}$$

holds. This leads to the following case distinction: If $x_j \notin \{x_1', ..., x_k'\}$ the equation $g_i$ stays within the formula, since it is an assumption stated for the free variable $x_j$. Following the argument that $g(\mathbf{x})$ can only be true in the two cases in (4) this shows that $f(\mathbf{x})$ is true if and only if $G_1(\mathbf{x}) \vee G_3(\mathbf{x}) \vee h(\mathbf{x})$ is true.

On the other hand if $x_j \in \{x'_1, ..., x'_k\}$ the quantified variable $x_j$ will be completely removed by substitution in $G_2(\mathbf{x})$ and the quantifier will vanish. In this case $f(\mathbf{x})$ is true if and only if $G_2(\mathbf{x}) \vee G_3(\mathbf{x}) \vee h(\mathbf{x})$ is true.                            □

Important for further computation is that this lemma should produce a normalized formula, but in general we don't have polynomial equations after substitution, because of the denominator $\pi(\mathbf{x})$.

To restore the normalization we remove denominators in the obvious way: If $d_j$ is the degree of $x_j$ in an given (in)equality prior to substitution, multiplying both sides by $\pi(\mathbf{x})^{d_j + (d_j \bmod 2)}$ after substitution produces a polynomial. Moreover, the relational operator stays the same, since $d_j + (d_j \bmod 2)$ is always even.

We will apply without loss of generalization the factorization within the next Lemma to the equation $g_1$.

**Lemma 2.** *Let $f(\mathbf{x})$ be a normalized formula with $f(\mathbf{x}) = g(\mathbf{x}) \vee h(\mathbf{x})$ as above where $g(\mathbf{x})$ is in the following form*

$$g(\mathbf{x}) := \exists(x'_1, ..., x'_k) [\underbrace{(0 = g_{11}(\mathbf{x}) \cdots g_{1n_1}(\mathbf{x}))}_{=:g_1(\mathbf{x})} \wedge g_2(\mathbf{x}) \wedge ... \wedge g_l(\mathbf{x})] \qquad (5)$$

*with $n_1, l \in \mathbb{N}$ so the following will hold*

$$f(\mathbf{x}) \equiv f'(\mathbf{x}) := \begin{cases} (\exists(x'_1, ..., x'_k)[(g_{11}(\mathbf{x}) = 0) \wedge g_2(\mathbf{x}) \wedge ... \wedge g_l(\mathbf{x})]) \vee \\ \vdots \\ \vee \\ (\exists(x'_1, ..., x'_k)[(g_{1n_1}(\mathbf{x}) = 0) \wedge g_2(\mathbf{x}) \wedge ... \wedge g_l(\mathbf{x})]) \\ \vee \\ h(\mathbf{x}) \end{cases} \qquad (6)$$

*Proof.* We use that $g_1(\mathbf{x})$ can be rephrased equivalently:

$$(0 = g_{11}(\mathbf{x}) \cdots g_{1n_1}(\mathbf{x})) \equiv (g_{11}(\mathbf{x}) = 0) \vee ... \vee (g_{1n_1}(\mathbf{x}) = 0) \qquad (7)$$

Using the distributive law for $\exists$ and $\vee$ and between $\wedge$ and $\vee$ we get

$$(\exists(x'_1, ..., x'_k)[((g_{11}(\mathbf{x}) = 0) \vee ... \vee (g_{1n_1}(\mathbf{x}) = 0)) \wedge g_2(\mathbf{x}) \wedge ... \wedge g_n(\mathbf{x})]$$
$$\equiv$$
$$\exists(x'_1, ..., x'_k)[((g_{11}(\mathbf{x}) = 0) \wedge g_2(\mathbf{x}) \wedge ... \wedge g_n(\mathbf{x}))]$$
$$\vee \cdots \vee$$
$$\exists(x'_1, ..., x'_k)[((g_{1n_1}(\mathbf{x}) = 0) \wedge g_2(\mathbf{x}) \wedge ... \wedge g_n(\boldsymbol{x}))]$$

Adding the disjunction of $h(\mathbf{x})$ concludes the proof.                            □

## 3   Algorithm for Choosing Substitutions

Our approach is to mimic search strategies, which have been introduced in the field of Artificial Intelligence, for calculating equivalent formulas. With this approach all possible equivalent formulas can be computed, or a smaller number of new formulas computed greedily.

## 3.1   Grading Function

Just as in A*-Search, introducing heuristics into our algorithm requires that we define a grading function, which assigns numerical values to formulas. A similar task has been discussed by Dolzmann, Seidel and Sturm in [12]. For the sample computations presented in section 4 we will use the grading functions discussed below. However, the algorithm presented in 3.2 can be used with other grading functions as well.

The natural grading function would be the time needed for quantifier elimination on the given formula by the target Q.E. program. This is, of course, impractical, so we need to be more heuristic. Several simple grading functions are possible. A first approach considers the number of quantifiers, equations, quantified and free variables, as well as the number of and's and or's within the formula. This induces the preliminary grading function $\eta_1$ for normalized formulas $f(\mathbf{x})$:

$$\eta_1(f) = w_a N(f, \wedge) + w_o N(f, \vee) + w_e N(f, \exists) +$$
$$w_q \sum_{x \in \{x'_1, \ldots, x'_k\}} (N(f, x)) + w_f \sum_{x \notin \{x'_1, \ldots, x'_k\}} (N(f, x))$$

Where $w_a, w_o, w_e, w_q, w_f \in \mathbb{R}$ are weighting parameters, the quantified variables in $f$ are given by $\{x'_1, \ldots, x'_k\} \subset \{x_1, \ldots, x_n\}$, $n, k \in \mathbb{N}$ and $N(f, o)$ denotes the number of occurrences of $o$ in $f$.

This grading function grades only linear in the length of the formula $f$, but the cost of quantifier elimination is at least polynomial, hence better approximation needs a more complex grading function. For computation of the results in section 4 we used the following grading function within our preprocessor.

$$\eta_2(f) = \begin{cases} w_{ex}\eta_2(g) & \text{if } f(\mathbf{x}) = (\exists y)[g(\mathbf{x})] \\ w_{con} \sum_{i=1}^{l} \eta_2(g_i) & \text{if } f(\mathbf{x}) = g_1(\mathbf{x}) \wedge \ldots \wedge g_l(\mathbf{x}) \\ w_{dis} \sum_{i=1}^{l} \eta_2(g_i) & \text{if } f(\mathbf{x}) = g_1(\mathbf{x}) \vee \ldots \vee g_l(\mathbf{x}) \quad (8) \\ w_{eq} \cdot \max_{x_i \in var\{f\}} (\deg x_i) + \eta_1(f) & \text{if } f(\mathbf{x}) \text{ is an equality} \\ w_{iq} \cdot \max_{x_i \in var\{f\}} (\deg x_i) + \eta_1(f) & \text{if } f(\mathbf{x}) \text{ is an inequality} \end{cases}$$

With further weighting parameters[1] $w_{ex}, w_{con}, w_{dis}, w_{eq}, w_{iq} \in \mathbb{R}$.

If the weighting parameters within the grading function can be chosen at runtime, one can improve the anticipation of the time needed for quantifier elimination with training of these parameters.

## 3.2   Algorithm

Using Lemmas 1 and  2, Algorithm 1 on page 94 computes an equivalent (and hopefully simpler) normalized formula for a given input. If the algorithm remem-

---
[1] Note that the parameters $w_a, w_o, w_e$ are no longer in use.

**Table 1.** Preprocessing algorithm

---

**Input:** Normalized formula $f(\mathbf{x})$
**Output:** Logical equivalent formula $f'(\mathbf{x})$

**Initialization:**
    Simplify $f(\mathbf{x})$ with the rules of boolean logic.
    set $p = \eta(f(\mathbf{x}))$, where $\eta$ is a grading function
    insert $p, f(\mathbf{x})$ into a priority queue

**Computation:**
    **while** the priority queue is not empty **do**
        dequeue $f(\mathbf{x})$ and factorize it as per Lemma 2
        simplify the factorized formula with the rules of boolean logic
        compute the substitution possibilities as per Lemma 1

        **for** all substitution possibilities **do**
            apply the substitution to $f(\mathbf{x})$ to get a new formula $g(\mathbf{x})$
            set $p = \eta(g(\mathbf{x}))$
            insert $p, g(\mathbf{x})$ into the priority queue
        **end do**
    **end do**

    **return** formula with the smallest grade

---

bers furthermore the equations which have been used for substitutions, then each substitution will be applied only once and the search tree has finite size due to the finiteness of a given input formula.

### 3.3 Occurring Problems

The main problem with the basic algorithm is the explosion in the number and complexities of formulas in the search tree. This section describes why these problems arise and what strategies are used to deal with them.

Formulas at each node grow more and more complex for two reasons: First, there is the case distinction between $\pi(\mathbf{x}) = 0$ and $\pi(\mathbf{x}) \neq 0$ in Lemma 1, which replaces one sub formula by two new formulas. So after application of $n$ substitutions the formula can be $2^n$ times larger then the initial version. Second is the factorization Lemma 2. If $g(\mathbf{x}) = g_1(\mathbf{x}) \cdots g_l(\mathbf{x})$, the new formula produced by factorization can be $l$ times larger then the initial formula.

Worse still is the large branching factor in the search space, such that even small input formulas produce huge search trees. If, at a given node, each equality $g_i(\mathbf{x})$ provides $s_i$ substitution possibilities, that node has $\sum_i s_i$ successors[2]. This

---

[2] Note that application of a substitution usually doesn't remove many other substitution possibilities, so that each succeeding node provides many possibilities as well, often even more than the parent node.

makes pruning and greedy strategies necessary. Furthermore one can observe that different nodes within the tree contain the same formulas, so that hashing to recognize duplication is recommend.

### 3.4   Combating the Increase in Formula Size

A human expert making the substitution $x := a/(b^2 + 1)$ would never construct a sub formula for the case in which the denominator vanishes. Moreover, for a substitution like $x := a/(b^3 + 1)$, an expert would spend some time seeing if the surrounding context in the formula constrained $b$ in such a way that the denominator could be determined to be non-zero. In certain cases our algorithm is able to do the same thing. It browses quickly through the formula and collects constraints concerning variables (e.g. $x > 0$) and checks whether these constraints trivially imply that the denominator in question is non-zero. For example, if the denominator is $a + 1$, and $a > 0$ is a constraint in the formula, the quick test determines that $a + 1 \neq 0$. A sub formula with the equation $a + 1 = 0$ would not be constructed.

During normalization, a certain degree of simplification along the same lines is done. If the constraints collected on the variables are easily seen to contradict an atomic formula, it is replaced with false and the surrounding formula is simplified at the boolean level.

### 3.5   Dealing with the Size of the Search Space

As described earlier, the size of the space in which we are searching usually makes it impractical to generate the entire search tree. Thus, we must employ some kind of pruning strategy to limit our search. When expanding a new node our preprocessor uses the following strategy: Successor nodes whose grades are $\alpha$-times larger than the grade of the parent node or $\beta$-times larger then the best known grade are thrown away. We have chosen values of $\alpha$ and $\beta$ within the interval $[1, 2] \subset \mathbb{R}$ which led to the satisfying results within the next section.

## 4   Results

This section examines several example problems and compares Redlog and Qepcad with and without the use of the preprocessor[3]. Our computations have been done with a Maple 10 implementation of the concepts above at a AMD 2400+ (1660 MHz) with 512 MB of Ram. For the quantifier elimination we have used on the one hand Redlog 3.0 (2004-04-15) with Reduce 3.8 (version of 2005-11-22), as well as the Qepcad B based simplification slfq [4] and on the other hand Qepcad B [3] (version of 2005-05-17). Redlog was free to choose a variable order within our experiments. For quantifier elimination with Qepcad B, variables were ordered by a simple heuristic based on degrees and number of appearances, and "equational constraints" as developed in [8, 15, 16, 7] were not used.

---

[3] The parameters for the grading function $\eta_2$ as introduced in section 3.1 have been chosen as follows: $w_q = 4, w_f = 3, w_{ex} = 40, w_{con} = 3, w_{dis} = 3, w_{eq} = 1, w_{iq} = 4$.

## 4.1   Edges Square Product

The following Problem has been stated by George Collins and provides two
equalities which can easily be used for linear or virtual substitution.

$$(\exists x_1)(\exists x_2)(\exists y_2) \, [ \, (x = x_1 x_2 - y_2) \wedge (y = x_1 y_2 + x_2) \, \wedge$$
$$(0 \leq x_1) \wedge (x_1 \leq 2) \wedge (2 \leq x_2) \wedge (x_2 \leq 4) \wedge (-1 \leq y_2) \wedge$$
$$(y_2 \wedge 1) \wedge (-1 \leq x) \wedge (x \leq 9) \wedge (-6 \leq y) \wedge (y \leq 6)]$$

With choice of $\alpha = \beta = 1$ our tool calculates a search tree of size 4 within
2.790 seconds and discovers one formula with only one quantified variable.

Our tool substitutes $x_2 = y - x_1 y_2$ and $y_2 = (-x + x_1 y)/(x_1^2 + 1)$, which leads
to the formula

$$(\exists x_1) \, [ \, (-x_1 \leq 0) \wedge (x_1 \leq 2) \wedge (0 \leq 1 + x) \wedge (x \leq 9) \, \wedge$$
$$(0 \leq y + 6) \wedge (y \leq 6) \wedge (0 \leq -(x_1^2 + 1)(-y - x_1 x + 2x_1^2 + 2)) \, \wedge$$
$$(((-x + x_1 y)(x_1^2 + 1) \leq (x_1^2 + 1)^2)) \wedge (x_1^2 + 1 \neq 0) \, \wedge$$
$$(0 \leq (x_1^2 + 1)(x_1^2 + 1 - x + x_1 y)) \wedge ((x_1^2 + 1)(y + x_1 x) \leq 4(x_1^2 + 1)^2)]$$

This formula has been expanded as the last node. Qepcad needs for its evaluation 0.973 seconds producing a formula consisting of 12 atoms if the variable
order is $(x, y, x_1)$. The overall time of both methods is 3.763 seconds.

Redlog took for the initial or the substituted formula 0.1 seconds and resulted
in both cases in a formula consisting of 231 atoms. Further simplification with
slfq led to a formula consisting of 9 atoms in 2.380 seconds.

## 4.2   Putnam Example

The following is the direct translation into a quantifier elimination problem of
Problem 2 from the 57th William Lowell Putnam Competition.

$$(\exists x_1)(\exists y_1)(\exists x_2)(\exists y_2) \, [ \, x_1^2 + y_1^2 - 1 = 0 \wedge (x_2 - 10)^2 + y_2^2 - 9 = 0 \, \wedge$$
$$x = (x_1 + x_2)/2 \wedge y = (y_1 + y_2)/2]$$

In this case we can eliminate up to two quantified variables which makes the
problem much easier for Qepcad to solve. With choice $\alpha = \beta = 1$ our tool
computed a tree of size 9 in 0.550 seconds and returned four formulas with
only two quantified variables as best results. For the following formula our tool
substituted $x_2 := 2x - x_1$ and $y_1 := 2y - y_2$

$$(\exists x_1)(\exists y_2) \, [ \, (x_1^2 + 4y^2 - 4yy_2 + y_2^2 - 1 = 0) \, \wedge$$
$$(4x^2 - 4xx_1 - 40x + x_1^2 + 20x_1 + 91 + y_2^2 = 0)]$$

The node containing this formula has been expanded as the 8th node and with
$(x, y, x_1, y_2)$ chosen as variable order this formula can be evaluated by Qepcad
in 0.256 seconds, which returns a formula with 2 atoms. Hence, computation of
this result with preprocessing and Qepcad took 0.806 seconds.

Redlog took for the initial formula and the substituted version less then 0.1 seconds and returned in both times a formula consisting of 53 atoms. Further simplification with slfq took 0.345 seconds which also resulted in a formula with 2 atoms.

### 4.3   YangXia

This problem comes from Yang Xia [19].

$$(\exists s)(\exists b)(\exists c) \, [ \, (a^2h^2 - 4s(s-a)(s-b)(s-c) = 0) \wedge (2Rh - bc = 0) \wedge$$
$$(2s - a - b - c = 0) \wedge (b > 0) \wedge (c > 0) \wedge (R > 0) \wedge (h > 0) \wedge$$
$$(a + b - c > 0) \wedge (b + c - a > 0) \wedge (c + a - b > 0)]$$

The computation with $\alpha = \beta = 1$ led to a search tree with 10 nodes and took 3.280 seconds. We had 4 different formulas with only one quantifier left (with pruning factor $\alpha = \beta = 2$ we discovered 7).

Substitution of $s := \frac{1}{2}(a + b + c)$, $c := 2Rh/b$ and cutting of the part $b = 0$, since $b > 0$ is assumed, produces the following equivalent formula

$$(\exists b) \, [ \, (-\frac{1}{2}b \neq 0) \wedge (0 < R) \wedge (0 < b) \wedge (0 < h) \wedge$$
$$(\frac{1}{16}a^2h^2b^4 - \frac{1}{32}a^2b^6 -$$
$$\frac{1}{8}a^2R^2h^2b^2 - \frac{1}{8}R^2h^2b^4 + \frac{1}{64}b^8 + \frac{1}{64}a^4b^4 + 1/4R^4h^4 = 0) \wedge$$
$$(0 < -1/4(-ab - b^2 + 2Rh)b) \wedge (0 < \frac{1}{2}Rhb) \wedge$$
$$(0 < \frac{1}{4}(2Rh + ab - b^2)b \wedge (0 < \frac{1}{4}(b^2 + 2Rh - ab)b)]$$

Qepcad took 4.339 seconds for the evaluation and returned a formula consisting of 6 atoms. We have chosen the order $(R, h, a, b)$ and assumed that $(0 < R) \wedge (0 < b) \wedge (0 < h)$. The overall time was 7.619 seconds.

Redlog was not able to compute a quantifier free formula equivalent to the initial formulation or to the substituted version due to internal errors whether we stated assumptions or not.

### 4.4   SEIT

The SEIT Model[4] introduced in epidemic modeling [17] has been investigated in [6] by quantifier elimination methods: Whereas it has been possible to derive

---

[4] The inability of Redlog and Qepcad B both individually or combined to give a satisfactory solution to the SEIT problem was the primary inspiration for the work in this paper. The fact that, with about 10 minutes of pencil and paper work, we could rewrite the original problem to something Qepcad B could easily solve seemed to indicate that there was a need for this kind of preprocessing tool. Thus, the reader may dismiss this example as a "rigged" experiment, as the tool was in some sense designed for it, but we hope it will be found to be especially illuminating instead.

threshold conditions by investigating the stability of the disease-free equilibrium, the following quantified formula asking for the existence of an endemic equilibrium could not be reduced to a quantifier-free equivalent by Redlog or Qepcad within one day of computation time.

$$(\exists s)(\exists F)(\exists J)(\exists T)[$$
$$(d - ds - b_1 J s = 0) \wedge (b_1 J s + b_2 J T - (d + v + r_1)F + (1 - q)r_2 J = 0) \wedge$$
$$(vF - (d + r_2)J = 0) \wedge (-dT + r_1 F + q r_2 J - b_2 T J = 0) \wedge$$
$$(F > 0) \wedge (J > 0) \wedge (T > 0) \wedge (s > 0) \wedge (b_1 > 0) \wedge (b_2 > 0) \wedge (d > 0) \wedge$$
$$(v > 0) \wedge (r_1 > 0) \wedge (r_2 > 0) \wedge (q > 0) \wedge (b_1 > b_2)]$$

Our tool took for traversing and expanding a tree consisting of 18 nodes 32.678 seconds and returned 3 formulas with only 1 quantified variable left. The following formula is one of them.

With substitution of $F := (d + r_2)J/v$, factorization and substitution of $J := 0$, $T := (-vb_1 s + dr_1 + vr_2 q + d^2 + dv + dr_2 + r_2 r 1)/(vb_2)$ and $v := 0$, as well as the substitution of $s := \frac{d}{d + Jb_1}$ and cutting of all contradicting sub formulas, this leads to

$$\exists J[$$
$$(0 < d) \wedge (0 < r_1) \wedge (0 < r_2) \wedge (0 < q) \wedge (0 < b_1 - b_2) \wedge (0 < v) \wedge$$
$$(0 < J) \wedge (0 < b_1) \wedge (0 < b_2) \wedge (d + Jb_1 \neq 0) \wedge (-v \neq 0) \wedge$$
$$(0 < (d + r_2)Jv) \wedge (vb_2 \neq 0) \wedge (0 < d(d + Jb_1)) \wedge$$
$$(0 < (d + Jb_1)b_2 v(-dvb_1 + d^2 v + d^2 r_2 + dvr_2 q + d^3 + d^2 r_1 + Jb_1 vr_2 q +$$
$$dr_2 r_1 + Jb_1 dv + Jb_1 dr_2 + Jb_1 r_2 r_1 + Jb_1 d^2 + Jb_1 dr_1)) \wedge$$
$$(-(d + Jb_1)b_2 v^3 d(-dvb_1 - Jb_1 vb_2 + d^2 v + d^2 r_1 + dvr_2 q +$$
$$d^3 + b_2 J d^2 + d^2 r_2 + dr_2 r_1 + b_2 J dv + b_2 J dr_2 + Jb_1 dv + Jb_1 dr_1 +$$
$$Jb_1 vr_2 q + Jb_1 d^2 + J^2 b_1 db_2 + Jb_1 dr_2 + Jb_1 r_2 r_1 + J^2 b_1 b_2 v + J^2 b_1 r_2 b_2) = 0)]$$

Redlog took for its quantifier elimination 0.1 seconds and returned a formula consisting of 113 atoms. Further simplification with slfq was impossible.

Qepcad B needed 4.496 seconds, when we chose $(b_1, d, v, b_2, r_2, q, r_1, J)$ as variable order [5] and assumed that $(0 < d) \wedge (0 < r_1) \wedge (0 < r_2) \wedge (0 < q) \wedge (b_2 < b_1) \wedge (0 < v) \wedge (b_1 > 0) \wedge (b_2 > 0)$. In this case a formula consisting of one atom was returned by Qepcad B. The overall time for preprocessing and Qepcad was 37.174 seconds.

## 5   Future Work

For improving the computation speed, we want to test a lazy evaluation approach for grading new formulas. By now, we have to apply a substitution for grading

---

[5] In this example the choice of the right variable order is crucial. The quantifier elimination with Qepcad for this formula with other variable orders needs much more time.

the resulting formula. But the substitution process is expensive and so it would be cheaper, if the grading would be based on the formula, associated with the parent node, and the substitution possibility.

It would be interesting to train the grading function by solving the minimization problem: Find parameters $p = (w_q, w_f, w_{ex}, w_{con}, w_{dis}, w_{eq}, w_{iq})$ so that for a set of given formulas $\{f_1, ..., f_k\}$ and computation times of Redlog or Qepcad B for this formulas $ct(f_i)$ the following holds

$$\sum_i (\eta_{2,p}(f_i) - ct(f_i))^2 = \min_{r \in \mathbb{R}^7} \sum_i (\eta_{2,r}(f_i) - ct(f_i))^2 \tag{9}$$

with $\eta_{2,p}$ the grading function introduced in section 3.1 and $p$ set as parameters.

The current work should be considered as a first step in developing new preprocessing methods for quantifier elimination. It seems to be possible to combine our approach with the general virtual substitution mechanism, making it applicable to variables of degree two and beyond (and also to other contexts, in which virtual substitution applies). We strongly suspect that a quick test for determining coefficients and factors to be non-vanishing *before* substitutions are made will prove to be far better than making the generic substitutions and trying to simplify afterwards. Also the investigation of the relationship to and a possible combination with other preprocessing methods applicable in special contexts will be the topic of future research, e.g. with the preprocessing method based on Gröbner bases computations used to generate first-order formulas describing the existence of Hopf bifurcation fixed points [13].

# References

1. Christopher W. Brown, *Simplification of truth-invariant cylindrical algebraic decompositions*, Proceedings of the 1998 international symposium on Symbolic and algebraic computation, ACM Press, 1998, pp. 295–301.
2. _____, *Simple CAD construction and its applications*, Journal of Symbolic Computation **31** (2001), no. 5, 521–547.
3. _____, *The QEPCAD B system*, 2002.
4. _____, *The SLFQ system*, 2002.
5. _____, *QEPCAD B: a program for computing with semi-algebraic sets using CADs*, ACM SIGSAM Bulletin **37** (2003), no. 4, 97–108.
6. Christopher W. Brown, M'hammed El Kahoui, Dominik Novotni, and Andreas Weber, *Algorithmic methods for computing threshold conditions in epidemic modelling*, Computer Algebra in Scientific Computing (CASC '04) (St. Petersburg, Russia) (V. G. Ganzha, E. W. Mayr, and E. V. Vorozhtsov, eds.), July 2004, pp. 51–60.
7. Christopher W. Brown and Scott McCallum, *On using bi-equational constraints in cad construction*, ISSAC '05: Proceedings of the 2005 international symposium on Symbolic and algebraic computation (New York, NY, USA), ACM Press, 2005, pp. 76–83.
8. G. E. Collins, *Quantifier elimination by cylindrical algebraic decomposition - 20 years of progress*, Quantifier Elimination and Cylindrical Algebraic Decomposition (B. Caviness and J. Johnson, eds.), Texts and Monographs in Symbolic Computation, Springer-Verlag, 1998.

9. G.E. Collins and H. Hong, *Partial cylindrical algebraic decomposition for quantifier elimination*, Journal of Symbolic Computation **12** (1991), no. 3, 299–328.

10. Georges Collins, *Quantifier elimination for real closed fields by cylindrical algebraic decomposition*, Second GI Conference on Automata Theory and Formal Languages, Lecture Notes in Computer Science, vol. 33, Springer-Verlag, 1975, pp. 134–183.

11. A. Dolzmann and T. Sturm, *Redlog: Computer algebra meets computer logic*, ACM SIGSAM BULLETIN **2-9** (1997), no. 31.

12. Andreas Dolzmann, Andreas Seidl, and Thomas Sturm, *Efficient projection orders for CAD*, ISSAC '04: Proceedings of the 2004 international symposium on Symbolic and algebraic computation (New York, NY, USA), ACM Press, 2004, pp. 111–118.

13. M. El Kahoui and A. Weber, *Deciding Hopf bifurcations by quantifier elimination in a software-component architecture*, Journal of Symbolic Computation **30** (2000), no. 2, 161–179.

14. Rüdiger Loos and Volker Weispfenning, *Applying linear quantifier elimination*, The Computer Journal **5** (1993), 450–462.

15. S. McCallum, *On projection in CAD-based quantifier elimination with equational constraint*, Proc. International Symposium on Symbolic and Algebraic Computation (Sam Dooley, ed.), 1999, pp. 145–149.

16. S. McCallum, *On propagation of equational constraints in CAD-based quantifier elimination*, Proc. International Symposium on Symbolic and Algebraic Computation (Bernard Mourrain, ed.), 2001, pp. 223–230.

17. P. van den Driessche and James Watmough, *Reproduction numbers and sub-threshold endemic equilibria for compartmental models of disease transmission*, Mathematical Biosciences **180** (2002), 29–48.

18. Volker Weispfenning, *The complexity of linear problems in fields*, Journal of Symbolic Computation **5** (1988), no. 1-2, 3–27.

19. Yang Xia, *Real solution classifications of parametric semi-algebraic systems*, Proceedings of A3L, 2005.

# Symbolic and Numerical Calculation of Transport Integrals for Some Organic Crystals

A. Casian[1], R. Dusciac[1], V. Dusciac[1], and V. Patsiuk[2]

[1] Technical University of Moldova, Av. Stefan cel Mare, 168,
MD-2004 Chisinau, Rep. of Moldova
[2] State University of Moldova, Str. A. Mateevici, 60,
MD-2009 Chisinau, Rep. of Moldova

**Abstract.** The kinetic integral equation for a quasi-one-dimensional organic crystal is solved exactly, and the expression for the electrical conductivity is presented as a transport integral. The latter has two singularities depending on crystal parameters. The possibilities to obtain analytic expressions in some particular cases and the general numerical calculations with pronounced singularities are analysed.

## 1 Introduction

Organic materials attract more and more attention of investigators as low cost replacements for conventional metals and inorganic semiconductors and as materials with much more diverse and often unusual physical properties [1]. These materials are now widely used as the active elements of organic-based devices such as light-emitting diodes and lasers for displays, photovoltaic cells, field-effect transistors, and real-time holographic optical recording and processing systems [2, 3, 4, 5]. Unusual high electrical conductivity [6, 7] and the thermoelectric efficiencies [8, 9, 10, 11] are also predicted in a special class of quasi-one-dimensional (Q1D) organic crystals under certain conditions. However, in spite of a great number of theoretical and experimental publications in this field, the mechanism of charge transport in organic materials is not completely understood yet. This is connected, on the one hand, with a rather large and diverse number of transport mechanisms in these materials, and on the other hand, each mechanism is more complicated than in ordinary inorganic materials.

We have studied a model of Q1D organic crystal which takes into account simultaneously two main interactions of conduction electrons with acoustic vibrations of crystalline lattice [6, 7], and also the scattering on impurities and defects. The first interaction is similar to deformation potential and is caused by the variation of the transfer energy of an electron from a molecule to the nearest one. The other interaction is polaron similar and is determined by the variation of the polarization energy of molecules surrounding the conduction electron. These interactions have been studied in the literature [12, 13, 14]. But we have considered these two interactions for the first time together in Q1D crystals. This is important because under certain conditions the interference between

V.G. Ganzha, E.W. Mayr, and E.V. Vorozhtsov (Eds.): CASC 2006, LNCS 4194, pp. 101–108, 2006.

these interactions can take place. As a result, these two interactions compensate each other significantly for a strip of states in the conduction band. This leads to the fact that the relaxation time as function of carrier energy has a maximum for these states, which may be rather sharp.

In this paper we analyse the solution of kinetic integral equation for the distribution function of carriers under the application of an electrical field and calculate numerically the transport integrals, which contain a function with two singularities. For a particular case of parameters, when an approximate analytical calculation is possible, the precision of this calculation is compared with that of numerical calculation.

## 2   Basic Equations

We will consider a simplified 1D crystal model the linear chains of which are described by the Hamiltonian [6, 7]:

$$H = \sum_k \varepsilon(k) a_k^+ a_k + \sum_q \hbar \omega_q b_q^+ b_q + \sum_{k,q} A(k,q) a_k^+ a_{k-q} \left( b_q + b_{-q}^+ \right) + H_{im}, \quad (1)$$

where the notations are the same as in [7]. In (1) the first term is the energy operator of electrons, the second term is the energy operator of acoustical vibrations (described by phonons), the third term is the operator of electron-phonon interaction, and the last term describes the scattering on impurity, $k$ and $q$ are the projections along the chain direction of the quasi-wave vectors of an electron and a longitudinal acoustic phonon, $a_k$ and $a_k^+$ are the creation and annihilation operators of an electron in the states with the wave vector projection $k$ and energy $\varepsilon(k) = 2w(1 - \cos ka)$, $b_q$ and $b_q^+$ are the respective operators for phonons with the frequency $\omega_q = 2v_s a^{-1} |\sin qa/2|$, $A(k,q)$ is the matrix element of the electron-phonon interaction

$$A(k,q) = 2i\hbar^{1/2} w' (2N_1 M \omega_q)^{-1/2} [\sin ka - \sin(k-q,a) + \gamma \sin qa]. \quad (2)$$

Here $M$ is the molecule mass, $N_1$ is the number of molecules in the basic region of the chain, $w'$ is the derivative with respect to the intermolecular distance of the transfer integral (or transfer energy) $w$ for an electron between the nearest molecules along the chains, $v_s$ and $a$ are the sound velocity and the lattice constant along the chains. In order to obtain the Hamiltonian of the 1D crystal, we need to sum up equation (1) on all chains in the basic region of the crystal. The Hamiltonian (1) takes into account simultaneously two main electron-phonon interaction mechanisms mentioned above. The parameter $\gamma$ in (2)

$$\gamma = 2e^2 \alpha_0 / a^5 w' \quad (3)$$

has the meaning of the ratio of amplitudes of these two interactions, $e$ is the electron charge, $\alpha_0$ is the average polarizability of the molecule.

## 3   Kinetic Equation

Let us now calculate the electrical conductivity $\sigma$ in a weak static electrical field directed along chains. Applying the method of two particle quantum Green functions [15, 16] we obtain

$$\sigma = \frac{i\pi e^2}{k_0 TV} \sum_{k,k'} v(k) v(k') G_{k;k'}, \tag{4}$$

where

$$G_{k;k'} \equiv \lim_{\varepsilon \to 0^+} << a_k^+ a_k \mid a_{k'}^+ a_{k'} >>_{i\varepsilon} \tag{5}$$

is the Fourier transformation of two-particle retarded Green function, $v(k) = 2\hbar^{-1} aw \sin ka$ is the velocity of the electron, $k_0$ is the Boltzmann constant, $T$ is the temperature, and $V$ is the volume of the crystal, $V = L_1 L_2 L_3$ ($L_1$, $L_2$, and $L_3$ are the basic lengths of the crystal along the chains and in transversal directions). It is convenient to introduce a new Green function $G(k)$

$$G(k) = i (L_2 L_3)^{-1} \sum_{k'} v(k') G_{k;k'}. \tag{6}$$

Then for conductivity $\sigma$ we will have

$$\sigma = \frac{\pi e^2}{k_0 T L_1} \sum_k v(k) G(k). \tag{7}$$

For the Green function $G(k)$ it is possible to write the first two equations of the infinite chain of equations and to decouple the chain in order to express the three-particle Green function through the two-particle one. For the degenerate carriers the scattering processes can be considered elastic up to temperatures of the order of 1K, and we obtain for $G(k)$ an integral equation of the kinetic equation type

$$(\pi bc)^{-1} v(k) n_k^0 \left(1 - n_k^0\right) = \sum_q [W_{k+q,k} G(k) - W_{k,k+q} G(k+q)], \tag{8}$$

$$\sum_q \ldots = \frac{L_1}{2\pi} \int_{-\pi/a}^{\pi/a} \ldots dq,$$

$$W_{k+q,k} = \frac{2\pi}{\hbar} \left[ |A(k+q,q)|^2 \left(1 + 2N_q^0\right) + \frac{n_{im} I^2 d^2}{L_1} \right] \delta [\varepsilon(k+q) - \varepsilon(k)], \tag{9}$$

where $W_{k+q,k}$ is the transition probability of electron from the state $k$ to $k+q$, $n_k^0$ and $N_q^0$ are equilibrium distribution functions for electrons and phonons. $I$ and $d$ characterize the effective height and width of the impurity potential, $n_{im}$ is the linear impurity concentration. The $\delta$-function from (9), which expresses the energy conservation law in the scattering process of electron, allows to solve exactly equation (8). Really

$$\delta\left[\varepsilon\left(k+q\right)-\varepsilon\left(k\right)\right]=\frac{1}{2wa\left|\sin ka\right|}\left[\delta\left(q\right)+\sum_{n=0}^{N_1/2}\delta\left(q+2k\pm\frac{2\pi n}{a}\right)\right]. \tag{10}$$

The first term on the right-hand side of (10) does not satisfy equation (8). Only the term with $n=0$, which corresponds to normal scattering processes with $q=-2k$, $-\pi/a\leq q\leq\pi/a$, $-\pi/2a\leq k\leq\pi/2a$, and with $n=1$, which corresponds to umklapp processes with $q=-2k\pm2\pi/a$, $-\pi/a\leq k\leq-\pi/2a$, $\pi/2a\leq k\leq\pi/a$ give contributions to (8). So as $G\left(k\right)$ is an odd function of $k$, we have $G\left(k+q\right)=G\left(-k\right)=-G\left(k\right)$. Then the kinetic equation (8) is solved exactly, and we obtain

$$G\left(k\right)=\left(\pi bc\right)^{-1}n_0^k\left(1-n_0^k\right)\upsilon\left(k\right)/M\left(k\right). \tag{11}$$

Here $M\left(k\right)$ has the meaning of mass-operator of Green's function $G\left(k\right)$

$$M\left(k\right)=\frac{4w'^2a^2k_0T_0}{\hbar Mv_s^2\left|w\right|\left|\sin ka\right|}\left[s\left|\sin ka\right|\coth\left(\frac{sT_0}{T}\left|\sin ka\right|\right)\left(1+\gamma\cos ka\right)^2+D_0^2\right], \tag{12}$$

where parameter $D_0^2$ takes into account the carriers scattering on impurities

$$D_0^2=n_{im}I^2d^2Mv_s^2/\left(4a^3k_0T_0w'^2\right) \tag{13}$$

and $s=\hbar v_s/\left(ak_0T_0\right)$, $T_0=300K$.

## 4   Electrical Conductivity

After introducing (11) and (12) in (7), we obtain for electrical conductivity

$$\sigma_{s,p}=\frac{\sigma_0T_0}{T}\int_0^{\pi}\frac{\sin^3 k\,n_k^0\left(1-n_k^0\right)dk}{\left[s\sin k\coth\left(\frac{sT_0}{T}\sin k\right)\left(1\pm\gamma\cos k\right)^2+D_0^2\right]}, \tag{14}$$

$$\sigma_0=2e^2z\left|w\right|^3 Mv_s^2/\left(\pi\hbar abcw'^2k_0^2T_0^2\right), \tag{15}$$

where the upper sign in the denominator of (14) is for an $s$-type band $\left(w>0\right)$, the lower is for a $p$-type band $\left(w<0\right)$, $\gamma$ means absolute value and $z$ is the number of chains through the transversal section of the elementary cell. It is more convenient to pass to another variable $\varepsilon=1-\cos k$, where $\varepsilon$ is the carrier energy in unities of $2w$. Then, we obtain

$$\sigma_{s,p}=\frac{\sigma_0T_0}{4T}\int_0^2\frac{\varepsilon\left(2-\varepsilon\right)d\varepsilon}{\left[s\sqrt{\varepsilon\left(2-\varepsilon\right)}\coth\left(\frac{sT_0}{T}\sqrt{\varepsilon\left(2-\varepsilon\right)}\right)\left(\gamma\varepsilon-\gamma\mp1\right)^2+D_0^2\right]}$$

$$\times\frac{1}{\cosh^2\left[\left(\varepsilon-\varepsilon_F\right)w/k_0T\right]}, \tag{16}$$

where $\varepsilon_F$ is the dimensionless Fermi energy (in units of $2w$).

Depending on physical parameters the integrand has two singularities: the first comes from $T^{-1}\cosh^{-2}\left[(\varepsilon - \varepsilon_F)\,w/k_0 T\right]$, which for $\varepsilon_F > 2k_0 T_0/w$ is very close to $\delta$-function, $(2k_0/w)\,\delta\,(\varepsilon - \varepsilon_F)$, especially at low temperatures, the second comes from the Lorentzian, which has a very sharp maximum at $\varepsilon = \varepsilon_0^{s,p} = (\gamma \pm 1)/\gamma$, when $\gamma > 1$ and $D_0 \ll 1$. If the first maximum is more pronounced and $D_0$ is not very small, applying the $\delta$-function approximation, we obtain

$$\sigma_{s,p} = \frac{\sigma_0 k_0 T_0}{2w}\,\frac{\varepsilon_F\,(2 - \varepsilon_F)}{s\sqrt{\varepsilon_F\,(2 - \varepsilon_F)}\coth\left(\frac{sT_0}{T}\sqrt{\varepsilon_F\,(2 - \varepsilon_F)}\right)\gamma^2\,(\varepsilon_F - \varepsilon_0^{s,p})^2 + D_0^2}.$$
(17)

Let us consider a crystal with parameters close to those of $s$-type TCNQ chains in the tetrathiofulvalene-tertracyanoquinodimethane (TTF-TCNQ) crystal [13]: $M = 3.5 \times 10^5 m_e$ ($m_e$ is the electron mass), $w = 0.125\mathrm{eV}$, $w' = 0.2\mathrm{eV\AA}^{-1}$, $v_s = 10^5 \mathrm{cm/s}$, $a = 12.3\mathrm{\AA}$, $b = 3.82\mathrm{\AA}$, $c = 18.47\mathrm{\AA}$, $z = 2$, $s = 0.066$, $\sigma_0 = 5.1 \times 10^3 \Omega^{-1}\mathrm{cm}^{-1}$. If $\gamma = 5$, $\varepsilon_F = 1$ (half filled band) and $D_0 = 1$, then from (17) follows that $\sigma_s = 265\Omega^{-1}\mathrm{cm}^{-1}$ at $T - 300\mathrm{K}$. Numerical calculation with Mathematica-5 gives $269\Omega^{-1}\mathrm{cm}^{-1}$, a very close result. If $D_0 = 10^{-1}$, (17) and (16) give $\sigma_s = 481\Omega^{-1}\mathrm{cm}^{-1}$ and $3.4 \times 10^3 \Omega^{-1}\mathrm{cm}^{-1}$, respectively. For $D_0 = 10^{-2}$ we have from (17) $\sigma_s = 529\Omega^{-1}\mathrm{cm}^{-1}$ and from the precise expression (16) $\sigma_s = 3.4 \times 10^4 \Omega^{-1}\mathrm{cm}^{-1}$. Thus, the approximate expression (17) is valid only when the Lorentzian from (16) is a rather smooth function. Otherwise it is necessary to calculate (16) numerically.

If the second maximum from the Lorentzian in (16) is more pronounced ($D_0 \ll 1$, $\gamma > 1$), it is possible to decompose the Lorentzian in series after $D_0$. The first term corresponds to the replacement of Lorentzian by $\delta$-function of $(\varepsilon - \varepsilon_0^{s,p})$, and we obtain

$$\sigma_s = \frac{\sigma_0 T_0}{4T}\,\frac{\pi\varepsilon_0^s\,(2 - \varepsilon_0^s)}{D_0 \gamma\left\{s\sqrt{\varepsilon_0^s\,(2 - \varepsilon_0^s)}\coth\left(\frac{sT_0}{T}\sqrt{\varepsilon_0^s\,(2 - \varepsilon_0^s)}\right)\right\}^{1/2}}$$

$$\times \frac{1}{\cosh^2\left[(\varepsilon_0^s - \varepsilon_F)\,w/k_0 T\right]}.$$
(18)

For $D_0 = 10^{-3}$, $\gamma = 5$ ($\varepsilon_0^s = 1.2$) and $\varepsilon_F = 0.2$, from (18), at $T = 300\mathrm{K}$, follows $\sigma_s = 205\Omega^{-1}\mathrm{cm}^{-1}$ and from (16) $\sigma_s = 216\Omega^{-1}\mathrm{cm}^{-1}$, a very close result. If $D_0$ is diminished up to $10^{-4}$ (extra pure crystals), we obtain from (18) and (16), respectively, $2.05 \times 10^3 \Omega^{-1}\mathrm{cm}^{-1}$ and $2.06 \times 10^3 \Omega^{-1}\mathrm{cm}^{-1}$, really the same result. Such small values for $D_0$ are necessary in (18) because in this case both maximums in the integrand from (16) are situated far one from the other, decrease quickly, and considerably suppress each other. If $\varepsilon_F$ is increased up to $\varepsilon_F = 1$, (16) and (18) give the same result already at $D_0 = 10^{-3}$.

In the case of $p$-type band, so as the maximums can be disposed close to each other, (16) and (18) give practically the same result even at $D_0 = 10^{-2}$.

However, when Mathematica-5 is applied directly to calculate (16) as a function of $T$ for $D_0$ smaller than $10^{-3}$, a notification appears that the integration is not convergent. Really, for some parameters the integrand in (16) has two pronounced singularities (see Fig.1). In order to overcome this difficulty, we have devided the region of integration in three parts: from zero to the first maximum, from the first maximum to the second one, and from the second maximum to the upper limit. Then the notification of non-convergency did not appear any more. In order to verify the achieved precision, a special program in Fortran has been developed, which calculates the contribution to the integral of each maximum and of the remaining regions separately. It was confirmed that Mathematica-5 ensures good precision in these cases.

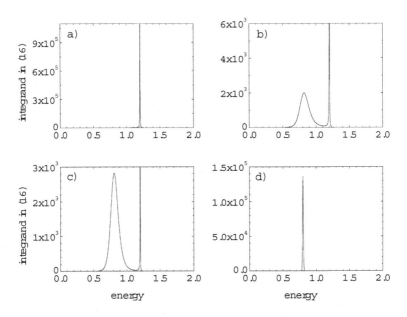

**Fig. 1.** The integrand in (16) for parameters of $s$-type TCNQ chains from the text and $\varepsilon_F = 0.8$, $\gamma = 5$, $D_0 = 10^{-2}$. $T$: a)300K; b)120K; c)100K; d)10K.

A special case arises when the high maximums are very close to each other, or even coincide. The integrand in (16) now represents the product of two almost $\delta$-functions. Mathematica-5 has managed this case too. Some results are presented in Figs. 2 and 3.

## 5    Conclusions

For quasi-one-dimensional organic crystals the kinetic equation is solved exactly, and the electrical conductivity is expressed as an ordinary integral named transport integral. The unusual situation is that for certain parameters the integrand expression has two singularities. When Mathematica-5 is directly applied, a notification appears that the calculation is not convergent. In order to exclude this

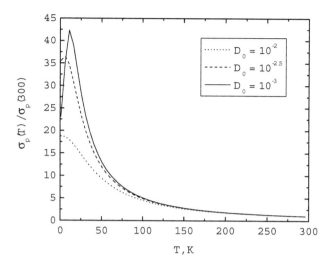

**Fig. 2.** The ratio of electrical conductivity $\sigma_p(T)$ to $\sigma_p(300)$ as a function of temperature $T$ for parameters from the text. $\varepsilon_F = 0.325$, $\gamma = 1.5$.

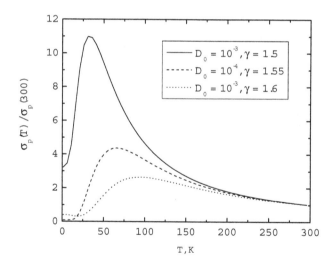

**Fig. 3.** The same as in Fig. 2 for $\varepsilon_F = 0.31$ and different $\gamma$ and $D_0$

difficulty, the domain of integration was divided into three parts: from initial position up to first singularity point, from this point up to the second singularity point, and from the second singularity point to the final limit. The results were verified by another program in Fortran developed to calculate the contributions to integral of singularity regions and of remaining regions separately. Thus, the needed precision of calculation with Mathematica-5 was confirmed. In particular cases approximate analytic expressions for electrical conductivity were obtained, the criteria of application were established, and their precision was verified by Mathematica-5. Even when the singularities of integrand are very close to each other or coincide, the numerical integration with Mathematica-5 can ensure the needed precision, if it is applied proficiently. This conclusion was verified by calculations with above mentioned additional program in Fortran.

# References

1. Pope, M., Swenberg, C.E.: Electronic Processes in Organic Crystals and Polymers. 2nd Ed., Oxford University Press, Oxford, 1999
2. Burroughes, J.H., Bradley, D.D.C., Brown, A.R., Marks, R.N., Mackay, K., Friend, R.H., Burn, P.L., Holmes, A.B.: Light-emitting Diodes Based on Conjugated Polymers. Nature **347**(1990) 539
3. Sariciftci, N.S., Smilowitz, L., Heeger, A.J., Wudl, F.: Photoinduced electron transfer from a conducting polymer to buckminsterfullerene. Science **258** (1992) 1474
4. Kippelen, B., Marder, S.R., Hendrickx, E., Maldonado, J.L., Guillemet, G., Volodin, B.L., Steele, D.D., Enami, Y., Sandalphon, L., Yao, Y.J., Wang, J.F., Rockel, H., Erskine, L., Peyghambarian, N.: Infrared photorefractive polymers and their applications for imaging. Science **279** (1998) 5347
5. Kubatkin, S., Danilov, A., Hjort, M., Cornil, J., Bredas, J.L., Stuhr-Hansen, N., Hedegard, P., Bjornholm, T.: Single-electron transistor of a single organic molecule with access to several redox states. Nature **425** (2003) 698–701
6. Casian, A., Dusciac, V., Coropceanu, Iu.: Huge carrier mobilities expected in quasi-one-dimensional organic crystals. Phys. Rev. B **66** (2002) 165404
7. Casian, A., Balandin, A., Dusciac, V., Coropceanu, Iu.: A mechanism for extremely high electrical conductivity in a quasi-one-dimensional organic crystal. Phys. Low-Dim. Struct. **9/10** (2002) 43–54
8. Casian, A., Balandin, A.A., Dusciac, V., Dusciac, R.: Modeling of the thermoelectric properties of quasi-one-dimensional semiconductors. In: Proc. 21st Intern. Conf. on Thermoel., Long Beach, USA. IEEE, Piscataway, NJ (2003) 310–313
9. Casian, A., Dashevsky, Z., Scherrer, H., Dusciac, V., Dusciac, R.: A possibility to realize a high thermoelectric figure of merit in quasi-one-dimensional organic crystals. In: Proc. 22nd Intern. Conf. on Thermoel., La Grande-Motte, France. IEEE, Piscataway, NJ (2004) 330–335
10. Dusciac, V.: Thermoelectric opportunities of the quasi-one-dimensional organic semiconductors. J. Thermoelectricity **1** (2004) 5–18
11. Casian, A.: Thermoelectric Properties of Organic Materials. Thermoelectric Handbook, Macro to Nano. CRC Press Inc. (2006) Chap. 36
12. Gosar, P., and Choi Sang-il: Linear response theory of the electron mobility in molecular crystals. Phys. Rev. **150** (1966) 529
13. Conwel, E.M.: Band transport in quasi-one-dimensional conductors in the phonon-scattering regime and application to Ttetrathiofulvalene-tertracyanoquinodimethane. Phys. Rev. B **22** (1980) 1761
14. Friedman, L.: Fluctuation of polarization energy as an important scattering mechanism in the quasi-one-dimensional organic solids. Sol. Stat. Comm. **40** (1) (1980) 41–44
15. Zubarev, D.N.: Nonequilibrium Statistical Thermodynamics. Nauka, Moscow, 1986 (in Russian)
16. Bonch-Bruevich, V.L., Tyablicov, S.V.: Method of Green Functions in Statistical Mechanics. Fizmatgiz, Moscow, 1961 (in Russian)

# On the Provably Tight Approximation of Optimal Meshing for Non-convex Regions

Dmytro Chibisov, Victor Ganzha, Ernst W. Mayr[1], and Evgenii V. Vorozhtsov[2]

[1] Institute of Informatics, Technical University of Munich, Garching 85748,
Boltzmannstr. 3, Germany
`chibisov@in.tum.de, ganzha@in.tum.de, mayr@in.tum.de`
[2] Institute of Theoretical and Applied Mechanics, Russian Academy of Sciences,
Novosibirsk 630090, Russia
`vorozh@itam.nsc.ru`

**Abstract.** Automatic generation of smooth, non-overlapping meshes on arbitrary regions is the well-known problem. Considered as optimization task the problem may be reduced to finding a minimizer of the weighted combination of so-called length, area, and orthogonality functionals. Unfortunately, it has been shown that on the one hand, certain weights of the individual functionals do not admit the unique optimizer on certain geometric domains. On the other hand, some combinations of these functionals lead to the lack of ellipticity of corresponding Euler-Lagrange equations, and finding the optimal grid becomes computationally too expensive for practical applications. Choosing the right functional for the particular geometric domain of interest may improve the grid generation very much, but choosing the functional parameters is usually done in the trial and error way and depends very much on the geometric domain. This makes the automatic and robust grid generation impossible. Thus, in the present paper we consider the way to compute certain approximations of minimizer of grid functionals independently of the particular domain. Namely, we are looking for the approximation of the minimizer of the individual grid functionals in the local sense. This means the functional has to be satisfied on the possible largest parts of the domain. In particular, we shall show that the so called method of envelopes, otherwise called the method of rolling circle, that has been proposed in our previous paper, guarantees the optimality with respect to the area and orthogonality functionals in this local sense. In the global sense, the grids computed with the aid of envelopes, can be considered as approximations of the optimal solution. We will give the comparison of the method of envelopes with well established Winslow generator by presenting computational results on selected domains with different mesh size.

## 1 Motivation and Introduction

Advanced computer technologies and parallel architectures allow one to solve time dependent problems with $10^9$ and more unknowns on rectangular regions in realistic time using hierarchical and adaptive approaches [2, 8]. In order to handle problems of such order of computational complexity on arbitrary regions and,

V.G. Ganzha, E.W. Mayr, and E.V. Vorozhtsov (Eds.): CASC 2006, LNCS 4194, pp. 109–128, 2006.
© Springer-Verlag Berlin Heidelberg 2006

in particular, with moving boundaries, we are interested to have efficient grid generation techniques, which would support hierarchical approach to computing and provide the possibility of adaptive mesh refinement as well as remeshing, due to the changes of boundaries, with minimal computational costs.

The problem of grid generation on an arbitrary region $\Omega$ in the $(x, y)$ plane can be solved by giving a map $x(\xi, \eta), y(\xi, \eta)$ from the unit square in the plane $(\xi, \eta)$ onto the $\Omega$. By choosing a uniform grid $(\xi_i, \eta_j)$ in the unit square, the map $x(\xi_i, \eta_j), y(\xi_i, \eta_j)$ would transform the grid $(\xi_i, \eta_j)$ to the region of interest. The required map may be computed in a number of ways. The variational grid generation is one of the most established approaches for this purpose, due to high quality of resulting grids. It provides the possibility to control the grid properties by choosing appropriate grid functionals to be minimized. The basic functionals are Length $(I_L)$, Area $(I_A)$, and Orthogonality $(I_O)$ functionals, which can be written in the form (see [6]):

$$I_L(x, y) = 1/2 \iint (x_\xi^2 + y_\xi^2 + x_\eta^2 + y_\eta^2) d\xi \, d\eta; \tag{1}$$

$$I_A(x, y) = 1/2 \iint (x_\xi^2 y_\eta^2 + y_\xi^2 x_\eta^2 - 2 x_\xi x_\eta y_\xi y_\eta) d\xi \, d\eta; \tag{2}$$

$$I_O(x, y) = 1/2 \iint (x_\xi^2 x_\eta^2 + 2 x_\xi x_\eta y_\xi y_\eta + y_\xi^2 y_\eta^2) d\xi \, d\eta. \tag{3}$$

The map $x(\xi, \eta), y(\xi, \eta)$ minimizing each of above functionals can be found by by solving corresponding Euler–Lagrange equations, which can be written in general form

$$\mathcal{T}_{1,1}\mathbf{x}_{\xi,\xi} + \mathcal{T}_{1,2}\mathbf{x}_{\xi,\eta} + \mathcal{T}_{2,2}\mathbf{x}_{\eta,\eta} + \mathcal{S} = 0,$$

where $\mathcal{T}_{i,j}$ are 2 x 2 matrices and $\mathcal{S}$ is a 2 x 1 vector. The terms in $\mathcal{T}_{i,j}$ and $\mathcal{S}$ depend on the particular functional and are nonlinear in the case of Area and Orthogonality Functionals. In the case of the Length functional $I_L$, $\mathcal{T}_{i,j}$ can be shown to be constant, and the Euler–Lagrange equations reduce to the simplest one:

$$x_{\xi,\xi} + x_{\eta,\eta} = 0, \qquad y_{\xi,\xi} + y_{\eta,\eta} = 0.$$

Minimizing $I_L$ by solving above equations leads to smooth grids. However, the intersections of grid lines may occur (Fig. 1). The folding of resulting grids by using the Length functional is inadmissible for practical applications. The Area functional leads to the following Euler–Lagrange equations, which produce unfolded but, unfortunately, nonsmooth grids:

$$x_{\xi,\xi} y_\eta^2 + y_\eta x_\xi y_{\xi,\eta} - y_\eta y_{\xi,\xi} x_\eta - 2 y_\eta y_\xi x_{\xi,\eta} - y_\xi x_\xi y_{\eta,\eta} + y_\xi y_{\xi,\eta} x_\eta + y_\xi^2 x_{\eta,\eta} = 0,$$
$$-x_\eta x_{\xi,\xi} y_\eta - 2 x_\eta x_\xi y_{\xi,\eta} + y_{\xi,\xi} x_\eta^2 + x_\eta y_\xi x_{\xi,\eta} + x_\xi x_{\xi,\eta} y_\eta + x_\xi^2 y_{\eta,\eta} - x_\xi y_\xi x_{\eta,\eta} = 0.$$

As described in [6], the further shortcoming of this method is that available numerical procedures for solving the above equations do not converge for certain domains. The Orthogonality functional produces orthogonal and sufficiently

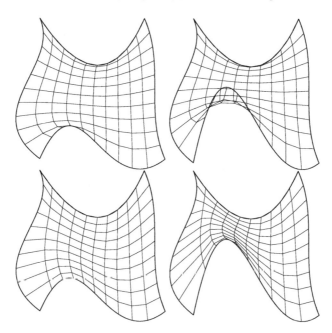

**Fig. 1.** Grid generation by minimizing the Length functional (top), and by minimizing the Winslow functional (bottom)

smooth grids on many domains, however, fails to converge in certain cases. Euler-Lagrange equations for the Orthogonality functional are:

$$x_{\xi,\xi}x_\eta{}^2 + 4\,x_\eta x_\xi x_{\xi,\eta} + x_\eta y_{\xi,\xi}y_\eta + x_\eta y_\xi y_{\xi,\eta} + 2\,x_{\xi,\eta}y_\xi y_\eta + x_\xi{}^2 x_{\eta,\eta} + x_\xi y_{\xi,\eta}y_\eta$$
$$+x_\xi y_\xi y_{\eta,\eta} = 0,$$
$$y_\eta x_{\xi,\xi}x_\eta + y_\eta x_\xi x_{\xi,\eta} + y_{\xi,\xi}y_\eta{}^2 + 4\,y_\eta y_\xi y_{\xi,\eta} + 2\,y_{\xi,\eta}x_\xi x_\eta + y_\xi x_{\xi,\eta}x_\eta + y_\xi x_{\xi,\eta}x_{\eta,\eta}$$
$$+y_\xi{}^2 y_{\eta,\eta} = 0.$$

In order to obtain smooth, orthogonal, and unfolded grids, the weighted combination of Length, Area, and Orthogonality functionals may be used:

$$I(x,y) = \omega_A I_A(x,y) + \omega_L I_L(x,y) + \omega_O I_I(x,y) \tag{4}$$

In particular, Area-Length combination overcomes the limitation of individual functionals because of avoiding grid folding produced by Length functional and producing smooth grids in contrast to the Area functional. However, the corresponding equations do not admit the continuous solution on many practically important domains like airfoil, backstep, and "C"-domains (see [6]). In order to preserve the advantages of the Length functional and avoid the grid foldings the famous Winslow grid generator has been proposed. The Winslow functional

$$I_W(x,y) = \iint \frac{x_\eta{}^2 + y_\eta{}^2}{(x_\xi y_\eta - x_\eta y_\xi)^2} + \frac{x_\xi{}^2 + y_\xi{}^2}{(x_\xi y_\eta - x_\eta y_\xi)^2}\,d\xi\,d\eta$$

leads to equations:

$$\left(x_\xi{}^2 + y_\eta{}^2\right) x_{\xi,\xi} - 2 \left(x_\xi x_\eta + y_\xi y_\eta\right) x_{\xi,\eta} + \left(x_\xi{}^2 + x_\eta{}^2\right) x_{\eta,\eta} = 0,$$
$$\left(x_\xi{}^2 + y_\eta{}^2\right) y_{\xi,\xi} - 2 \left(x_\xi x_\eta + y_\xi y_\eta\right) y_{\xi,\eta} + \left(x_\xi{}^2 + x_\eta{}^2\right) y_{\eta,\eta} = 0.$$

The Winslow generator inherits the grid smoothness from the Length functional and tends to produce smooth non-folded grids (see Fig. 1). However, the lack of orthogonality may lead, for example, to high truncation errors by using the Winslow grids for numerical solution of PDE's. Further modifications of the presented functionals may be found in the literature (see [6]), which tend to produce good meshes in certain cases and fail to admit the solution in other cases. Choosing the right functional for a certain geometric domain, or, in particular, choosing optimal weights in (4) may improve the resulting grids significantly. The optimal choosing, however, depends on the particular domain very much and is usually performed in the trial-and-error way. All this makes the automatic and robust grid generation impossible. Thus, in the present paper we study the possibility of overcoming this difficulty by considering a domain independent approach for the approximation of the minimizers of grid functional in the local sense. This means, we are interested in satisfying the corresponding Euler–Lagrange equations on the possible large part of the domain. We admit the discontinuities in the resulting mapping $x(\xi, \eta), y(\xi, \eta)$ between certain parts of the geometric domain and study the approximation of the minimizers of functionals (1) – (3) in discrete form, which will be derived in Section 2. In Section 3 we shall describe the method of envelopes, or, otherwise called the method of rolling circle, and show the quality of the approximation of the discrete optimization problem. Section 4 is devoted to the comparison of computational performance of the method of rolling circle with Winslow grid generators.

## 2    Variational Grid Generation: Discrete Optimization Formulation

In order to admit discontinuous mappings from the unit square onto the geometric region of interest we formulate the Area, Length, and Orthogonality conditions in the discrete sense. We consider the discretized map $x(i, j), y(i, j)$ from uniform grid in the unit square onto the arbitrary region in the $(x, y)$ plane. As can be seen in Fig. 2, the square of the length of two grid segments (horizontal and vertical) intersecting in the common grid vertex $(i, j)$ is given by the following polynomial:

$$L_{i,j,i-1,j-1} = \left(x_{i,j} - x_{i-1,j}\right)^2 + \left(y_{i,j} - y_{i-1,j}\right)^2 + \left(x_{i,j} - x_{i,j-1}\right)^2 + \left(y_{i,j} - y_{i,j-1}\right)^2.$$

Summation over the grid vertices leads to the discrete form of the Length functional:

$$I_L(x_{1,1}, y_{1,1}, ..., x_{N,N}, y_{N,N}) = \sum_{i,j=2}^{N} \left(x_{i,j} - x_{i-1,j}\right)^2 + \left(y_{i,j} - y_{i-1,j}\right)^2 +$$
$$\left(x_{i,j} - x_{i,j-1}\right)^2 + \left(y_{i,j} - y_{i,j-1}\right)^2.$$

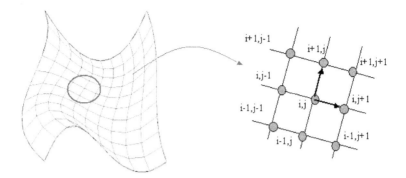

**Fig. 2.** Grid discrete function

The orthogonality condition between two intersecting line segments can be expressed in a similar way as polynomial by using scalar product of vectors $(x_{i+1,j} - x_{i,j}, y_{i+1,j} - y_{i,j})$ and $(x_{i,j+1} - x_{i,j}, y_{i,j+1} - y_{i,j})$:

$$O_{i,j,i+1,j+1} = ((x_{i+1,j} - x_{i,j})(x_{i,j+1} - x_{i,j}) + (y_{i+1,j} - y_{i,j})(y_{i,j+1} - y_{i,j}))^2.$$

The orthogonality condition for four angles in each cell becomes:

$$O_{i,j}^{cell} = O_{i,j,i+1,j+1} + O_{i+1,j,i-1,j+1} + O_{i+1,j+1,i-1,j-1} + O_{i,j+1,i+1,j-1}.$$

Summation over all cells then leads to the discrete orthogonality functional:

$$I_O(x_{1,1}, y_{1,1}, ..., x_{N,N}, y_{N,N}) = \sum_{i,j=1}^{N-1} O_{i,j}^{cell}.$$

The squared grid cell area can be expressed as follows

$$A_{i,j}^{cell} = (-\left(x_{i,j} - x_{i,j-1}\right)\left(y_{i,j-1} - y_{i+1,j-1}\right) + \left(x_{i,j-1} - x_{i+1,j-1}\right)\left(y_{i,j} - y_{i,j-1}\right)$$
$$+ \left(x_{i+1,j} - x_{i,j}\right)\left(y_{i+1,j-1} - y_{i+1,j}\right) - \left(x_{i+1,j-1} - x_{i+1,j}\right)\left(y_{i+1,j} - y_{i,j}\right))^2.$$

Finally, the Area functional is given as

$$I_A(x_{1,1}, y_{1,1}, \ldots, x_{N,N}, y_{N,N}) = \sum_{i,j=1}^{N-1} A_{i,j}^{cell}.$$

Similarly to the Euler–Lagrange equations for the continuous Area, Length, and Orthogonality functionals, we obtain the system of $2N^2$ algebraic equations necessary for the function $I(x_{1,1}, y_{1,1}, \ldots, x_{N,N}, y_{N,N})$ to reach a minimum in $(x_{1,1}, y_{1,1}, ..., x_{N,N}, y_{N,N})$:

$$\frac{\partial I}{\partial x_{i,j}} = 0, \qquad \frac{\partial I}{\partial y_{i,j}} = 0. \tag{5}$$

Similarly to the continuous case, equations (5) for the Length functional $I_L$ are linear

$$8\,x_{i,j} - 2\,x_{i-1,j} - 2\,x_{i,j-1} - 2\,x_{i+1,j} - 2\,x_{i,j+1} = 0,$$
$$8\,y_{i,j} - 2\,y_{i-1,j} - 2\,y_{i,j-1} - 2\,y_{i+1,j} - 2\,y_{i,j+1} = 0$$

and equations (5) for $I_A$ and $I_O$ are cubic and are similar to the corresponding discretized Euler–Lagrange equations.

Please note, in our approach we do not optimize the meshing with respect to conditions (5) directly. It can be shown that existing global optimization approaches like branch and bound strategy, would become computationally too expensive, especially in the case when mesh size decreases. Instead of applying computationally expensive direct optimization techniques, we use the method of envelopes, which has been introduced in [4] and will be described in Section 3. We will, namely, show that this method leads to satisfaction of (5) on the most part of the region for the Length, Area, and Orthogonality functionals.

## 3    Approximation of Grid Functionals by the Method of Envelopes

In the present section, we describe the so-called method of rolling circle that has been introduced in [4]. We shall show that this method produces nearly optimal grids in the sense that equations (5) are satisfied locally. Let the region $\Omega \subset R^2$ be bounded by the roots of polynomials $f_i(x, y)$. The so-called Tarski formula describing the set of points, which belong to this region can be written as follows:

$$\Omega(x, y) \equiv \bigwedge_i f_i(x, y) \geq 0$$

We propose to calculate the lines of the curvilinear grid in the following way. We contact a circle $C(x, y) = x^2 + y^2 - r^2 = 0$ with $\Omega$ and move $C$ along the boundary of $\Omega$ keeping them in contact. The motion of a circle can be produced by shifting it by $x_0, y_0$ units:

$$C(x - x_0, y - y_0) = 0.$$

The circle moving along some boundary curve $f_i(x, y) = 0$ describes a curve $g_i(x, y) = 0$ called *envelope* (Fig. 3). More precisely, the envelope in our case is a curve, whose tangent at each point coincides with the tangent of a moving circle at each time of its motion. In our grid generation approach the envelopes correspond to grid lines parallel to the boundary (Fig. 5, left at the top). As will be shown in Section 3.2, connecting the intersection points of the circle $C$ and envelope $g_i$ on the one hand and the circle $C$ and boundary $f_i$ on the other hand produces the line segment which is orthogonal to both curves (Fig. 5, right at the top) and satisfies (5) for Length, Area and Orthogonality Functionals.

The contact of $C$ and $f_i$ can be expressed in terms of common roots of bounding polynomials. The envelope $g_i$ corresponds also to such shifts $x_0, y_0$ of $C$,

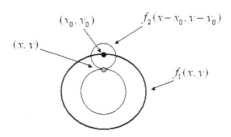

**Fig. 3.** Calculating of envelopes by quantifier elimination

where polynomials $f_i$ and $C$ have common roots and coinciding tangents. This can be formalized using polynomial equations as follows:

$$h : \{(x_0, y_0)|\exists x, y : f(x, y) = 0 \wedge C(x - x_0, y - y_0) = 0 \wedge$$
$$-\frac{\partial f(x,y)}{\partial x}\frac{\partial C(x-x_0,y_0 0)}{\partial y} + \frac{\partial f(x,y)}{\partial y}\frac{\partial C(x-x_0,y_0 0)}{\partial x} = 0\}.$$

Alternatively, if the boundary curve $f(x, y) = 0$ is given parametrically ($x = x(t), y = y(t)$), the envelope may be defined by (see [1]):

$$h : \{(x_0, y_0)|\exists t : C(x(t) - x_0, y(t) - y_0, r) = 0 \wedge \frac{\partial}{\partial t}C(x(t) - x_0, y(t) - y_0, r) = 0.$$

Eliminating $\exists$-quantifiers with existing methods described below produces the point set, which corresponds to the envelope $g$ (Fig. 3). After $h(x, y)$ is calculated for different values of radius $r$ of the circle, the grid points distributed along them should be connected with those of $f(x, y)$ in such a way the resulting grid satisfies equations (5) for the Length, Area, and Orthogonality functionals on the possible large part of the region. This construction will be presented in the next Section after introducing the method to calculate $h(x, y)$.

### 3.1   Elimination of Variables Using Resultants

In this section we shall describe how the well known approach to the elimination of variables from the following first-order formulas (so-called existential first-order theory over the reals) with the aid of resultants can be used to calculate the grid lines parallel to the boundaries. Assume the geometric region is bounded by $N$ parametric curves $[x_{(j)}(t), y_{(j)}(t)], t \in [0, 1]$ of degree $deg \leq M$:

$$
\begin{aligned}
x_{(1)}(t) &= \sum_{i=1}^{M} a_i^{(1)}t^i, & y_{(1)}(t) &= \sum_{i=1}^{M} b_i^{(1)}t^i, \\
&\dots \\
x_{(N)}(t) &= \sum_{i=1}^{M} a_i^{(N)}t^i & y_{(N)}(t) &= \sum_{i=1}^{M} b_i^{(N)}t^i.
\end{aligned}
\tag{6}
$$

The envelope described by a circle rolling along parametric curves (6) can be described with the following formula:

$$\exists t : \bigvee_{i=1}^{N} C(x_{(i)}(t)-x_c, y_{(i)}(t)-y_c, r) = 0 \wedge \frac{\partial}{\partial t} C(x_{(i)}(t)-x_c, y_{(i)}(t)-y_c, r) = 0, \quad (7)$$

where $C$ is the circle equation with indeterminate radius $r$ and center position $x_c, y_c$ on the curve $(x^{(i)}(t), y^{(i)}(t))$. Given a polynomial $f(x)$ of degree $n$ with roots $\alpha_i$ and a polynomial $g(x)$ of degree $m$ with roots $\beta_j$, the resultant is defined by

$$\rho(f, g) = \prod_{i,j}(\alpha_i - \beta_j).$$

$\rho(f, g)$ vanishes iff $\exists a : f(a) = 0 \wedge g(a) = 0$. The resultant can be computed as the determinant of the so-called Sylvester Matrix [3]. In the multivariate case, the computation of resultant can be reduced to the univariate one by considering the polynomials $f, g \in \mathbb{K}[x_1, ..., x_N]$ as univariate polynomials in $\mathbb{K}[x_1]$ with unknown coefficients in $\mathbb{K}[x_2, ..., x_N]$ ( denoted by $\mathbb{K}(x_2, ..., x_N)[x_1]$). In the following we call the resultant of $f, g \in \mathbb{K}(x_1, ..., x_{i-1}, x_{i+1}, ..., x_N)[x_i]$ as $res_{x_i}(f, g)$.

In order to eliminate $t$ from (7) and find the envelope $h$ of the circle and the curve $(x(t), y(t))$ we may calculate

$$h(x_0, y_0, r) = res_t \left( C(x(t) - x_0, y(t) - y_0, r), \frac{\partial}{\partial t} C(x(t) - x_0, y(t) - y_0, r) \right).$$
(8)

In this way the first family of grid lines, namely parallel to the boundary, can be calculated in analytic form. A Maple calculation shows that the resultant for the envelope to the curve $(x(t), y(t))$ with indeterminate coefficients has already 2599 terms even if the degree of the original curve is equal to 2. This makes direct computation with resulting envelopes impossible. Bellow we shall describe the way how to avoid such computations for by the generation of grid lines perpendicular to the envelopes.

In order to generate the second family of grid lines, which are perpendicular to the first family, we discretize each of the curves (6) with step size $\Delta t = \frac{1}{M}$ and obtain a number of points $p_j = (x_j, y_j), j = 1..M$. This can be done with symbolic $M$ by substitution of $t = \frac{j}{M}, j = 1, ..., N$ in (1). We place the circle center in each $p_j$ and compute the intersection of $C(x - x_j, y - y_j)$ with $h(x, y, r)$. Let us denote the common roots of the both polynomials as $\mathbf{V}(C(x - x_j, y - y_j), h(x, y, r))$. Thus, the second family of grid lines $v(j, r)$ perpendicular to $h(x, y, r) = 0$ can be obtained by computing

$$v(j, r) = \mathbf{V}(C(x - x_j, y - y_j), h(x, y, r)).$$
(9)

Since $h(x, y, r)$ is a large symbolic expression, as mentioned previously, computing (9) in a direct way by elimination of variables using resultants becomes a very expensive task.

**GenVertex**$(x_i, y_i, j, r)$
(* The procedure computes grid vertex corresponding to the boundary curve
$[x_i(j), y_i(j)]$ using the circle of radius r*)

(*generate new grid vertex*)

$$x_i^{j,r} \leftarrow x_i(j/N) - \frac{r \frac{dy_i}{dt}|_{x_i(j/N)}}{\sqrt{\frac{dx_i}{dt}|^2_{x_i(j/N)} + \frac{dx_i}{dt}|^2_{x_i(j/N)}}};$$

$$y_i^{j,r} \leftarrow y_i(j/N) + \frac{r \frac{dx_i}{dt}|_{y_i(j/N)}}{\sqrt{\frac{dx_i}{dt}|^2_{y_i(j/N)} + \frac{y_i(j/N)}{dt}|^2_{y_i(j/N)}}};$$

**return** $x_i^{j,r}, y_i^{j,r}$;

**Fig. 4.** Calculating the grid vertex corresponding to the given boundary curve and radius of the circle according to the Proposition 1

Therefore, we use the following simple result, which gives the intersection of a circle with middle point $(x_j, y_j) \in [x(t), y(t)]$ and radius $r$ and $h(x, y, r)$ given by (8):

**Proposition 1.** *Let $h(x, y, r)$ be envelope of a family of circles $C(x - x(t), y - y(t), r)$ with radius $r$ given by (8). Then for any $t_j \in \mathbb{R}$ the following is satisfied:*

$$\mathbf{V}(C(x - x(t_j), y - y(t_j), r), h(x, y, r)) =$$

$$\left( x(t_j) \pm \frac{r \frac{dy}{dt}|_{t_j}}{\sqrt{\left(\frac{dy}{dt}|_{t_j}\right)^2 + \left(\frac{dx}{dt}|_{t_j}\right)^2}}, \quad y(t_j) \mp \frac{r \frac{dx}{dt}|_{t_j}}{\sqrt{\left(\frac{dx}{dt}|_{t_j}\right)^2 + \left(\frac{dx}{dt}|_{t_j}\right)^2}} \right).$$

*Proof.* According to (8) $\mathbf{V}(C(x - x(t_j), y - y(t_j)), h(x, y, r)) = \mathbf{V}(C(x - x(t), y - y(t), r), \frac{\partial}{\partial t} C(x - x(t), y - y(t), r))$ for some $t$. Note that

$$\frac{\partial}{\partial t} C(x - x(t), y - y(t), r) = \frac{\partial C}{\partial x} \frac{dx}{dt} + \frac{\partial C}{\partial y} \frac{dy}{dt} = -2(x - x(t)) \frac{dx}{dt} - 2(y - y(t)) \frac{dy}{dt}.$$

This means that all solutions of $\mathbf{V}(C(x - x(t_j), y - y(t_j), r), h(x, y, r))$ lie on a line $x \frac{dx}{dt}|_{t_j} + y \frac{dy}{dt}|_{t_j} - x(t_j) \frac{dx}{dt}|_{t_j} - y(t_j) \frac{dy}{dt}|_{t_j} = 0$ independently of $r$. Thus, we are interested to find the intersections of circle $C(x - x_j, y - y_j, r)$ and this line going through the middle point of $C$. Using a bit of elementary mathematics we obtain the statement of this proposition. $\diamond$

Now we are able to find the points $(x_h, y_h)$ on envelope $h(x, y, r)$, which correspond to the particular position $(x_b, y_b)$ of a circle on the boundary of the

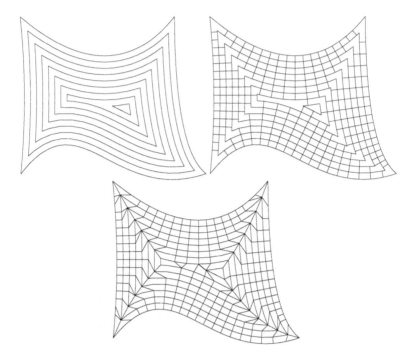

**Fig. 5.** Grid generation by the method of rolling circle: 1) calculating the first family of grid lines parallel to the boundary; 2) calculating grid lines orthogonal to the first family; 3) connecting "hanging" grid vertices

region. For example, when the bounding curve (1) is of degree 3 with unknown coefficients $a_1, ..., a_4, b_1, ..., b_4$ we obtain using Proposition 1:

$$
\begin{aligned}
x_b &= a_1 + a_2 t_j + a_3 t_j{}^2 + a_4 t_j{}^3, \\
y_b &= b_1 + b_2 t_j + b_3 t_j{}^2 + b_4 t_j{}^3, \\
x_h &= a_1 + a_2 t_j + a_3 t_j{}^2 + a_4 t_j{}^3 + \frac{r\left(b_2 + 2\,b_3 t_j + 3\,b_4 t_j^2\right)}{\sqrt{b_2{}^2 + 4\,b_2 b_3 t_j + 4\,b_3{}^2 t_j{}^2 + a_2{}^2 + 4\,a_2 a_3 t_j + 4\,a_3{}^2 t_j{}^2}}, \quad (10) \\
y_h &= b_1 + b_2 t_j + b_3 t_j{}^2 + b_4 t_j{}^3 - \frac{r\left(a_2 + 2\,a_3 t_j + 3\,a_4 t_j^2\right)}{\sqrt{b_2{}^2 + 4\,b_2 b_3 t_j + 4\,b_3{}^2 t_j{}^2 + a_2{}^2 + 4\,a_2 a_3 t_j + 4\,a_3{}^2 t_j{}^2}}.
\end{aligned}
$$

Using (10) we are able to calculate the spatial positions of the individual grid nodes lying on envelopes dependent on the distance $r$ from the boundary ($r$ is a radius of the circle, which produces corresponding envelope). Because of Proposition 2 (see below) the line segments induced by the calculated nodes are orthogonal. For example, the vertices of the grid shown in Figs. 5 and 6 have been generated in this way.

So far we have considered successive generation of grid cells starting from an individual boundary curve by computing two families of grid lines: perpendicular and parallel to this curve. Since the given region is bounded by several trimmed curves, it is convenient to provide a method guaranteeing that the edges of grid cells generated for individual curves do not intersect or even coincide in their

**GenGrid**$(x_1, ...., x_n, y_1, ...., y_n)$
(* The procedure computes grid vertices for the region with a boundary given by parametric polynomials $[x_i(t), y_i(t)]$*)

(*preprocessing: calculating envelopes *)
**for** i **from** 1 **to** n **do**
(*calculating the envelopes with distance $r$ from boundary*)
$h_i(x, y, r) := res_t(C(x - x_i(t), y - y_i(t), r), \frac{\partial}{\partial(t)} C(x - x_i(t), y - y_i(t), r))$
**od:**

(*calculating grid vertices*)
$nodes \leftarrow Empty;$
**for** r **from** 1 **to** m; i **from** 1 **to** n; j **from** 1 **to** N **do**
(*generate new grid vertex*)
$(x_i^{j,r}, y_i^{j,r}) \leftarrow GenVertex(x_i, y_i, j, r)$

(* check the intersection with already generated grid edges using equation of envelopes and append the new vertex to the list *nodes*)
**if** $h_k(x_i^{j,r}, y_i^{j,r}) \leq 0$ **for all** $k \neq i$ **then** $nodes \leftarrow (x_i^{j,r}, y_i^{j,r})$;
**od:**
**return** *nodes*;

**Fig. 6.** Grid generation algorithm: Calculating the grid vertices. The generated vertices may use the algorithm shown in Fig. 7.

nodes. As can easily be seen by considering, for example, Fig. 5, a certain initial distribution of points $(x_b, y_b)$ on boundary curves could produce the coincidence of the grid nodes generated by (10) separately for each curve. However, the calculation of the boundary point distribution is computationally very expensive because of involving the solution of nonlinear equations like (10). With regard to needed CPU time this computational task can be compared with solving equations (5) themselves, which we are looking to approximate. Thus, instead of calculating the initial distribution of boundary points, we start with any given distribution. Because of possible intersection of grid edges generated by (10) in the case of arbitrary boundary points distribution, we check at each step of our algorithm the intersection. Please note, we do not need to check intersection with already calculated grid edges. Instead of it we may use the equations of the envelopes by substituting the coordinates of grid nodes $(x_h, y_h)$ calculated using (10) in the equations $h_i(x, y) = 0$. Depending on the sign of $h_i(x_h, y_h)$ the new generated grid edge intersects already generated edges or does not intersect. In this way the grid vertices shown in Fig. 5 on the top, right are produced. The description of this part of the algorithm is given in Figs. 4 and 6 The arising "hanging" nodes are eliminated by connecting them to the closest nodes using

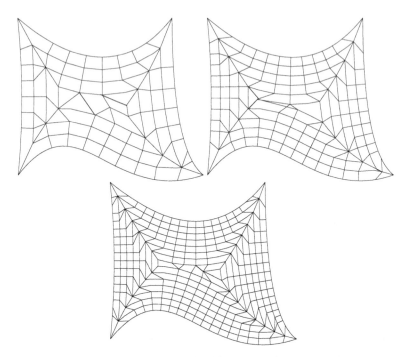

**Fig. 7.** Grids with different mesh size

the algorithm shown in Fig. 9. The resulting grid is depicted in Fig. 5, at the bottom.

As can be seen the grid edges induced by grid vertices generated by the algorithm shown in Figs. 4, 6, and 9 are orthogonal to the boundary curves since the grid vertices lye on the normals to the boundary curves (Proposition 1). In the same way it can easily be shown that most grid edges induced by (10) are orthogonal to all the envelopes:

**Proposition 2.** *Let $h(x, y, r)$ be envelope of a family of circles $C(x - x(t), y - y(t), r)$ with radius $r$ given by (8). Then for any $t_j \in \mathbb{R}$ the line segment given by $P_1$, $P_2 \in \mathbb{R}^2$, where $P_1 = (x(t_j), y(t_j))$ and $P_2$ lies on the envelope $h(x, y, r)$ and is given by*

$$P_2 = \left( x(t_j) \pm \frac{r\frac{dy}{dt}|_{t_j}}{\sqrt{\left(\frac{dy}{dt}|_{t_j}\right)^2 + \left(\frac{dx}{dt}|_{t_j}\right)^2}}, \quad y(t_j) \mp \frac{r\frac{dx}{dt}|_{t_j}}{\sqrt{\left(\frac{dx}{dt}|_{t_j}\right)^2 + \left(\frac{dx}{dt}|_{t_j}\right)^2}} \right)$$

*and intersects $h(x, y, r)$ orthogonally.*

*Proof.* By the result of Proposition 1 the envelope $h(x, y, r) = 0$ may be parameterized by

$$\left( x - \frac{r y_t}{\sqrt{x_t^2 + y_t^2}}, y + \frac{r x_t}{\sqrt{x_t^2 + y_t^2}} \right).$$

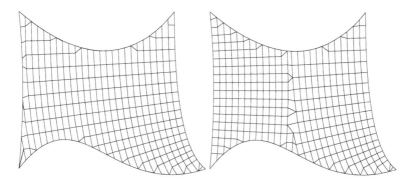

**Fig. 8.** Grids generated by rolling a circle along one boundary curve (left) and along two boundary curves (right)

Differentiation yields the tangent vector $\mathbf{t}_h$ to the envelope:

$$\mathbf{t}_h = \left( x_t - \frac{r y_{t,t}}{\sqrt{x_t^2 + y_t^2}} + 1/2 \, \frac{r y_t \, (2 \, x_t x_{t,t} + 2 \, y_t y_{t,t})}{(x_t^2 + y_t^2)^{3/2}}, \right.$$
$$\left. y_t + \frac{r x_{t,t}}{\sqrt{x_t^2 + y_t^2}} - 1/2 \, \frac{r x_t \, (2 \, x_t x_{t,t} + 2 \, y_t y_{t,t})}{(x_t^2 + y_t^2)^{3/2}} \right).$$

Then the inner product of vectors $P_2 - P_1$ and $\mathbf{t}_h$ is given by the following expression. After reducing the expression to the common denominator, we obtain the inner product to be equal to 0:

$$- \left( x_t - \frac{r y_{t,t}}{\sqrt{x_t^2 + y_t^2}} + 1/2 \, \frac{r y_t \, (2 \, x_t x_{t,t} + 2 \, y_t y_{t,t})}{(x_t^2 + y_t^2)^{3/2}} \right) r y_t \, \frac{1}{\sqrt{x_t^2 + y_t^2}}$$
$$+ \left( y_t + \frac{r x_{t,t}}{\sqrt{x_t^2 + y_t^2}} - 1/2 \, \frac{r x_t \, (2 \, x_t x_{t,t} + 2 \, y_t y_{t,t})}{(x_t^2 + y_t^2)^{3/2}} \right) r x_t \, \frac{1}{\sqrt{x_t^2 + y_t^2}}$$
$$= \frac{r^2 y_t y_{t,t}}{x_t^2 + y_t^2} - \frac{r^2 y_t^2 x_t x_{t,t}}{(x_t^2 + y_t^2)^2} - \frac{r^2 y_t^3 y_{t,t}}{(x_t^2 + y_t^2)^2} + \frac{r^2 x_t x_{t,t}}{x_t^2 + y_t^2}$$
$$- \frac{r^2 x_t^3 x_{t,t}}{(x_t^2 + y_t^2)^2} - \frac{r^2 x_t^2 y_t y_{t,t}}{(x_t^2 + y_t^2)^2} = 0.$$

In this way, we have shown that grid edges generated by our algorithm are orthogonal to boundary as well as to envelopes.                                    ◇

## 3.2   Satisfaction of the Local Optimality Conditions

In the previous sections we have described the calculation of the grid nodes starting from each boundary curve. The algorithm shown in Fig. 6 generates grid nodes iteratively till no more grid nodes can be generated because of intersections. As can be seen in Fig. 5, on the top, right, the hanging nodes are produced. At the next step we connect the hanging nodes in such a way as

**ConnectVertices**$(x_1, ...., x_n, y_1, ...., y_n)$
(* The procedure connects grid vertices computed by *GenGrid* *)

(*calculating intersections of envelopes, which correspond to the adjacent boundary curves*)
**for all** i,j,r **if** $adjacent(h_i, h_j)$ **do**
$intersections[i, j, r] \leftarrow solve(h_i(x, y, r) = 0, h_j(x, y, r) = 0)$
**od:**

$line\_segments \leftarrow Empty;$
(*calculating grid line segments*)
(*connect grid vertices corresponding to the particular position on the boundary curve and the radius of the circle*)
**for all** $r, i, j$ **do**

(*calculating grid line segments parallel to the boundary: if both vertices could be generated, then connect them*)
**if** $(x_i^{j-1,r}, y_i^{j-1,r}), (x_i^{j,r}, y_i^{j,r}) \in nodes$ **then**
$line\_segments \leftarrow [(x_i^{j-1,r}, y_i^{j-1,r}), (x_i^{j,r}, y_i^{j,r})]$

(*if the left vertex could not be generated, then connect the right vertex to the intersection of envelopes on the left hand side*)
**if** $(x_i^{j-1,r}, y_i^{j-1,r}) \notin nodes$ **then**
$line\_segments \leftarrow [intersections[i, i - 1, r + 1], (x_i^{j,r}, y_i^{j,r})]$

(*if the right vertex could not be generated, then connect the left vertex to the intersction of envelopes on the right hand side*)
**if** $(x_i^{j,r}, y_i^{j,r}) \notin nodes$ **then**
$line\_segments \leftarrow [intersections[i, i + 1, r + 1], (x_i^{j-1,r}, y_i^{j-1,r})]$

(*calculating grid line segments perpendicular to the boundary: if both vertices could be generated, then connect them*)
**if** $(x_i^{j,r}, y_i^{j,r}), (x_i^{j,r+1}, y_i^{j,r+1}) \in nodes$ **then**
$line\_segments \leftarrow [(x_i^{j,r}, y_i^{j,r}), (x_i^{j,r+1}, y_i^{j,r+1})]$
**od:**

(*if the top vertex could not be generated, then connect the bottom vertex to the closest intersection point of envelopes*)
**if** $(x_i^{j,r+1}, y_i^{j,r+1}) \notin nodes$ **then**
$line\_segments \leftarrow [find\_closest\_intersection(), (x_i^{j,r}, y_i^{j,r})]$

**return** $line\_segments;$

**Fig. 9.** Grid generation algorithm: Connecting the grid vertices

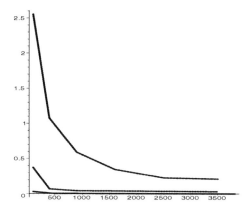

**Fig. 10.** Values of the Orthogonality Functional $I_O$ in dependence on the number of mesh nodes for the Winslow Generator (dash-dot), the Length Generator (dot), and, the method of envelopes (solid)

to obtain the valid meshing. In this section we shall consider our method with respect to local minimization of Orthogonality and Length functionals. As has been mentioned in the Introduction, we are interested in satisfying the corresponding Euler–Lagrange equations locally. This means the following. Let us fix the hanging nodes and consider above functionals for all the nodes in between. First, consider the Orthogonality functional:

**Proposition 3.** *The grid generated by*

$$x(\xi, \eta) = x(\xi) \pm \frac{\eta \frac{dy}{d\xi}}{\sqrt{\left(\frac{dx}{d\xi}\right)^2 + \left(\frac{dy}{d\xi}\right)^2}};$$

$$y(\xi, \eta) = y(\xi) \mp \frac{\eta \frac{dx}{d\xi}}{\sqrt{\left(\frac{dx}{d\xi}\right)^2 + \left(\frac{dy}{d\xi}\right)^2}}. \tag{11}$$

*minimizes $I_O$.*

*Proof.* Substituting (11) into each of Euler–Lagrange Equations for the Orthogonality functional $I_O$ given by

$$x_{\xi,\xi} x_\eta{}^2 + 4 x_\eta x_\xi x_{\xi,\eta} + x_\eta y_{\xi,\xi} y_\eta + x_\eta y_\xi y_{\xi,\eta} + 2 x_{\xi,\eta} y_\xi y_\eta + x_\xi{}^2 x_{\eta,\eta} + x_\xi y_{\xi,\eta} y_\eta$$
$$+ x_\xi y_\xi y_{\eta,\eta} = 0,$$
$$y_\eta x_{\xi,\xi} x_\eta + y_\eta x_\xi x_{\xi,\eta} + y_{\xi,\xi} y_\eta{}^2 + 4 y_\eta y_\xi y_{\xi,\eta} + 2 y_{\xi,\eta} x_\xi x_\eta + y_\xi x_{\xi,\eta} x_\eta + y_\xi x_\xi x_{\eta,\eta}$$
$$+ y_\xi{}^2 y_{\eta,\eta} = 0,$$

we obtain with the aid of Maple a large differential expression containing 27 terms. After reducing this expression to the common denominator we obtain the expression equal to 0:

$$-8 \frac{\eta\, y_{\xi,\xi}{}^2 y_\xi}{(x_\xi{}^2 + y_\xi{}^2)^{3/2}} - 2 \frac{x_\xi{}^4 x_{\xi,\xi}}{(x_\xi{}^2 + y_\xi{}^2)^2} + 2 \frac{\eta\, y_\xi{}^3 x_\xi x_{\xi,\xi,\xi}}{(x_\xi{}^2 + y_\xi{}^2)^{5/2}} + \cdots$$

$$2\,\frac{\eta\,y_{\xi,\xi,\xi}\,x_\xi{}^2 y_\xi{}^2}{(x_\xi{}^2+y_\xi{}^2)^{5/2}}+6\,\frac{\eta\,y_{\xi,\xi}x_\xi{}^3 x_{\xi,\xi}}{(x_\xi{}^2+y_\xi{}^2)^{5/2}}+2\,\frac{x_\xi{}^2 x_{\xi,\xi}}{x_\xi{}^2+y_\xi{}^2}=0.$$

The Euler–Lagrange equations for the Orthogonality functional are also satisfied.                                                                              ◇

In this way we have proved that grid calculated using our algorithm is locally optimal with respect to the Orthogonality functional. In the global sense the computational comparison for the Winslow generator and our method has been performed and is shown in Fig. 10.

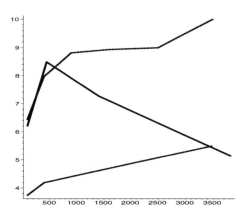

**Fig. 11.** Values of the Length Functional $I_L^v$ in dependence on the number of mesh nodes for the Winslow generator (dash-dot), the Length generator (dot), and, the method of envelopes (solid)

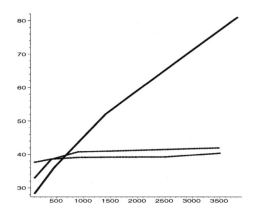

**Fig. 12.** Values of the Length Functional $I_L^h$ in dependence on the number of mesh nodes for the Winslow generator (dash-dot), the Length generator (dot), and, the method of envelopes (solid)

Furthermore, by the construction, the method does not produce grid foldings. Since the meshing is required to cover the whole area of the domain, the method is optimal with respect to the area. Let us consider the presented method with respect to the Length functional. Let us write the Length functional as a sum of two functionals $I_L = I_L^v + I_L^h$ corresponding to grid edges, which are perpendicular and parallel to the boundary:

$$I_L^h(x,y) = 1/2 \iint (x_\xi^2 + y_\xi^2) d\xi \, d\eta; \qquad I_L^v(x,y) = 1/2 \iint (x_\eta^2 + y_\eta^2) d\xi \, d\eta.$$

The following Proposition shows that our method optimizes $I_L^v$.

**Proposition 4.** *The grid generated by (11) minimizes $I_L^v$.*

*Proof.* The proof is similar to the proof of Proposition 3. Substituting (11) into the Euler–Lagrange equations corresponding to $I_L^v$ yields the statement of the proposition.                                                                                  ◇

From the geometric point of view, the statement of Proposition 4 holds because of the following reasons. Consider the fixed boundary point $(x(t_j), y(t_j))$. By our construction (as shown in Proposition 1) increasing $r$ produces grid nodes $(x_h(j,r), y_h(j,r))$, which lie on the straight line perpendicular to the boundary in $(x(t_j), y(t_j))$. Of course, a straight line produces the minimal length among all the curves, which may connect the points $(x_h(j,r), y_h(j,r))$. On the other hand, the statement does not hold for $I_L^h$, because the points $(x_h(j,r), y_h(j,r))$ by increasing $j$ lie on the envelope, which is not the curve of minimal length between these grid vertices, in contrast to the straight line. Our method allows to minimize $I_L^h$ as well, namely, by appropriate changing of the radius of the rolling circle in such a way as the points $(x_h(j,r), y_h(j,r))$ lie on the straight line. However, as can be easily seen from the presented considerations, minimizing $I_L^h$ would destroy the orthogonality of grid edges. The computational comparsion for our method, the Winslow generator, and the Length generator with respect to $I_L^h$ and $I_L^v$ is shown in Figs. 11 and 12.

## 4   Computational Experiments

In order to compare the presented method of envelopes with well-established grid generation methods, we describe the numerical procedure implementing a finite-difference method for solving the Euler–Lagrange equations corresponding to the individual functionals. For this purpose we use the *Alternating Direction Implicit* (ADI) method introduced in [7].

For instance, consider the Winslow grid generator, which is based on the solution of the following system of nonlinear coupled PDE's:

$$\left(x_\xi{}^2 + y_\eta{}^2\right) x_{\xi,\xi} - 2 \left(x_\xi x_\eta + y_\xi y_\eta\right) x_{\xi,\eta} + \left(x_\xi{}^2 + x_\eta{}^2\right) x_{\eta,\eta} = 0,$$
$$\left(x_\xi{}^2 + y_\eta{}^2\right) y_{\xi,\xi} - 2 \left(x_\xi x_\eta + y_\xi y_\eta\right) y_{\xi,\eta} + \left(x_\xi{}^2 + x_\eta{}^2\right) y_{\eta,\eta} = 0.$$

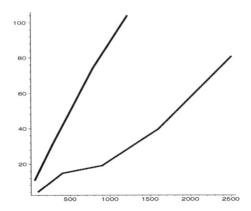

**Fig. 13.** CPU time versus the number of mesh nodes for the Length generator (dot) and the method of envelopes (solid)

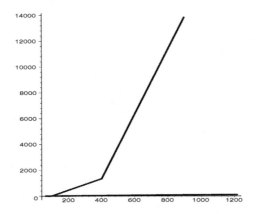

**Fig. 14.** CPU time versus the number of mesh nodes for the Winslow generator (dash-dot) and the method of envelopes (solid)

As described in [5], we use the following second-order approximation for the partial derivatives of the function $f(\xi, \eta)$:

$$(f_\xi)_{i,j} = 1/2(f_{i+1,j} - f_{i-1,j}),$$
$$(f_\eta)_{i,j} = 1/2(f_{i,j+1} - f_{i,j-1}), \quad (f_{\xi,\xi})_{i,j} = (f_{i+1,j} - 2f_{i,j} + f_{i-1,j}),$$
$$(f_{\eta,\eta})_{i,j} = (f_{i,j+1} - 2f_{i,j} + f_{i,j-1}),$$
$$(f_{\eta,\xi})_{i,j} = 1/4(f_{i+1,j+1} - f_{i+1,j-1} - f_{i-1,j+1} + f_{i-1,j-1}).$$

Let us introduce the following difference operators:

$$\Lambda_\xi^n f_{i,j} = \left[(x_\eta^2)_{i,j}^n + (y_\eta^2)_{i,j}^n\right](f_{i+1,j} - 2f_{i,j} + f_{i-1,j}),$$

$$\Lambda_\eta^n f_{i,j} = \left[(x_\xi^2)_{i,j}^n + (y_\xi^2)_{i,j}^n\right](f_{i,j+1} - 2f_{i,j} + f_{i,j-1}),$$

$$\Lambda_{\eta,\xi}^n f_{i,j} = -1/2\left[(x_\xi x_\eta)_{i,j}^n + (y_\xi y_\eta)_{i,j}^n\right](f_{i+1,j+1} - f_{i+1,j-1} - f_{i-1,j+1} + f_{i-1,j-1}).$$

The superscript denotes the number of iterations. Then the ADI difference scheme, which converges to the solution of Winslow equations using pseudo-time steps $\tau$, may be written as follows:

$$\frac{\tilde{x}_{i,j} - x_{i,j}^n}{0.5\tau} = \Lambda_\xi^n \tilde{x}_{i,j} + \Lambda_{\xi,\eta}^n x_{i,j}^n + \Lambda_\eta^n x_{i,j}^n \, ,$$

$$\frac{x_{i,j}^{n+1} - \tilde{x}_{i,j}}{0.5\tau} = \Lambda_\xi^n \tilde{x}_{i,j} + \Lambda_{\xi,\eta}^n \tilde{x}_{i,j} + \Lambda_\eta^n x_{i,j}^{n+1} \, ,$$

$$\frac{\tilde{y}_{i,j} - y_{i,j}^n}{0.5\tau} = \Lambda_\xi^n \tilde{y}_{i,j} + \Lambda_{\xi,\eta}^n y_{i,j}^n + \Lambda_\eta^n y_{i,j}^n \, ,$$

$$\frac{y_{i,j}^{n+1} - \tilde{y}_{i,j}}{0.5\tau} = \Lambda_\xi^n \tilde{y}_{i,j} + \Lambda_{\xi,\eta}^n \tilde{y}_{i,j} + \Lambda_\eta^n y_{i,j}^{n+1} \, .$$

Using this scheme, for example, the grid in Fig. 1 has been obtained. In Figs. 10, 11, and 12 the values of the discrete Length and Orthogonality functionals, which have been introduced in Section 2, are compared for the Winslow Generator and our method of envelopes for different sizes of the mesh. We have used the region shown in Fig. 5. The proposed method of envelopes proves to be efficient in terms of the CPU time needed for its computer implementation. As shown in Figs. 13 and 14, increasing mesh size produces almost linear time growth in the case of our method, whereas the CPU time for the Winslow generator tends to grow much faster.

## 5   Conclusion

In the present paper we have presented a method of envelopes, called otherwise the method of rolling circle, which allows us to obtain the tight approximation of optimal meshing on nonconvex regions. As has been shown, Area and Orthogonality functionals, which have to be minimized in order to obtain optimal unfolded meshing are minimized by the method of envelopes in the local sense. This means, the corresponding Euler–Lagrange equations are satisfied locally. For the split Length functional (in the sense of Proposition 4) one part is completely minimized, and the value of another part depends on the curvature of the boundary. Increasing curvature leads to the worse approximation of the minimizer. Minimizing the second part of Length functional is also possible using our method. However, as can easily be derived from the presented consideration, minimizing the second part of the split Length functional would destroy the orthogonality. The obtained results show the tight approximation of the minimizers of individual mesh functionals by the method of envelopes. The main advantage of the method is its algorithmic simplicity and efficiency. Furthermore, the method is domain independent and can be applied to domains for which many classical iterative procedures do not converge, or require the manual choosing of weights of individual functionals.

# References

1. Bruce, J.W., Giblin, P.J.: Curves and Singularities, Cambridge University Press, 1984
2. Bungartz, H.-J.: Dünne Gitter und deren Anwendung bei der adaptiven Lösung der dreidimensionalen Poisson-Gleichung. Dissertation, Institut für Informatik, Technische Universitait München, 1992
3. Cox, D.A., Little, J.B., O'Shea, D.: Ideals, Varieties, and Algorithms, Springer-Verlag, Berlin, 1996
4. Chibisov, D., Ganzha, V. G., Mayr, E. W., Vorozhtsov, E.V.: Generation of orthogonal grids on curvilinear trimmed regions in constant time. In: Proc. CASC'2005, LNCS 3718, Springer-Verlag, Berlin, Heidelberg, 2005, 105–114
5. Ganzha, V. G., Vorozhtsov, E.V.: Numerical Solution for Partial Differential Equations: Problem Solving Using *Mathematica*, CRC Press, Boca Raton, Ann Arbor, 1996
6. Knupp, P., Steinberg, S.: Fundamentals of Grid Generation. CRC Press, Boca Raton, Ann Arbor, 1994
7. Peaceman, D. W. and Rachford, H.H., jr.: The numerical solution of parabolic and elliptic differential equations, J. of SIAM **3**(1955) 28–41
8. Zenger, C.: Sparse grids. In: Parallel Algorithms for Partial Differential Equations, Proc. Sixth GAMM-Seminar, Kiel, 1990, Hackbusch, W. (ed.), Vol. 31 of Notes on Num. Fluid Mech. Vieweg-Verlag, Braunschweig/ Wiesbaden (1991) 241–251

# Providing Modern Software Environments to Computer Algebra Systems

Svetlana Cojocaru, Ludmila Malahova, and Alexander Colesnicov

Institute of Mathematics and Computer Science,
Academy of Sciences of Moldova, Chişinău

**Abstract.** Many computer algebra systems lack modern user-friendly software environment. Poorly designed interface depreciates rich mathematical ideas implemented in calculation engine. It obstructs extensive usage of such systems because of requiring special knowledge and skills, e.g., in programming, to use them. Another problem of computer algebra systems is multitude of data formats and the implied difficulty in simultaneous usage of different systems. We discuss basics of and requirements to interfaces for computer algebra systems and techniques of their implementation. Modern software engineering approaches permit to provide a toolkit for semi-automated development of software environments for computer algebra systems.

## 1 Introduction

We will strictly distinguish between the computational and interface aspects of computer algebra (CA) systems. We suppose the existence of a program executing algebraic computations that we will refer to as an *engine*. Multitude of solved problems makes investigators to create specialized CA engines in the cases when general purpose systems are inefficient, or the necessary functionality is not provided even by commercial systems. As a rule, the creator of a specialized CA system has not enough time, resources, and qualification to develop a software environment for it. It is not unusual that rich mathematical ideas implemented in a CA engine are enveloped in poorly designed interface. The absence of the user-friendly standard software environment does not permit the extensive usage of such system. Another problem of CA engines is multitude of data formats and the implied difficulty in communication between different engines at their simultaneous usage.

The paper is organized as follows:

- General problems of interfaces to CA systems are discussed in Sec. 2. We refer to several existing systems to illustrate our observations.
- Two CA systems, Bergman and Singular, and their interfaces are discussed in Sec. 3.
- We discuss and motivate desired features of interface to a computer algebra system (CAS) in Sec. 4.

V.G. Ganzha, E.W. Mayr, and E.V. Vorozhtsov (Eds.): CASC 2006, LNCS 4194, pp. 129–140, 2006.

– The result of previous discussions finds its implementation during an INTAS project "Interface Generating Toolkit for Symbolic Computation Systems" (grant ref. $05 - 104 - 7553$). We discuss this project in Sec. 5.

## 2    Interfaces to Computer Algebra Systems

We divided all functionality of a CAS into two parts: the engine features (the computations that the CAS can execute) and the interface features. It is obvious that the latter are external relative to the former and are almost independent of them.

Development of interfaces for CAS was and remains an object of long-time investigations [1, 2].

The problem of CAS interface has the following aspects:

1. Interaction with text editors;
2. Graphics;
3. Interaction with numerical calculation systems;
4. Interaction between different CAS (including interaction through computer networks);
5. Testing support;
6. Interaction with end users.

Investigations show that CAS interface development should solve the following problems:

– 2-D representation of mathematical expressions,
– Editing of mathematical expressions that includes sub-expression manipulation,
– Windows that model sheets of paper and combine texts, formulae, and graphics,
– Processing and presentation of long expressions,
– Simultaneous use of several CASs, which implies the necessity to solve problems of data conversion, configuration management, and communication protocols,
– Satisfaction of special needs for teaching systems, (in particular, the possibility to show intermediate results and explications of processes applied to obtain them; elaboration of electronic manuals, and especially interactive ones),
– Interface extensibility providing additions of new menus, new fragments of on-line documentation, etc.,
– Guiding of the user during the whole period of his/her problem solving,
– The system should be self-explanatory; its operational mode should be understandable directly from the experience of interaction with the system,
– Control over problem formulation correctness and over information necessary to solve it.

The primary scope of an interface is creation of a comfortable environment for a mathematician or another specialist that uses mathematical apparatus. It would be preferable for these users to input data and to obtain mathematical results in their natural 2-dimensional form. The linear form of input that can be also used in this mode is faster but it imposes additional conventions to enter powers, indices, fractions, etc., or uses additional characters. It is necessary also to provide possibilities to edit expressions, integrate them with a usual text, and obtain results in a form suitable for publication of an article (e.g., LaTeX) or in Internet (e.g., MathML).

The syntactic check of entered mathematical expressions and the spelling check of accompanying text would be also desired features.

The following three categories of CAS can be identified:

1. Systems or packages that do not have a special interface,
2. Interfaces based on a (specialized) programming language,
3. Graphical interfaces.

This division is not strict: e.g., most systems with graphical interface possess their own programming language also.

There are many systems that have command line interface only, e.g., Singular [9], Bergman [3, 4, 5, 6, 7], and Yacas [10].

Absence of graphical interfaces is compensated partially by integration in an existing editor like Emacs or its derivatives (e.g., such are Macaulay [11] and Singular), in Scientific Workplace, Scientific Word, or Scientific Notebook [12] (e.g., Maple [13] and MuPAD [14]). MuPAD has the interface to Java but their approach is opposite to ours: MuPAD itself is regarding as the shell, and Java programs are treated as applets or plug-ins expanding MuPAD. Services provided from these editors may be not too sophisticated but create much more comfortable environment than operating in ASCII from the command line.

We can mention also specialized editors developed to serve as interfaces to CASs. One such editor is TeXmacs by Joris van der Hoeven [15]. It combines elements of TeX and Emacs and was successfully applied for Macaulay 2, Reduce [16], MuPAD, Maxima [17]. Another front-end product with graphical interface is FrontMan [18]. It offers a small but useful set of possibilities (syntactic coloring of input information and results, export of sessions in HTML format, integrated document visualization through a Web browser, multiple simultaneous sessions, safe transmission through the network using the SSH protocol). An important fact is that these editors permit creation of user's own style. It is a kind of personalization created by a user.

Most of features mentioned above can be found in systems with graphical interfaces. Examples of such systems are Derive [19], Mathematica [20], etc. In general, most of these systems provide:

- Visualization of mathematical formulae in 2-dimensional format,
- Sub-expression manipulation,
- Separate windows for data input and results output,
- Separate windows for graphical operations,

- Export of results in a printable format (RTF, PDF, LaTeX, etc.)
- Comfortable navigation with on-line help,
- Integration with an existing or specially developed editor that facilitates editing of mathematical texts,
- Demonstration of intermediate steps to explain processes of expression transformation.

In addition, so-called "notebooks" [12] support operations over text, mathematical formulae, and graphics. Most of them can be adapted to user's preferences individualizing menus and toolbars, and assigning hot keys to actions.

Interaction between different systems is to be supported by use of specially developed unified formats for mathematical formulae, like OpenMath [21] or OpenXM [22]. Most modern systems lack supports of such formats because these formats were developed after these systems were created.

# 3   Case Studies: Bergman and Singular

## 3.1   Bergman

Bergman [23] is a CAS for symbolic calculations in non-commutative and commutative algebra. It calculates Gröbner basis and related functions. The current version does not provide general procedures for polynomial calculations, but it is planned for the future.

Bergman is written in Lisp. In the beginning the users were to communicate with Bergman through its underlying Lisp console. This was found unsuitable for most users, and a graphical shell was developed in Java. The system and its interface shell were described elsewhere [3, 4, 5, 6, 7].

## 3.2   Singular

Singular [9] is positioned as a specialized CAS for polynomial computations in commutative algebra, algebraic geometry, and singularity theory.

Singular works with ideals and modules over rings. These base rings are polynomial rings over fields (e.g., finite fields, the rationals, floats, algebraic extensions, transcendental extensions), or quotient rings with respect to an ideal. Singular's calculation engine includes several various algorithms for computing Gröbner bases, and polynomial calculations like factorizations, GCD, syzygy computations, etc.

Singular has an interactive shell that is in fact an ASCII terminal, or it can be run within Emacs. It has its own C-like programming language, that includes variables of different types (e.g., integers or rings) and other usual features, e.g., procedures. It has so-called *communication links* that permit to call its functions from other programs.

Singular execution can be interrupted, but only at certain places, namely, after finishing the current *kernel command*.

The Emacs interface provides everything that is available at the ASCII-terminal interface, and offers many additional features, e.g., color highlighting, parentheses matching, running interactive demo sessions, etc. The Emacs interface can be widely customized.

### 3.3    Inferences from Experience of Bergman and Singular

To implement the Bergman shell, we selected Java [24] as the *implementation language*. The resulting shell is highly portable. We developed the Bergman shell in Intel PC under Linux and Windows, and in Sun Sparc under Solaris, and tested the CLISP port of Bergman and the shell in all these environments. The selection of Java should be classified as very successful. Some small problems like distortion of screen forms under different environments were solved by careful design and obligatory testing under all available systems.

If the calculation engine is separated from interface, the inefficiency of Java interpretive nature can be ignored. The time for typing data is small comparatively to the calculation time. For the interface, the ruling consideration is its comfortability for the user.

Both Bergman and Singular permit to save *working sessions* in files and restore them later. For Bergman, we generalized sessions to *environments*. A session is a set of parameters that fully defines the problem to be solved. An environment is a partial set of data common for several sessions. It corresponds to the group of mathematical problems the user investigates during several sessions. E.g., the environment "commutative" fixes a single parameter, the commutativity. All new sessions are created using the current default environment. Inversely, when a new environment is created, it is based on the parameters of the current session. To save a session as an environment, the user selects parameters that are to be fixed, and drops other parameters.

Like Singular, the user of Bergman inputs list of variables and polynomials as texts. A lot of data may be represented as switches (flags) and selections. Bergman has approx. 30 input fields the user should set. We provided a new solution that permitted to concentrate all information in a relatively small panel. Our solution uses drop-down menus and variable labels that shows the current selection. The drop-down menu is activated when the corresponding label is clicked. We call such panel a *parameter panel*.

Like Singular, Bergman has some suitable points where the user can interfere in the process. We call them *internal hooks* and can supply variable modules to be called at these points. There are many possible uses of such hooks, e.g., stop of calculations or progress indicators.

## 4    Desired Features of Interface to Computer Algebra System

We can conclude from all aforementioned that CAS interfaces have a lot of features and functions in common.

Discussions in Sec. 2 and 3 show that all discussed problems can be reviewed as general problems of graphical shells to CASs. We can look at Bergman or Singular as at the **calculation engine** which is called from the dialog shell. Inversely, we can look from the shell side and treat Bergman or Singular as the **calculation plug-in** for the shell.

The separation of calculation engine and interface shell means that a toolkit can be developed for semi-automatic generation of interface shell for a CAS. Such toolkit should consist of an interface generator and ready made adaptable interface modules, e.g., in Java.

The adaptable interface modules should provide the following functionality:

– To implement selected features of CAS interface at the generalized level permitting automated production of interface, taking into account specifics of this area.
– To implement cross-process interaction with a symbolic computation engine.
– To implement inside the interface tools for smart user personalization and intellectual adaptation to his/her preferences.
– To implement tools for interaction with other existing CAS.
– To adapt, implement, and integrate tools for auxiliary development tasks, e.g., help and documentation tools, extensibility support, testing support, etc.
– To adapt, implement and integrate tools for visual presentation of data and results.

## 4.1   Principles of Interface Creation

At present, a lot of principles were proposed for interface design. The following ten principles were formulated by Jakob Nielsen [25]. We would apply these principles in our research and development of the CAS interface construction toolkit.

Mathematicians and computer scientists from our group estimated several CA systems against these principles. We selected well-known and widely used commercial systems Mathematica and Maple, and took for comparison Bergman with its Java shell. The results are shown in Tab. 1.

These estimations show that even widely used systems with a good interface can be improved in many aspects. E.g., the following complaints were made for Maple 9.5 and 10:

1. The red and blue coloring of the Maple working document irritates eyes and can't be changed.
2. The system does not keep its status between runs.
3. It is difficult to create aliases for frequently used commands.
4. Clearing of system memory by an internal command (does not exist in menu).
5. Help is less understandable and usable than in Mathematica; no immediate suggestion at wrong command spelling.
6. It is difficult to manipulate quickly with different working windows (no functional keys provided).

The following complaints were made for Matematica 5.0:

1. Conversion in the LATEX format is made by an internal command (does not exist in menu).
2. It is difficult to manipulate quickly with different working windows.
3. Non-convenient interface made from several different objects (base menu, symbol list, working window).
4. It is almost impossible to return to the preceding step (difficult undo).
5. The system does not keep the style of the preceding formatting for the new input.
6. No immediate suggestion at wrong command spelling.
7. Quick editing panel permits to enter a matrix but it is not internalized for further calculations.

These complaints are of course subjective but they permit to motivate the notes from Tab. 1, and we can use them as inspirations in our work. One of main conclusions we made after analyzing opinions of CAS users is as follows: a CAS should adapt to each user's needs and habits without user's efforts. The customization tools are an important part of graphical software environment but they took user's time and may request additional qualification. The adaptation problem can be successfully solved in a system with an intelligent interface that we discuss right below.

## 4.2   Intelligent Interfaces

As to program products, the last decade is characterized by increasing of their complexity. This implies the quick increasing of area where computers are used, and makes the problem of human-computer interfaces extremely important. On the one hand, the number of non-professional users is increasing; on the other hand, programs became more and more complicated. This situation led to the interface intellectualization: a process where the interface intercepts some of user's functions and even makes some decisions for the user.

No exact definition of an intelligent interface exists, and it seems impossible to provide one.

Generalizing principles proposed by different authors [8, 26, 27], we can formulate the following four aspects of an intelligent interface.

- Adaptability of the interface: techniques permitting human-computer interaction to be adapted to different users and to different situations at usage.
- Modelling of the user: techniques permitting to operate in the system with knowledge on the user.
- Natural language technology: techniques permitting interpretation and generation of utterances in natural language in the form of text or speech.
- Intelligent help in complex systems: techniques permitting system usage from the very first run and providing help at moments and situations when the user needs them most probably.

**Table 1.** Ten heuristic principles in user interfaces of some CAS

| | Principle | Mathematica 5 | Maple 9.5 | Bergman (with shell) |
|---|---|---|---|---|
| 1 | Visibility of system status | − | − | − |
| 2 | Match between system and the real world | +− | +− | +− |
| 3 | User control and freedom | +− | −+ | −+ |
| 4 | Consistency and standards | +− | −+ | −+ |
| 5 | Error prevention | +− | +− | −+ |
| 6 | Recognition rather than recall | −+ | −+ | − |
| 7 | Flexibility and efficiency of use | +− | +− | −+ |
| 8 | Aesthetic and minimalist design | −+ | −+ | +− |
| 9 | Help users recognize, diagnose, and recover from errors | + | ++ | − |
| 10 | Help and documentation | ++ | + | −+ |

Legend: − not implemented; + implemented;
+− implemented but need modifications; −+
not implemented but is alike; ++ a very good
implementation

These four characteristics are mentioned by the majority of authors. We will remark another two found in different sources and being, in our opinion, rather important.

– Generation of explications: techniques that permit to explain to the user how the result was obtained.
– Problem interception from the user: techniques that permit to understand user's actions and intentions and to propose him a suitable variant of further actions making it possible to concentrate on other problems.

All these characteristics are not absolutely necessary or even sufficient to declare that a given interface is an intelligent one. Some of them can be manifested more accentuated, to be absent or mutually dependent. Moreover, it seems that not all of them are necessary for our project. We selected for the implementation the following:

– to provide possibilities to formulate the problem, not requiring the description of solution;

- to make it possible to formulate the problem in user's professional (sub) language;
- adaptation to the user.

## 5  Implementation

Now a project is financed by INTAS (grant ref. $05 - 104 - 7553$) for implementation of a toolkit to develop interfaces for CASs. The toolkit will use modern software technologies and provide semi-automated creation of the necessary interface. The toolkit is supposed to consist of several Java packages (Java class libraries) and an interface generator that will be used to construct the interface from these libraries. The toolkit will also provide the support of communication between different symbolic computation engines. This support will be based on use of XML standard formats for mathematical data.

The use of the implemented toolkit will be exemplified by using it to develop interfaces for two symbolic computation systems: Singular [9] and Bergman [23]. Singular works in commutative mode, while Bergmans powerful part are non-commutative computations. Both Singular and Bergman lacks modern graphical interface. Last but not least, using the toolkit with Singular and Bergman means a kind of their integration.

Preliminary researches include:

- Classification and technological description of features and functions specific to interfaces for CAS;
- Clarification and classification of cross-process interactions between interface modules and symbolic computation engines;
- Generalization of selected features to the level permitting automated interface generation;
- Description of adaptive behavior of interface elements, and techniques to implement it;
- As the result, the necessary basis will be created to automated development of interfaces for CAS.

The main part of our project is the Interface Generator. It will semi-automatically produce a graphic interface shell for a given CAS from adaptable Java modules.

These modules form the following functional groups: sessions and environments; input; processes; output; data convertors and inter-system communications; configuration; help; testing; other modules.

Some of functionality from this list is available as ready-made open source modules, but we do not implement a single shell for a specific CAS. We implement a graphic software environment generator. It implies that even ready freely available modules should be included in our toolkit after some generalization and parameterization.

Now let us discuss some of these functional groups of modules.

Sessions and environments were discussed above.

The input modules will implement different input techniques. For example, editors and pretty-printing were the subject of a lot of investigations and implementations, and we would, most probably, generalize some of existing free solutions. There exist also ready-made open source modules for 2-D input and sub-expression manipulation. However, there is no good implementation of wizards. Wizard mode is necessary for novice users, and also to input very complex data. Parameter panels we used in Bergman shell also need to be worked over as we constructed them ad hoc from raw dialog elements. Such aspects as input data analysis, checking, reactions to erroneous data, etc., need attention also. We should provide external checking of input data, and we can also interact with the shell at input checking.

Running processes also includes many aspects. As with wizards, there is no good implementation of consoles and process management. In our experience, we found that the console should provide some additional features, e.g., additional button to open a file and to input its contents as if it were typed in console. There are also logging, safety checks during run to detect external intervention during calculation, disk overflow, processor overloading, etc. Cross-process interaction includes running of several calculations simultaneously, or running in parallel several different engines under the same shell, or running parts of the same calculation in parallel and to exchange data between them, etc.

In many cases it is necessary to interact with the engine in some critical points of calculations, e.g., after terminating a step of calculation to show progress indicator. The standard technique for this is inclusion in the engine code special "hooks" that are calls of external procedures with preliminary fixed parameters. Before calculation, the shell transmits to the engine entry points of such functions that interrupt calculations when hooks are met. We are to formulate a set of requests for such hooks and places where they are to be inserted, and to implement components that interact through them. Stopping and continuation of calculations and progress indicators are the obvious aspects that can be implemented using hooks.

For output, we need components that manage presentation of calculation results, keep them, send them by e-mail, and convert them into different formats. We should generate a lot of standard formats (LaTeX, PostScript, RTF, PDF, MathML, OpenMath, OpenXM), and formats of different existing CAS.

We would provide components that compare results of several calculations and reveal differences. This is necessary, e.g., to find differences in results obtained with varied input data, or to compare test calculation results with standard ones.

We will provide components that manage configurations of the shell. More general, we will support smart adaptation of the shell to user's needs and preferences.

Help files are produced by compiling from a specially prepared (marked) sources using a compiler. For modern CHM format, the corresponding compiler by Microsoft is free. It is also possible to have HTML help that is browsed by standard browsers but needs special indexation module. In any case, some preliminary operations are necessary before compilation.

Testing components manage test examples and analyze results of test runs of the engine, e.g., compare them with their standards.

There are also some tools that were not listed under previous topics. For example, we need components that manage list of external tools in the shell. External tools are some user selected programs called directly from the shell, e.g., an additional text editor. They can be run in console or in window, and can have parameters. It is necessary to insert tools in the list, to delete them from the list, and reorder the list.

## 6   Conclusions

CAS is a very useful tool that simplifies formula manipulation and handling of mathematical models for engineering applications, for mathematical research, for education, and for many other areas. Our approach can be applied in all these cases and these areas will gain time and efforts for interface development. Universality of the proposed solution will be guaranteed if we will use Java as the technology for its implementation. The developed packages and applications will permit investigators in different areas to concentrate efforts on symbolic computation engines and to use ready-made interface solutions.

## References

1. Kajler, N., Soiffer, N.: A survey of user interfaces for computer algebra systems. Journal of Symbolic Computation **25** (2) (1998) 127–159
2. Computer-Human Interaction in Symbolic Computation, N. Kajler (ed.). Springer-Verlag, Wien, 1998
3. Colesnicov, A.: Implementation and usage of the Bergman package shell. Computer Science Journal of Moldova **4**, No. 2 (11) (1996) 260–276
4. Backelin, J., Cojocaru, S., Ufnarovski, V. BERGMAN. In: Computer Algebra Handbook. Grabmeier, J., Kaltofen, E., Weispfenning, V. (eds.). Springer-Verlag (2003) 349–352
5. Backelin, J., Cojocaru, S., Ufnarovski, V.: The Computer Algebra Project Bergman: Current State. In: "Commutative algebra, Singularities and Computer Algebra", J. Herzog and V. Vuletescu (eds.) (2003) 75–101 Series II. Mathematics, Physics and Chemistry. **Vol. 115,** Kluwer Academic Publishers
6. Backelin, J., Cojocaru, S., Colesnicov, A., Malahova, L., Ufnarovski, V.: Problems in interaction with the Computer Algebra System Bergman. In: "Computational Commutative and Non-Commutative Algebraic Geometry", **196,** NATO Science Series: Computer & Systems Sciences. S. Cojocaru et al. (eds.). IOS Press (2005) 185–198
7. Backelin, J., Cojocaru, S., Ufnarovski, V.: Mathematical Computations Using Bergman. Lund University, Centre for Mathematical Science, 2005
8. Ehlert, P.: Intelligent User Interfaces: Introduction and survey. Research Report DKS03–01/ICE 01, February, 2003, Delft University of Technology

## Web References

9. Singular: *http://www.singular.uni-kl.de/*
10. Yacas: *http://yacas.sourceforge.net/*
11. Macaulay: *http://www.math.uiuc.edu/Macaulay2/*
12. Scientific Workplace, Scientific Word, Scientific Notebook: *http://www.mackichan.com/*
13. Maple: *http://www.maplesoft.com/*
14. MuPAD: *http://www.mupad.de/*
15. TEXmacs: *http://www.math.upsud.fr/ anh/TeXmacs/TeXmacs.html*
16. Reduce: *http://www.reduce-algebra.com/*
17. Maxima: *http://maxima.sourceforge.net/*
18. FrontMan: *http://rpmfind.net/linux/RPM/sourceforge/r/rp/rpmsforsuse/ frontman-0.3.4-1.i386.html* *http://www.eleceng.ohio-state.edu/ ravi/kde/frontman.html*
19. Derive: *http://www.chartwellyorke.com/derive.html*
20. Mathematica: *http://www.wolfram.com/products/mathematica/index.html*
21. OpenMath: *http://www.openmath.org/*
22. OpenXM: *http://www.math.kobe-u.ac.jp/OpenXM/*
23. Bergman: *http://www.math.su.se/bergman/*
24. Java: *http://java.sun.com/*
25. *http://www.useit.com/papers/heuristic/heuristic_list.html*
26. E. Ross. Intelligent User Interfaces: Survey and Research directions. Technical report, CSTR-00-004, March, 2000, Department of Computer Science, University of Bristol. (*http://www.cs.bris.ac.uk/Publications/Papers/1000447.pdf* )
27. A. Waern. What is an Intelligent Interface? *http://www.sics.se/~annika/papers/intint.html*

# The Instability of the Rhombus-Like Central Configurations in Newton 9-Body Problem

D. Diarova[1] and N.I. Zemtsova[2]

[1] Institute of Oil and Gas, Atyrau, Kazakhstan
ddiarova@mail.ru
[2] Computing Center of RAS, Moscow
zemni@ccas.ru

**Abstract.** E.A.Grebenikov and A.N.Prokopenya proved that rhombus-like central configuration in Newton 5-body problem is unstable. In this article, the problem of existence and stability of the rhombus-like central configurations in Newton 9-body problem, which consists of two homothetic rhombuses, is studied. It is proved that these central configurations are unstable. All computations are executed by means of computer algebra system Mathematica.

## 1   Necessary and Sufficient Conditions of Existence of Central Configurations

The differential equations of the new restricted problems of cosmic dynamics [1] are not integrable. For search of exact particular solutions of such equations A. Wintner has developed the theory of homographic solutions, which is based on

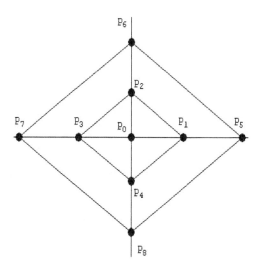

**Fig. 1.**

V.G. Ganzha, E.W. Mayr, and E.V. Vorozhtsov (Eds.): CASC 2006, LNCS 4194, pp. 141–148, 2006.
© Springer-Verlag Berlin Heidelberg 2006

the concept of central configuration [2]. Development of new computer technologies and computer algebra systems, for example, such as Mathematica [3], allows
one to make a search for new central configurations, and, hence, for new exact
particular solutions, rather effective.

We analyse the Newton 9-body problem. The bodies $P_1$, $P_2$, $P_3$, $P_4$, $P_5$, $P_6$,
$P_7$, $P_8$ with masses $m_1$, $m_2$, $m_3$, $m_4$, $m_5$, $m_6$, $m_7$, $m_8$ are mutually attracted
according to Newton gravitational law and are in vertices of two similar homo-
thetical rhombuses, which rotate at a constant angular speed around the central
body $P_0$ with mass $m_0$. All bodies are located in the same plane (Fig. 1).

We analyse the problem of existence of the central configurations [1] for this
model in the non-inertial Cartesian coordinate system $P_0xy$. Existence of the
central configurations does not depend on orientation and the sizes of rhombuses
[1], therefore, for convenience we will choose the elementary variant: $P_1(\alpha, 0, 0)$,
$P_2(0, 1, 0)$, $P_3(-\alpha, 0, 0)$, $P_4(0, -1, 0)$, $P_5(\beta, 0, 0)$, $P_6(0, \beta/\alpha, 0)$, $P_7(-beta, 0, 0)$,
$P_8(0, \beta/\alpha, 0)$.

The necessary conditions of existence of non-inertial central configurations for
any model (the number of bodies $n$ is arbitrary) look like [1]:

$$
\begin{cases}
\sum_{s=1}^{n}{}' m_s(x_s y_k - x_k y_s) \left( \dfrac{1}{\Delta_{ks}^3} - \dfrac{1}{r_s^3} \right) = 0, \\[2mm]
\sum_{s=1}^{n}{}' m_s(y_s z_k - y_k z_s) \left( \dfrac{1}{\Delta_{ks}^3} - \dfrac{1}{r_s^3} \right) = 0, \\[2mm]
\sum_{s=1}^{n}{}' m_s(x_s z_k - x_k z_s) \left( \dfrac{1}{\Delta_{ks}^3} - \dfrac{1}{r_s^3} \right) = 0, \\[2mm]
k = 1, 2, \ldots, n, \ s \neq k. \\[2mm]
\Delta_{ks}^2 = (x_k - x_s)^2 + (y_k - y_s)^2 + (z_k - z_s)^2, \\[2mm]
r_s^2 = x_s^2 + y_s^2 + z_s^2.
\end{cases}
\tag{1}
$$

For our model the system (1) will be written in the form

$$
\begin{cases}
\dfrac{\alpha(1+\alpha^2)^{3/2} - \alpha}{(1+\alpha^2)^{3/2}}(m_2 - m_4) + \dfrac{\alpha^3(\alpha^4+\beta^2)^{3/2} - \alpha^3\beta^3}{\beta^3(\alpha^4+\beta^2)^{3/2}}(m_6 - m_8) = 0 \\[3mm]
\dfrac{\alpha^3 - (1+\alpha^2)^{3/2}}{(1+\alpha^2)^{3/2}}(m_1 - m_3) + \dfrac{\beta^3 - (1+\beta^2)^{3/2}}{\beta^2(1+\beta^2)^{3/2}}(m_5 - m_7) = 0 \\[3mm]
\dfrac{\beta(1+\beta^2)^{3/2} - \beta}{(1+\beta^2)^{3/2}}(m_2 - m_4) + \dfrac{\alpha^2(1+\alpha^2)^{3/2} - \alpha^2}{\beta(1+\alpha^2)^{3/2}}(m_6 - m_8) = 0 \\[3mm]
\dfrac{\beta\alpha^6 - \beta(\alpha^4+\beta^2)^{3/2}}{\alpha^3(\alpha^4+\beta^2)^{3/2}}(m_1 - m_3) + \dfrac{\alpha^3 - (1+\alpha^2)^{3/2}}{\alpha\beta(1+\alpha^2)^{3/2}}(m_5 - m_7) = 0
\end{cases}
\tag{2}
$$

If one considers system (2) as a linear system, in which masses are unknown, then it is possible to show that it is satisfied only in case

$$m_1 = m_3, \quad m_2 = m_4, \quad m_5 = m_7, \quad m_6 = m_8, \tag{3}$$

In order words, a necessary condition of existence of central configurations in our model is equality of masses in pairs of symmetric vertices of rhombuses.

The necessary and sufficient conditions of existence of non-inertial central configurations are [1]:

$$
\begin{cases}
I\left(\displaystyle\sum_{s=1}^{n}{}' \left(\dfrac{x_s - x_k}{\Delta_{ks}^3} - \dfrac{x_s}{r_s^3}\right) - \dfrac{(m_0 + m_k)x_k}{r_k^3}\right) = -W x_k, \\[3mm]
I\left(\displaystyle\sum_{s=1}^{n}{}' \left(\dfrac{y_s}{\Delta_{ks}^3} - \dfrac{y_k}{r_s^3} - \dfrac{y_s}{r_s^3}\right) - \dfrac{(m_0 + m_k)y_k}{r_k^3}\right) = -W y_k, \\[3mm]
I\left(\displaystyle\sum_{s=1}^{n}{}' \left(\dfrac{z_s - z_k}{\Delta_{ks}^3} - \dfrac{z_s}{r_s^3}\right) - \dfrac{(m_0 + m_k)z_k}{r_k^3}\right) = -W z_k, \\[3mm]
k = 1, 2, \ldots, n, \ s \neq k.
\end{cases}
\tag{4}
$$

where $I$ is the total moment of inertia of body system, $W$ is the analogue of potential function in non-inertial coordinate system:

$$I = \sum_{k=1}^{n} m_k (x_k^2 + y_k^2 + z_k^2)^2,$$

$$
W = \frac{1}{2}\sum_{k=1}^{n}\sum_{s=1}^{n}{}' \frac{m_k m_s}{\Delta_{ks}} + \frac{1}{2}\sum_{k=1}^{n}\sum_{s=1}^{n}{}' m_k m_s (x_k x_s + y_k y_s + z_k z_s)\left(\frac{1}{r_k^3} + \frac{1}{r_s^3}\right)
$$
$$
+ \sum_{k=1}^{n} \frac{(m_0 + m_k)m_k}{r_k}.
$$

After substitution in system (4) of the coordinate values of points $P_1$, $P_2$, $P_3$, $P_4$, $P_5$, $P_6$, $P_7$, $P_8$ and necessary simplifications we obtain system (5).

We now study this system, in which $m_0=1$, and $m_1$, $m_2$, $m_5$, $m_6$ are unknowns. Then system (5) is a linear system of three equations in four unknowns $m_1$, $m_2$, $m_5$, $m_6$.

$$
\begin{cases}
\dfrac{\beta^3 - \alpha^3}{\alpha^2 \beta^2} m_0 + \dfrac{\beta(\beta^2 - \alpha^2)^2 - 8\alpha^3(\alpha^2 + \beta^2)}{4\alpha^2(\beta^2 - \alpha^2)^2} m_1 + \\[2mm]
+\dfrac{2\alpha\beta((1 + \beta^2)^{3/2} - (1 + \alpha^2)^{3/2})}{(1 + \alpha^2)^{3/2}(1 + \beta^2)^{3/2}} m_2 - \dfrac{16\alpha\beta^4 - +\alpha(\beta^2 - \alpha^2)^2}{4\alpha^2(\beta^2 - \alpha^2)^2} m_5 + \\[2mm]
+\dfrac{2\alpha^4\beta(\beta^3(1 + \alpha^2)^{3/2} - (\alpha^4 + \beta^2)^{3/2})}{\beta^3(1 + \alpha^2)^{3/2}(\alpha^4 + \beta^2)^{3/2}} m_6 = 0, \\[3mm]
\dfrac{1 - \alpha^3}{\alpha^2} m_0 + \dfrac{(1 + \alpha^2)^{3/2} - 8\alpha^3}{4\alpha^2(1 + \alpha^2)^{3/2}} m_1 + \dfrac{8\alpha - \alpha(1 + \alpha^2)^{3/2}}{4(1 + \alpha^2)^{3/2}} m_2 - \\[2mm]
-\dfrac{4\alpha\beta(1 + \beta^2)^{3/2} - 2\alpha\,(\beta^2 - \alpha^2)^2}{(\alpha + \beta)^2(-\alpha + \beta)^2(1 + \beta^2)^{3/2}} m_5 + \\[2mm]
+\dfrac{\alpha^3(4\alpha\beta(\alpha^4 + \beta^2)^{3/2} + 2\alpha(\beta^2 - \alpha^2)^2)}{(\beta - \alpha)^2(\alpha + \beta)^2(\alpha^4 + \beta^2)^{3/2}} m_6 = 0, \\[3mm]
\dfrac{\beta^3 - \alpha^6}{\alpha^2 \beta^2} m_0 + \dfrac{\beta((\alpha^4 + \beta^2)^{3/2} - 8\alpha^6)}{4\alpha^2(\beta^2 + \alpha^4)^{3/2}} m_1 + \\[2mm]
+\dfrac{2\alpha\beta(\beta^2 - \alpha^2)^2 - 2\alpha^4(1 + \alpha^2)^{3/2}(\alpha^2 + \beta^2)}{(1 + \alpha^2)^{3/2}(\beta - \alpha)^2(\alpha + \beta)^2} m_2 - \\[2mm]
-\dfrac{4\alpha\beta^4(\alpha^2 + 1)^{\frac{3}{2}} + 2\alpha^4(\beta^2 - \alpha^2)^2}{\beta^2(\alpha^2 + 1)^{3/2}(\alpha + \beta)^2(\beta - \alpha)^2} m_5 + \dfrac{\alpha^4(8\beta^3 - (\alpha^4 + \beta^2)^{\frac{3}{2}})}{4\beta^2(\alpha^4 + \beta^2)^{3/2}} m_6 = 0,
\end{cases} \tag{5}
$$

Solving system (5), we obtain for $m_2$, $m_5$, $m_6$ the analytical expressions:

$$
\begin{cases}
m_2 = f_{20}(\alpha, \beta)m_0 + f_{21}(\alpha, \beta)m_1, \\
m_5 = f_{50}(\alpha, \beta)m_0 + f_{51}(\alpha, \beta)m_1, \\
m_6 = f_{60}(\alpha, \beta)m_0 + f_{61}(\alpha, \beta)m_1.
\end{cases} \tag{6}
$$

The coefficients $f_{20}(\alpha, \beta), \ldots, f_{61}(\alpha, \beta)$ have very complex analytical expressions, and they are more conveniently defined by methods of computer algebra. We give here only the expression for $f_{20}$:

$$
\begin{aligned}
f_{20} = &(4(1 + \alpha^2)^{3/2}(\alpha - \beta)^2(\alpha + \beta)^2(1 + \beta^2)^{3/2}(4\alpha(1 + \alpha^2)^{3/2}(\alpha - \beta)^2 \\
&(\alpha + \beta)^2(-2\alpha(\alpha^2 - \beta^2)^2 - 4\alpha\beta(1 + \beta^2)^{3/2})((1 + \alpha^2)^{3/2}(-\alpha^3 + \beta^3) \\
&(8\beta^3 - (\alpha^4 + \beta^2)^{3/2}) - 8(-\alpha^6 + \beta^3)((1 + \alpha^2)^{3/2}\beta^3 - (\alpha^4 + \beta^2)^{3/2})) + \\
&+16(1 + \beta^2)^{3/2}(-4\alpha(1 + \alpha^2)^{3/2}\beta^4 - 2\alpha^4(\alpha^2 - \beta^2)^2)(2\alpha(1 - \alpha^3)(\beta^2 - \alpha^2)^2 \\
&((1 + \alpha^2)^{3/2}\beta^3 - (\alpha^4 + \beta^2)^{3/2}) - 2\alpha(1 + \alpha^2)^{3/2}(-\alpha^3 + \beta^3)((\alpha^2 - \beta^2)^2 + \\
&+3\beta(\alpha^4 + \beta^2)^{3/2})) - \alpha(1 + \alpha^2)^3(1 + \beta^2)^{3/2}(\alpha^4 - 2\alpha^2\beta^2 + 17\beta^4)
\end{aligned}
$$

$$(-\alpha(1-\alpha^3)(\alpha-\beta)^2(\alpha+\beta)^2(8\beta^3-(\alpha^4+\beta^2)^{3/2})+8\alpha(-\alpha^6+\beta^3)$$
$$((\alpha^2-\beta^2)^2+2\beta(\alpha^4+\beta^2)^{3/2}))))/(\alpha^3(\alpha(8-(1+\alpha^2)^{3/2})(\alpha-\beta)^4$$
$$(\alpha+\beta)^4(1+\beta^2)^3(\alpha(1+\alpha^2)^3(\alpha^4-2\alpha^2\beta^2+17\beta^4)(8\beta^3-(\alpha^4+\beta^2)^{3/2})+$$
$$32(-4\alpha(1+\alpha^2)^{3/2}\beta^4-2\alpha^4(\alpha^2-\beta^2)^2)((1+\alpha^2)^{3/2}\beta^3-(\alpha^4+\beta^2)^{3/2}))+$$
$$+32(1+\alpha^2)^{3/2}\beta^2(1+\beta^2)^{3/2}(2\alpha\beta(\alpha^2-\beta^2)^2-2\alpha^4(1+\alpha^2)^{3/2}(\alpha^2+\beta^2))$$
$$(-4(\alpha-\beta)^2(\alpha+\beta)^2(-2\alpha(\alpha^2-\beta^2)^2-4\alpha\beta(1+\beta^2)^{3/2})((1+\alpha^2)^{3/2}\beta^3-$$
$$-(\alpha^4+\beta^2)^{3/2})\alpha(1+\alpha^2)^{3/2}(1+\beta^2)^{3/2}(\alpha^4-2\alpha^2\beta^2+17\beta^4)((\alpha^2-\beta^2)^2+$$
$$2\beta(\alpha^4+\beta^2)^{3/2}))+32((1+\alpha^2)^{3/2}(\alpha-\beta)^2\beta^3(\alpha+\beta)^2(-(1+\alpha^2)^{3/2}+$$
$$+(1+\beta^2)^{3/2})(\alpha(1+\alpha)^{3/2}(\alpha-\beta)^2(\alpha+\beta)^2(-2\alpha(\alpha^2-\beta^2)^2-4\alpha\beta$$
$$(1+\beta^2)^{3/2})(8\beta^3-(\alpha^4+\beta^2)^{3/2})-8\alpha(1+\beta^2)^{3/2}(-4\alpha(1+\alpha^2)^{3/2}\beta^4-$$
$$-2\alpha^4(\alpha^2-\beta^2)^2)((\alpha^2-\beta^2)^2+2\beta(\alpha^4+\beta^2)^{3/2})))),$$

It was proved in [1] that rotational angular speed $\omega$ of configurations is no arbitrary value, it should satisfy eight analytical relations

$$\omega^2 = \frac{1}{x_k}\left(\sum_{s=1}^{n}{}' \frac{m_s(x_k-x_s)}{\Delta_{ks}^3} + \frac{m_0 x_k}{r_k^3}\right) = \frac{1}{y_k}\left(\sum_{s=1}^{n}{}' \frac{m_s(y_k-y_s)}{\Delta_{ks}^3} + \frac{m_0 y_k}{r_k^3}\right) \quad (7)$$
$$k = 1,\ldots,8.$$

These rations define all dependences between dynamic and geometrical parameters of model for existence of the central configuration. For our model we have

$$\omega^2 = \frac{1}{\alpha^3}m_0 + \frac{1}{4\alpha^3}m_1 + \frac{2}{(1+\alpha^2)^{3/2}}m_2 + \frac{4\beta}{(\beta^2-\alpha^2)^2}m_5 + \frac{2\alpha^3}{(\alpha^4+\beta^2)^{3/2}}m_6. \quad (8)$$

Parameters $m_2$, $m_5$, $m_6$, and $\omega^2$ at various values of positive parameters $m_1$, $\alpha$, and $\beta$ should satisfy obvious inequalities

$$m_2 \geq 0, \quad m_5 \geq 0, \quad m_6 \geq 0, \quad \omega^2 \geq 0. \quad (9)$$

Dependencies of $m_2$, $m_5$, $m_6$, $\omega^2$ of parameters $m_1$, $\alpha$, and $\beta$ are complex enough, therefore, it is impossible to determine in analytical form the family of rhombuses satisfying all inequalities (9). We verified conditions (9) for specific values of $m_1$, $\alpha$, and $\beta$. Some computed results are presented in table 1.

**Table 1.** The calculated values $m_2$, $m_5$, $m_6$, $\omega^2$ for different $m_1$, $\alpha$, and $\beta$

| $m_1$ | $\alpha$ | $\beta$ | $m_2$ | $m_5$ | $m_6$ | $\omega^2$ |
|-------|----------|---------|-------|-------|-------|------------|
| 0.01 | 0.1 | 0.2 | -127.122... | 0.936... | 44.852... | -68.864... |
| 0.01 | 0.7 | 1.6 | 17.089... | 10.675... | 4.085... | 6.369... |
| 0.01 | 0.7 | 1.7 | -10.555... | -4.999... | -0.281... | -2.817... |
| 0.01 | 1. | 1.1 | 0.01 | -0.07... | -0.07... | 1.664... |
| 0.01 | 1. | 1.2 | 0.01 | 0.08... | 0.08... | 0.799... |
| 0.1 | 0.5 | 1. | -13.618... | -0.386... | 5.383... | -7.319... |
| 0.1 | 0.8 | 1.2 | -0.339... | 0.134... | -0.016... | 1.313... |
| 0.1 | 0.8 | 1.3 | 0.414... | 0.299... | 0.126... | 1.030... |

**Table 1.** (*continued*)

| $m_1$ | $\alpha$ | $\beta$ | $m_2$ | $m_5$ | $m_6$ | $\omega^2$ |
|-------|----------|---------|-------|-------|-------|------------|
| 1. | 0.9 | 1.5 | 1.571... | 0.059... | -0.249... | 2.760... |
| 1. | 0.9 | 1.6 | 1.651... | 0.453... | 0.138... | 2.159... |
| 10. | 0.8 | 1.5 | 28.337... | 3.582... | -4.716... | 24.419... |
| 10. | 0.8 | 1.6 | 32.685... | 9.539... | 0.124... | 24.424... |

This table shows that the set of values of parameters $m_1$, $\alpha$, and $\beta$, for which inequalities (9) are satisfied, is not empty.

The following statement holds.

*Necessary and sufficient conditions of existence of the rhombus-like central configurations located in a rotating coordinate system $P_0 xy$ are satisfaction of conditions (3), (6), (8), and (9).*

## 2   Linear Stability of Rhombus-Like Configurations

Using the Lyapunov's first method [5], we investigate linear stability of the found configuration.

The differential equations of movement of plane Newton 9-body problem in the Cartesian non-inertial coordinate system $P_0 xy$ are [4]

$$
\begin{cases}
\dfrac{d^2 x_k}{dt^2} = -\dfrac{(m_0 + m_k) x_k}{r_k^3} + \sum_{s=1}^{n}{}' m_s \left( \dfrac{(x_s - x_k)}{\Delta_{ks}^3} - \dfrac{x_s}{r_s^3} \right), \\[2ex]
\dfrac{d^2 y_k}{dt^2} = -\dfrac{(m_0 + m_k) y_k}{r_k^3} + \sum_{s=1}^{n}{}' m_s \left( \dfrac{(y_s - y_k)}{\Delta_{ks}^3} - \dfrac{y_s}{r_s^3} \right), \\[2ex]
k = 1, 2, \ldots, n, \ s \neq k.
\end{cases}
\tag{10}
$$

It is necessary to linearize the system of differential equations (10) in neighbourhood of each point $P_1$, $P_2$, ..., $P_8$ and to study the properties of eigenvalues of the matrix of obtained linearized systems.

For this purpose we shall transform system (10) to normal form [6] using the formulas

$$
x = x, \quad y = y, \quad u = \frac{dx}{dt}, \quad v = \frac{dy}{dt}.
$$

Then instead of system (10) we will have

$$
\begin{cases}
\dfrac{dx_k}{dt} = u_k, \quad \dfrac{dy_k}{dt} = v_k, \\[2ex]
\dfrac{du_k}{dt} = -\dfrac{(m_0 + m_k) x_k}{r_k^3} + \sum_{s=1}^{8}{}' m_s \left( \dfrac{(x_s - x_k)}{\Delta_{ks}^3} - \dfrac{x_s}{r_s^3} \right), \\[2ex]
\dfrac{dv_k}{dt} = -\dfrac{(m_0 + m_k) y_k}{r_k^3} + \sum_{s=1}^{8}{}' m_s \left( \dfrac{(y_s - y_k)}{\Delta_{ks}^3} - \dfrac{y_s}{r_s^3} \right).
\end{cases}
\tag{11}
$$

We denote coordinates of the rhombus vertices $P_1, P_2, \ldots, P_8$ by $x_k^*, y_k^*$, and $X$ is the vector of small 32 values

$$X = (u_1 - u_1^*, \ldots, u_8 - u_8^*, v_1 - v_1^*, \ldots, v_8 - v_8^*, x_1 - x_1^*, \ldots, x_8 - x_8^*, y_1 - y_1^*, \ldots, y_8 - y_8^*).$$

In these notations $u_k^* = v_k^* = 0, \quad k = 1, \ldots, 8$. These values correspond to the found configuration.

We executed the procedure of linearization of system (11) in the neighbourhood of each point $P_1, P_2, \ldots, P_8$ by system Mathematica [3]. As a result, we obtain the system of linear differential equations

$$\frac{dX}{dt} = AX \tag{11}$$

where $A$ is a $32 \times 32$ matrix of the form

$$A = \begin{bmatrix} 0 & B \\ E & 0 \end{bmatrix}.$$

Here 0 is a $16 \times 16$ zero matrix, $+$ is the $16 \times 16$ unit matrix, $B$ is a $16 \times 16$ matrix

$$B = \begin{bmatrix} a_{1,17} & a_{1,18} & \cdots & a_{1,32} \\ a_{2,17} & a_{2,18} & \cdots & a_{2,32} \\ \cdots & \cdots & \cdots & \cdots \\ a_{16,17} & a_{16,18} & \cdots & a_{16,32} \end{bmatrix}.$$

Coefficients $a_{1,17}, \ldots, a_{16,32}$ depend on $m_0, m_1, m_2, m_5, m_6$, and their expressions have quite a bulky form. For example, the first element is

$$a_{1,17} = \frac{2}{\alpha^3} m_0 + \frac{9}{4\alpha^3} m_1 + \frac{4\alpha^2 - 2}{(1 + \alpha^2)^{5/2}} m_2 + \frac{12\alpha^2\beta + 4\beta^3}{(\beta^2 - \alpha^2)^3} m_5 + \frac{4\alpha^7 - 2\alpha^3\beta^2}{(\alpha^4 + \beta^2)^{5/2}} m_6,$$

and the last element is

$$a_{16,32} = \frac{2\alpha^3}{\beta^3} m_0 + \frac{4\beta^2\alpha^3 - 2\alpha^7}{(\alpha^4 + \beta^2)^{5/2}} m_1 + \frac{4\alpha^3(3\alpha^2\beta + \beta^3)}{(\beta - \alpha)^3(\beta + \alpha)^3} m_2 + \frac{2\alpha^3(2 - \alpha^2)}{\beta^3(\alpha^2 + 1)^{5/2}} m_5 + \frac{9\alpha^3}{4\beta^3} m_6,$$

where $m_2, m_5$, and $m_6$ are defined by formulas (6).

Matrix $A$, as a matrix of a linear Hamilltonian system, is symplectic matrix [7], therefore, in order that our configuration be stable, it is necessary that all 32 eigenvalues of matrix $A$ are purely imaginary. We could not obtain analytical dependencies for eigenvalues of matrix $A$ on parameters $\alpha, \beta$, and $m_1$ because its elements have very complex analytical expressions. Therefore, we used possibilities of CAS "Mathematica", the instruction "Eigenvalues", for calculation of its eigenvalues. We have executed calculations for "practically visible" set of values $m_1$, $\alpha$, and $\beta$ satisfying conditions (9). These calculations have shown that eigenvalues of matrix $A$ are not all purely imaginary, therefore, the following statement holds:

*The rhombus-like 9-body configuration is linearly unstable for "admissible" values of dynamic and geometrical parameters $m_1$, $\alpha$, and $\beta$.*

As the set of real numbers has the power of a continuum, this statement has probability character.

# References

1. Grebenikov, E.A., Kozak-Skoworodkin, D., Jakubiak, M.: Methods of Computer Algebra in Many-Body Problem (in Russian). Published by UFP, Moscow (2002)
2. Wintner, A.: The Analytical Foundations of Celestial Mechanics. Princeton Univ. Press, Princeton (1941)
3. Wolfram, S.: The Mathematica Book. 4th edn. Wolfram Media/ Cambridge University Press (1999)
4. Abalakin, V.K., Aksenov, E.P., Grebenikov, E.A., Demin, V.G., Ryabov, Yu.A.: Handbook on Celestial Mechanics and Astrodynamics (in Russian). Nauka, Moscow (1976)
5. Lyapunov, A.M.: General Problem on Stability of Motion (in Russian), The USSR Academy of Sciences, Moscow (1954)
6. Stepanov, V.V.: Course of Differential Equations (in Russian). Nauka, Moscow (1968)
7. Arnold, V.I.: About stability of equilibrium positions of Hamiltonian systems in general elliptic case (in Russian). DAN USSR **137** (1961) 255–257
8. Grebenikov, E.A., Prokopenya, A.N.: About instability of rhombus-like homographical solutions of Mewton five-body problem (in Russian). The Problem of Security Theory and Stability Systems. Published by Computing Center of the Russian Acad. Sci., Moscow (2005)

# Algorithmic Invariants for Alexander Modules

Jesús Gago-Vargas[1], Isabel Hartillo-Hermoso[2],
and José María Ucha-Enríquez[1],[*]

[1] Depto. de Álgebra, Univ. de Sevilla, Apdo. 1160, 41080 Sevilla, Spain
{gago, ucha}@us.es
[2] Dpto. de Matemáticas, Univ. de Cádiz, Apdo. 40, 11510 Puerto Real, Cádiz, Spain
isabel.hartillo@uca.es

**Abstract.** Let $G$ be a group given by generators and relations. It is possible to compute a presentation matrix of a module over a ring through Fox's differential calculus. We show how to use Gröbner bases as an algorithmic tool to compare the chains of elementary ideals defined by the matrix. We apply this technique to classical examples of groups and to compute the elementary ideals of Alexander matrix of knots up to 11 crossings with the same Alexander polynomial.

## 1 Introduction

Let $G = \langle \mathbf{x} : \mathbf{r} \rangle$ be a group given by generators and relations, where $\mathbf{x} = (x_1, \ldots, x_n)$ is a base of the free group $F$ and $\mathbf{r} = (r_1, \ldots, r_m)$ are the relations. Through Fox's differential calculus [Crowell et al.(1977)] it is possible to compute the presentation matrix of the Alexander module of the group. We review briefly these concepts.

We build from $G$ the ring of the group $\mathbb{Z}G$. A derivation over the group ring is a map $D : \mathbb{Z}G \to \mathbb{Z}G$ such that

$$D(\nu_1 + \nu_2) = D\nu_1 + D\nu_2,$$
$$D(\nu_1 \nu_2) = (D\nu_1)\mathsf{t}(\nu_2) + \nu_1 D\nu_2,$$

where $\mathsf{t}$ is the trivializer and $\nu_1, \nu_2 \in \mathbb{Z}G$. For elements in $G$, the second condition is

$$D(g_1 g_2) = Dg_1 + g_1 Dg_2.$$

Then a derivation can be seen as the unique linear extension to $\mathbb{Z}G$ of a map $D : G \to \mathbb{Z}G$ that verifies the previous condition.

It is known that each generator $x_j$ in the group $G$ defines a unique derivation $D_j = \partial/\partial x_j$ in $\mathbb{Z}G$, such that

$$\frac{\partial x_i}{\partial x_j} = \delta_{ij}.$$

Let $H$ be the abelianized group of $G$. Considering the group rings we have the composition of maps

$$\mathbb{Z}F \xrightarrow{D_j} \mathbb{Z}F \xrightarrow{\gamma} \mathbb{Z}G \xrightarrow{\mathsf{a}} \mathbb{Z}H,$$

---

[*] All the authors partially supported by FQM-333 and MTM2004-01165.

V.G. Ganzha, E.W. Mayr, and E.V. Vorozhtsov (Eds.): CASC 2006, LNCS 4194, pp. 149–154, 2006.

where $\gamma$ is the projection and $\mathbf{a}$ is the abelianizer. The Alexander matrix from $G$ is $A = (a_{ij})$, where

$$a_{ij} = \mathbf{a}\gamma \left( \frac{\partial r_j}{\partial x_i} \right).$$

Note that this matrix is the transposed of the matrix defined by [Crowell et al.(1977)]. The Alexander matrix presents a module over the ring $\mathbb{Z}H$. If two groups are isomorphic then the modules are isomorphic.

A finite presentation for $M$ is an exact sequence

$$R^n \xrightarrow{\alpha} R^m \xrightarrow{\Phi} M \to 0$$

where $R^n$ and $R^m$ are free $R$-modules with respective bases $\mathbf{f}_1, \ldots, \mathbf{f}_n$ and $\mathbf{e}_1, \ldots \mathbf{e}_m$. If $\alpha$ is represented by the matrix $A$ with respect to these bases then the $m \times n$ matrix $A$ is a presentation matrix for $M$.

**Theorem 1.** *[Lickorish(1998), Thm. 6.1] If $A_1$ and $A_2$ are presentation matrices of a module $M$ then they are related by a sequence of matrix transformations of the following form and their inverses:*

1. *Permutation of rows and columns.*
2. *Replacement of the matrix $A_1$ by $\begin{pmatrix} A_1 & 0 \\ 0 & 1 \end{pmatrix}$.*
3. *Addition of an extra column of zeros to the matrix $A_1$.*
4. *Addition of a scalar multiple of a row (column) to another row (column).*

We say that $A_1$ and $A_2$ are Fitting equivalents.

**Definition 2.** *Let $M$ be a $R$ module, with an $m \times n$ presentation matrix $A$. The $r$-th elementary ideal $F_r$ of $M$ is the ideal generated by all the $(m - r + 1) \times (m - r + 1)$ minors of $A$.*

By convention, $F_r(M) = R$ when $r > m$ and $F_r(M) = 0$ if $r \leq 0$. They form an ascending chain $F_k(M) \subset F_{k+1}(M)$. The elementary ideals are independent of the presentation matrix chosen to evaluate them.

## 2   Algorithms in the Ring Group

The ring $\mathbb{Z}H$ is commutative, because $H$ is an abelian group, and it has a special form.

**Proposition 3.** *The ring $\mathbb{Z}H$ is isomorphic to $\mathbb{Z}[x_1^{\pm}, \ldots, x_n^{\pm}]/J$, where $J$ is the ideal generated by the relations $r_1, \ldots, r_m$ under commutativity.*

*Proof.* Through the abelianizer, all the relations have the form $\prod x_i^{e_i} = 1$, so $J$ is generated by the elements $\prod x_i^{e_i} - 1$.

**Corollary 4.** *There is an algorithm to compare ideals in $\mathbb{Z}H$.*

*Proof.* Through the bijection between ideals in $\mathbb{Z}[x_1^{\pm}, \ldots, x_n^{\pm}]/J$ and ideals in $R = \mathbb{Z}[x_1^{\pm}, \ldots, x_n^{\pm}]$ that contains $J$, the problem is reduced to compare ideals in $R$. In this ring we can compute Gröbner bases [Sims(1994), Pauer et al.(1999)], or by the isomorphism $R \simeq \mathbb{Z}[x_1, \ldots, x_n, w]/\langle x_1 \cdots x_n w - 1 \rangle$ [Adams et al.(1994)].

There is no known algorithm to decide whether two matrices present isomorphic modules. There are other invariants as the ideal row (column) class [Fox et al.(1964)] or the Nakanishi index [Kawauchi(1996)]. However we do not know algorithms to compute them and *ad hoc* arguments are needed to give their values for specific matrices [Fox et al.(1964), Kearton et al.(2003)].

*Example 5.* Let consider the groups given by the presentations

$$D_8 = \langle x, y | x^4 = 1, y^2 = 1, yxy^{-1} = x^{-1} \rangle, Q_8 = \langle x, y | x^4 = 1, x^2 = y^2, y^{-1}xy = x^{-1} \rangle.$$

$D_8$ is the dihedral group of order 8 (symmetry group of the square) and $Q_8$ is the quaternion group. A classical exercise in group theory is to show that these two groups are not isomorphic. Let see how can this be accomplished with elementary ideals. Let $r_i$ be the the relations in $D_8$:

$$r_1 : x^4 = 1, r_2 : y^2 = 1, r_3 : yxy^{-1}x = 1.$$

Then

$$\frac{\partial r_1}{\partial x} = 1 + x + x^2 + x^3, \quad \frac{\partial r_2}{\partial x} = 0, \quad \frac{\partial r_3}{\partial x} = y + x^{-1},$$

$$\frac{\partial r_1}{\partial y} = 0, \quad \frac{\partial r_2}{\partial y} = 1 + y, \quad \frac{\partial r_3}{\partial y} = 1 - x^{-1}$$

In the abelianized group we add the relation $xy = yx$, so $x^2 = 1, y^2 = 1$. The Alexander module of the group has a presentation matrix

$$M(D_8) = \begin{pmatrix} 2 + 2x & 0 & x + y \\ 0 & y + 1 & 1 - x \end{pmatrix},$$

over the ring $\mathbb{Z}[x^{\pm}, y^{\pm}]/\langle x^2 - 1, y^2 - 1 \rangle$.

We proceed in an analogous way with $Q_8$. We write the relations

$$s_1 : x^4 = 1, s_2 : x^2 y^{-2} = 1, s_3 : xy^{-1}xy = 1,$$

and

$$\frac{\partial s_1}{\partial x} = 1 + x + x^2 + x^3, \quad \frac{\partial s_2}{\partial x} = 1 + x, \quad \frac{\partial s_3}{\partial x} = 1 + xy^{-1},$$

$$\frac{\partial s_1}{\partial y} = 0, \quad \frac{\partial s_2}{\partial y} = -x^2(1 + y), \quad \frac{\partial s_3}{\partial y} = -xy^{-1} + y^{-1}$$

As before, in the abelianized group the relations are reduced to $x^2 = 1, y^2 = 1$ and a presentation matrix of the Alexander module is

$$M(Q_8) = \begin{pmatrix} 2 + 2x & 1 + x & 1 + xy \\ 0 & -1 - y & -xy + y \end{pmatrix}$$

over the ring $\mathbb{Z}[x^{\pm}, y^{\pm}]/\langle x^2 - 1, y^2 - 1 \rangle$.

We compute a Gröbner basis in $\mathbb{Z}[x^{\pm}, y^{\pm}]$ of the second elementary ideal. Adding the polynomials $x^2 - 1, y^2 - 1$, we get

$$F_2(M(D_8)) = \langle 4, 1 + y, 1 - x \rangle, F_2(M(Q_8)) = \langle 2, 1 + x, 1 + y \rangle.$$

They are different so the groups are not isomorphic.

*Example 6.* In [Kanenobu(1986)] it is defined a class of knots $K_{p,q}$, with $p, q \in \mathbb{N}$, that has the Alexander matrix

$$A_{p,q} = \begin{pmatrix} t^2 - 3t + 1 & (p - q)t \\ 0 & t^2 - 3t + 1 \end{pmatrix}.$$

Lemma 2 of [Kanenobu(1986)] asserts that $K_{p,q}$ and $K_{p',q'}$ have isomorphic Alexander modules if and only if $|p - q| = |p' - q'|$. Let us show how to apply our approach to give a new proof of this lemma. If the modules are isomorphic then the second elementary ideals $F_2$ must coincide. A Gröbner basis of the ideal is equal to $\{t^2 - 3t + 1, p - q\}$, so $F_2(A_{p,q}) = F_2(A_{p',q'})$ if and only if $|p - q| = |p' - q'|$.

# 3    An Application to Knot Theory

One of the main invariants in knot theory is the fundamental group of the knot complement. The Alexander matrix can be computed from the Seifert matrix, and with the tables listed in [Burde et al.(1985)] and [Livingston(2004)] we can get the Alexander matrix of knots up to 11 crossings. As an application of the algorithm described before we give a list of knots with the same Alexander polynomial (grouped by boxes) and where the elementary ideals give more information to distinguish knots (see Table 1). For example, from Table 1 we deduce that $11a_{102}$ and $11a_{181}$ are different, but we cannot say anything about $11a_{102}$ and $11a_{199}$.

The first step was to compute the Alexander matrix of the knot and reduce it through the transformations given by Theorem 1. In all cases we have got at most a $2 \times 2$ presentation matrix, so $F_3$ is always equal to $R$. The Gröbner bases were computed over the ring $\mathbb{Z}[t, w]$, adjoining to the ideals the polynomial $tw - 1$.

**Table 1.** Elementary ideals

| Name | Elem. ideal $F_2$ | Name | Elem. ideal $F_2$ | Name | Elem. ideal $F_2$ | Name | Elem. ideal $F_2$ |
|---|---|---|---|---|---|---|---|
| $11n_{100}$ | $R$ | $10_{140}$ | $\langle 2, t^2 - t + 1\rangle$ | $10_{60}$ | $R$ | $11n_{162}$ | $\langle 2, t^2 + t + 1\rangle$ |
| $9_{37}$ | $\langle 3, t - 2\rangle$ | $11n_{73}$ | $\langle t^2 - t + 1\rangle$ | $11n_{165}$ | $\langle 2, t^2 + t + 1\rangle$ | $9_{39}$ | $R$ |
| | | $11n_{74}$ | $\langle t^2 - t + 1\rangle$ | | | | |
| $11n_{97}$ | $R$ | | | $11a_{223}$ | $R$ | $11a_{31}$ | $R$ |
| $6_1$ | $R$ | $11n_{116}$ | $R$ | $11n_{148}$ | $\langle 5, t^2 + 2t + 1\rangle$ | $11a_{317}$ | $\langle t + 1, 5\rangle$ |
| $9_{46}$ | $\langle 3, t + 1\rangle$ | $11n_{49}$ | $\langle 2, t^2 + t + 1\rangle$ | | | | |
| | | | | $11a_{108}$ | $R$ | $10_{67}$ | $R$ |
| $11a_{102}$ | $R$ | $11n_1$ | $R$ | $11a_{139}$ | $R$ | $10_{74}$ | $\langle 3, t + 1\rangle$ |
| $11a_{181}$ | $\langle 3, t - 2\rangle$ | $9_{48}$ | $\langle 3, t + 1\rangle$ | $11a_{231}$ | $\langle t^2 - t + 1\rangle$ | $11n_{68}$ | $R$ |
| $11a_{199}$ | $R$ | | | $11a_{57}$ | $\langle t^2 - t + 1\rangle$ | | |
| | | $10_{155}$ | $\langle t + 1, 5\rangle$ | $11a_{88}$ | $R$ | $11a_{157}$ | $\langle 2, t^4 + t^2 + 1\rangle$ |
| $10_{113}$ | $R$ | $11n_{37}$ | $R$ | | | $11a_{264}$ | $R$ |
| $11a_{107}$ | $\langle 2, t^2 + t + 1\rangle$ | $8_9$ | $R$ | $11a_{109}$ | $R$ | $11a_{305}$ | $R$ |
| $11a_{347}$ | $\langle 2, t^2 + t + 1\rangle$ | | | $11a_{44}$ | $\langle t^2 - t + 1\rangle$ | $11a_{80}$ | $R$ |
| | | $11n_{164}$ | $\langle t^2 - t + 1\rangle$ | $11a_{47}$ | $\langle t^2 - t + 1\rangle$ | | |
| $11a_{187}$ | $R$ | $11n_{85}$ | $R$ | | | $10_{63}$ | $\langle 2, t^2 + t + 1\rangle$ |
| $11a_{249}$ | $\langle 3, t - 2\rangle$ | $8_{18}$ | $\langle t^2 - t + 1\rangle$ | $10_{123}$ | $\langle t^4 - 3t^3 + 3t^2 - 3t + 1\rangle$ | $9_{38}$ | $R$ |
| $11a_{38}$ | $R$ | $9_{24}$ | $R$ | | | | |
| $11a_8$ | $R$ | | | $11a_{28}$ | $R$ | $11a_{277}$ | $\langle t + 1, 3\rangle$ |
| | | $10_{163}$ | $\langle 2, t^2 + t + 1\rangle$ | | | $11a_{99}$ | $R$ |
| $11a_{132}$ | $\langle 2, t^2 + t + 1\rangle$ | $11n_{87}$ | $R$ | $10_{87}$ | $R$ | | |
| $11a_{352}$ | $\langle 3, t^2 - t + 1\rangle$ | $9_{28}$ | $R$ | $10_{98}$ | $\langle 1 - t + t^2\rangle$ | $11a_{196}$ | $\langle 7, t + 1\rangle$ |
| $11a_6$ | $R$ | $9_{29}$ | $R$ | $11a_{165}$ | $\langle 2, 1 - t + t^2\rangle$ | $11a_{216}$ | $R$ |
| | | | | $11a_{58}$ | $R$ | $11a_{286}$ | $R$ |
| $10_{65}$ | $\langle 2, t^2 + t + 1\rangle$ | $10_{59}$ | $R$ | $11n_{72}$ | $\langle 1 - t + t^2\rangle$ | | |
| $10_{77}$ | $R$ | $11n_{66}$ | $R$ | | | | |
| $11n_{71}$ | $\langle -t^2 + t - 1\rangle$ | $9_{40}$ | $\langle t^2 - 3t + 1\rangle$ | $10_{144}$ | $\langle 2, 1 - t + t^2\rangle$ | | |
| $11n_{75}$ | $\langle -t^2 + t - 1\rangle$ | | | $11n_{99}$ | $R$ | | |
| | | $10_{42}$ | $R$ | | | | |
| $10_{103}$ | $\langle 5, t + 1\rangle$ | $10_{75}$ | $\langle 3, t + 1\rangle$ | $11n_{83}$ | $\langle 2, t^2 + t + 1\rangle$ | | |
| $10_{40}$ | $R$ | | | $9_{41}$ | $\langle 7, 1 + t\rangle$ | | |

# References

[Adams et al.(1994)] W.W. Adams, P. Loustaunau, *An introduction to Gröbner bases*, volume 3 of *Graduate Studies in Mathematics*, American Mathematical Society, Providence, RI, 1994.

[Burde et al.(1985)] G. Burde, H. Zieschang, *Knots*, volume 5 of *de Gruyter Studies in Mathematics*, Walter de Gruyter, Berlin and New York, 1985.

[Crowell et al.(1977)] R.H. Crowell, R.H. Fox, *Introduction to Knot Theory*, volume 57 of *Graduate Texts in Mathematics*, Springer-Verlag, New York, 1977.

[Fox et al.(1964)] R.H. Fox, N. Smythe, An ideal class invariant of knots, Proc. Amer. Math. Soc., 15:707–709, 1964.

[Kanenobu(1986)] T. Kanenobu, Infinitely many knots with the same polynomial invariant, Trans. Amer. Math. Soc., 97:158–162, 1986.

[Kawauchi(1996)] A. Kawauchi, *A survey of knot theory*, Birkhäuser Verlag, Basel, 1996.

[Kearton et al.(2003)]  C. Kearton, S.M.J. Wilson, Knot modules and the Nakanishi index, Proc. Amer. Math. Soc., 131:655–663, 2003.

[Lickorish(1998)]  W.B.R. Lickorish, *An introduction to knot theory*, volume 175 of *Graduate Texts in Mathematics*, Springer-Verlag, New York, 1998.

[Livingston(2004)]  C. Livingston, *Table of knot invariants* at `http://www.indiana.edu/~knotinfo/`.

[Pauer et al.(1999)]  F. Pauer, A. Unterkircher, Gröbner Bases for Ideals in Laurent Polynomials Rings and their Application to Systems of Difference Equations, Appl. Algebra Engrg. Comm. Comput., 9:271–291, 1999.

[Sims(1994)]  C.C. Sims, *Computation with finitely presented groups*, volume 48 of *Encyclopedia of Mathematics and its Applications*, Cambridge University Press, Cambridge, 1994.

# Sudokus and Gröbner Bases: Not Only a *Divertimento*

Jesús Gago-Vargas[1], Isabel Hartillo-Hermoso[2],
Jorge Martín-Morales[3], and José María Ucha-Enríquez[1,*]

[1] Dpto. de Álgebra, Univ. de Sevilla, Apdo. 1160, 41080 Sevilla, Spain
{gago, ucha}@us.es
[2] Dpto. de Matemáticas, Univ. de Cádiz, Apdo. 40, 11510 Puerto Real, Cádiz, Spain
isabel.hartillo@uca.es
[3] Depto. de Matemáticas, Univ. de Zaragoza, Zaragoza, Spain
jorge@unizar.es

**Abstract.** Sudoku is a logic-based placement puzzle. We recall how to translate this puzzle into a 9-colouring problem which is equivalent to a (big) algebraic system of polynomial equations. We study how far Gröbner bases techniques can be used to treat these systems produced by Sudokus. This general purpose tool can not be considered as a good solver, but we show that it can be useful to provide information on systems that are —in spite of their origin— hard to solve.

## 1 Introduction

During the last years some games called of 'Number place' type have become very popular. The target is to put some numbers or pieces on a board starting from some information given by other numbers. We have analyzed some of these puzzles and reduced them to equivalent algebraic systems of polynomial equations. We think that this modelling is itself a good motivation for students.

We have found that these systems —particularly in the case of Sudoku— are a good source of non-trivial examples to:

1. Study the limits and applicability of the available solving methods.
2. Compare the methods.

This work is a report on what can be expected of Gröbner bases as the natural first approach to this study. The reader is referred to the classical bibliography as [1], [4], [11] or [3] as excellent introductions to this subject.

## 2 Describing and Modelling Sudoku

Sudoku is a puzzle that became very popular in Japan in 1986 and all around the world in 2005, although its origin happened in New York, under the name

---

* All the authors partially supported by FQM-333 and MTM2004-01165.

V.G. Ganzha, E.W. Mayr, and E.V. Vorozhtsov (Eds.): CASC 2006, LNCS 4194, pp. 155–165, 2006.

'Number Place'. You have to fill in a $9 \times 9$ board divided in 9 regions of size $3 \times 3$ with the digits 1 to 9, starting from some numbers given on the board in such a way that two numbers cannot be repeated in any row, column or $3 \times 3$ region. A proper Sudoku has only one solution.

Sudoku can be expressed as a graph colouring problem:

- The graph has 81 vertices, one for each cell.
- You need 9 colours, one for each number.
- The edges are defined by the adjacency relations of Sudoku: where we want different numbers (taking into account rows, columns and regions) we need different colours.

|   | 9 |   |   | 4 |   |   | 7 |   |
|---|---|---|---|---|---|---|---|---|
|   |   |   |   | 7 | 9 |   |   |   |
| 8 |   |   |   |   |   |   |   |   |
| 4 |   | 5 | 8 |   |   |   |   |   |
| 3 |   |   |   |   |   |   |   | 2 |
|   |   |   |   | 9 | 7 |   |   | 6 |
|   |   |   |   |   |   |   |   | 4 |
|   |   | 3 | 5 |   |   |   |   |   |
| 2 |   |   | 6 |   |   |   | 8 |   |

**Fig. 1.** A typical Sudoku

The resulting graph $\mathcal{G}$ is a regular graph with degree 20, so the number of edges of $\mathcal{G}$ is equal to $\frac{81 \cdot 20}{2} = 810$.

We can solve the colouring problem through a polynomial system ([2]. cf. [1], [11]) described by an ideal $\mathcal{I}$ of $\mathbb{Q}[x_1, \ldots, x_{81}]$ —a variable for each vertex— with the following generators $F(x_j), j = 1, \ldots, 81$ and $G(x_i, x_j), 1 \leq i < j \leq 81$:

- We will consider the colours numbered from 1 to 9. For each vertex $x_j$ we consider the polynomial $F(x_j) = \prod_{i=1}^{9}(x_j - i)$.
- If two vertexes are adjacent then $F(x_i) - F(x_j) = (x_i - x_j)G(x_i, x_j) = 0$, so the condition about different colours is given by adding the polynomial $G(x_i, x_j)$.

We number the cells in a Sudoku as in Figure 2.

In addition, all the initial information of the Sudoku must be included. For example, if we want to solve the Sudoku in Figure 1, we have to add the following polynomials to the ideal $\mathcal{I}$:

$$x_2 - 9, \ x_6 - 4, \ x_9 - 7, \ x_{15} - 7, \ x_{16} - 9, \ x_{19} - 8,$$
$$x_{28} - 4, \ x_{30} - 5, \ x_{31} - 8, \ x_{37} - 3,$$
$$x_{45} - 2, \ x_{51} - 9, \ x_{52} - 7, \ x_{56} - 6, \ x_{63} - 4,$$
$$x_{66} - 3, \ x_{67} - 5, \ x_{73} - 2, \ x_{76} - 6, \ x_{80} - 8.$$

| 1 | 2 | 3 | 4 | 5 | 6 | 7 | 8 | 9 |
|---|---|---|---|---|---|---|---|---|
| 10 | 11 | 12 | 13 | 14 | 15 | 16 | 17 | 18 |
| 19 | 20 | 21 | 22 | 23 | 24 | 25 | 26 | 27 |
| 28 | 29 | 30 | 31 | 32 | 33 | 34 | 35 | 36 |
| 37 | 38 | 39 | 40 | 41 | 42 | 43 | 44 | 45 |
| 46 | 47 | 48 | 49 | 50 | 51 | 52 | 53 | 54 |
| 55 | 56 | 57 | 58 | 59 | 60 | 61 | 62 | 63 |
| 64 | 65 | 66 | 67 | 68 | 69 | 70 | 71 | 72 |
| 73 | 74 | 75 | 76 | 77 | 78 | 79 | 80 | 81 |

**Fig. 2.** Cells enumeration

All the information about the solutions of a given Sudoku is contained in the set of zeros of $\mathcal{I}$ noted by $V(\mathcal{I})$:

$$V(\mathcal{I}) = V_{\mathbb{Q}}(\mathcal{I}) = \{(s_1, \ldots, s_{81}) \in \mathbb{Q}^{81} \text{ such that } H(s_1, \ldots, s_{81}) = 0, \text{ for any } H \in \mathcal{I}\}.$$

*Remark 1.* Once we have added the polynomials corresponding to initial data, it is easy to see that the polynomials $F(x_i)$ are redundant so we can delete them. The system of equations has 810 equations, one for each edge of the graph.

The following are elementary results:

**Proposition 2.** *A Sudoku has solution if and only if $\mathcal{I} \neq \mathbb{Q}[x_1, \ldots, x_{81}]$ if and only if any reduced Gröbner basis of $\mathcal{I}$ with respect to any term ordering is not $\{1\}$.*

**Proposition 3.** *If we start from a proper Sudoku then any reduced Gröbner basis of $\mathcal{I}$ with respect to any term ordering has the form $G = \{x_i - a_i \mid i = 1, \ldots, 81\}$ where every $a_i$ are numbers from 1 to 9. The numbers $a_i$ describe the solution.*

*Example 4.* Let us consider the problem from Figure 3. We have written the set

| 9 |   |   |   |   |   |   |   | 8 |
|---|---|---|---|---|---|---|---|---|
| 5 |   |   | 2 |   | 8 |   | 6 |   |
|   |   | 3 | 7 | 1 |   |   |   | 9 |
|   |   |   |   | 7 | 3 |   | 5 |   |
| 2 |   |   |   |   |   |   |   | 4 |
|   | 5 |   | 1 | 6 |   |   |   |   |
| 8 |   |   |   | 2 | 7 | 3 |   |   |
|   | 4 |   | 3 |   | 9 |   |   | 1 |
| 7 |   |   |   |   |   |   |   | 2 |

**Fig. 3.** Sudoku with 28 numbers

of 810 equations and a program in a file available on the Internet[1] called `sudoku` that runs under SINGULAR [7]. The syntax is

---

[1] http://www.us.es/gmcedm/

```
<"sudoku";
intmat A[9][9] =
        9,0,0,0,0,0,0,0,8, 5,0,0,2,0,8,0,6,0,
        0,0,3,7,1,0,0,0,9,
        0,0,0,0,7,3,0,5,0, 2,0,0,0,0,0,0,0,4,
        0,5,0,1,6,0,0,0,0,
        8,0,0,0,2,7,3,0,0, 0,4,0,3,0,9,0,0,1,
        7,0,0,0,0,0,0,0,2;
def G = sudoku(A); vdim(G);
//used time: 1.65 sec
//-> 1
```

This Sudoku has a unique solution and is encoded in the reduced Gröbner basis $G$.

| 9 | 2 | 6 | 5 | 3 | 4 | 7 | 1 | 8 |
|---|---|---|---|---|---|---|---|---|
| 5 | 7 | 1 | 2 | 9 | 8 | 4 | 6 | 3 |
| 4 | 8 | 3 | 7 | 1 | 6 | 5 | 2 | 9 |
| 1 | 9 | 8 | 4 | 7 | 3 | 2 | 5 | 6 |
| 2 | 6 | 7 | 9 | 8 | 5 | 1 | 3 | 4 |
| 3 | 5 | 4 | 1 | 6 | 2 | 9 | 8 | 7 |
| 8 | 1 | 9 | 6 | 2 | 7 | 3 | 4 | 5 |
| 6 | 4 | 2 | 3 | 5 | 9 | 8 | 7 | 1 |
| 7 | 3 | 5 | 8 | 4 | 1 | 6 | 9 | 2 |

**Fig. 4.** Solution for Figure 3

Unfortunately, in general the systems produced by Sudokus are not so friendly. Backtracking solvers (more or less guided by logic) are all over the web and are usually very fast. An interesting alternative method (which admits an algebraic approach too that has to be considered in the future) is that of the *dancing links* ([10]).

We think that writing down the equations and computing Gröbner bases is not a good solving method in general[2]. Nevertheless this approach has some advantages: if the Sudoku has many solutions the Gröbner bases allow us to obtain the number of solutions, as we will see in the next section.

## 3    Counting Solutions

It is well known that, as ideals $\mathcal{I}$ produced by Sudokus are radical (cf. [4, Ch. 2, Prop. 2.7.]) the number of elements in $V(\mathcal{I})$ is equal to the dimension of the

---

[2] This is something that perhaps could be expected: when you solve Sudokus by hand, you consider proper subsets of the initial data of the Sudoku that produce new values in the cells. The polynomial approach in principle take into account *all the system at the same time* and does not take advantage of the subsystems, unless you choose ad hoc term orderings for each Sudoku.

$\mathbb{Q}$-vector space $\mathbb{Q}[x_1, \ldots, x_{81}]/\mathcal{I}$, and that this number can be computed with *any* Gröbner basis $G$ with respect to any term ordering $<$:

**Proposition 5.** *(cf. [1, Prop. 2.1.6.]) A basis of the $\mathbb{Q}$-vector space $\mathbb{Q}[x_1, \ldots, x_{81}]/\mathcal{I}$ consists of the cosets of all the power products that are not divisible by $lp_<(g_i)$ for every $g_i \in G$.*

The Singular command ([7]) to obtain this invariant for a given ideal of a ring is `vdim`.

*Example 6.* Suppose now that we start from the Sudoku of Figure 3 but cells number 64 and 82 are empty.

```
A[6,4]=0; A[8,2]=0;
G=sudoku(A);
vdim(G);
//used time: 127.71 sec
//-> 53
```

Then there are 53 different solutions. To compute all of them

```
LIB "solve.lib";
def S = solve(G,5,0,"nodisplay");
setring S; size(SOL);
//-> 53
SOL[1];  //First solution in the list
```

In an analogous way we can see that if cell 26 is empty too there are 98 solutions.

*Example 7.* We have easily obtained that the number of different Sudokus $4 \times 4$ that has the same rules that the $9 \times 9$ but only four colours and four $2 \times 2$ regions is 288. The ideal to be considered is the one corresponding with no initial data.

The number of possible configurations for the case $9 \times 9$ is known ([6]) and it is a work in progress to apply a mixed approach between brute force and Gröbner bases computation to obtain the number of configurations. It should be pointed out that mixed volume might be another interesting approach.

*Example 8.* If the initial configuration of Figure 3 has the number 1 in cell number 4 then the problem has no solution: any reduced Gröbner basis is equal to $\{1\}$. To obtain that a given Sudoku has no solution is often in practice reasonably fast. So, although solving a given Sudoku can be very hard, it is not so hard in general to guess the value of a given cell trying the set of possible values.

# 4   Modelling More Fashionable Games

There exist many variants of the previous game. We briefly overview some of them and give their mathematical modelling with an algebraic system, above all because of their pedagogical interest. In general they are not colouring problems and in all cases Gröbner bases count the number of possible solutions.

## 4.1  Variants of Sudoku

The following games are variants of the classical Sudoku:

1. *Killer Sudoku*. Instead of being given the values of a few individual cells, the sum of groups of cells are given. No duplicates are used within the groups. The algebraic system is built by adding to the 810 equations those that define the linear relations coming from the sums of groups of cells. For example, in Figure 5 the cells $x_1$ and $x_2$ give us the polynomial $x_1 + x_2 - 3$.

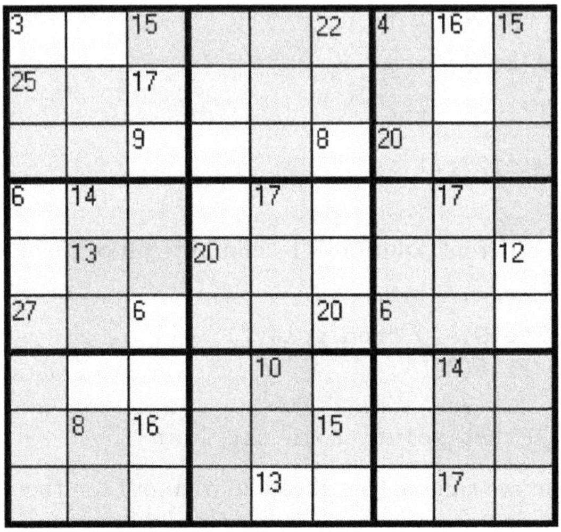

**Fig. 5.** Killer Sudoku

2. *Even-Odd Sudoku*. Fill in the grid so that every row, column, $3 \times 3$ box, contains the digits 1 through 9, with gray cells even, white cells odd. The grey cells bring out a polynomial of the form $\prod_{i \text{ even}} (x_j - i)$.
3. *1-way Disallowed number place*. All the places where orthogonally adjacent cells are consecutive numbers have been specially marked. If two cells $x_i$ y $x_j$ are adjacent we have to add the following equations. If they are specially marked we have to write an equation of the form $(x_i - x_j - 1)(x_i - x_j + 1)$. If they don't then it is the negative proposition, so we can write it as $z(x_i - x_j - 1)(x_i - x_j + 1) - 1$, where $z$ is a new variable.
4. *Greater than Sudoku*. It only appears "greater than" or "less than" signs in adjacent cells. The board is empty: we have not any data. Any relation of the form $x_i > x_j$ can be written as $x_i - x_j = b_{ij}$, where $b_{ij} \in \{1, 2, \ldots, 8\}$.
5. *Geometry Sudoku*. The board is not rectangular, it can even be a torus. We only have to change the adjacency relations.
6. *Factor Rooms*. It is similar to Killer Sudoku, but now with products and without $3 \times 3$ blocks.

## 4.2  Kakuro

The rules are

1. Place a number from 1 to 9 in each empty cell.
2. The sum of each vertical or horizontal block equals the number at the top or on the left of that block.
3. Numbers may only be used once in each block.

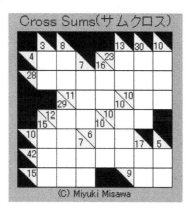

**Fig. 6.** Kakuro

The equations are of the form

1. $F(x_j) = \prod_{i=1}^{9}(x_j - i)$ for each cell.
2. $G(x_j, x_k) = \frac{F(x_j) - F(x_k)}{x_j - x_k}$ for cells $(j, k)$ in the same block.
3. Linear relations defined by the sums in each block.

For example, in Figure 6 we have the following linear relations for the first cell:

$$x_1 + x_2 - 4, x_1 + x_6 - 3.$$

## 4.3  Bridges or Hashiwokakero

This is another popular game in Japan. The rules are

1. The number of bridges is the same as the number inside the island.
2. There can be up to two bridges between two islands.
3. Bridges cannot cross islands or other bridges.
4. There is a continuous path connecting all the islands.

The unknowns are the bridges that cross from one island to other. For example, from the top left island in Figure 7, with value 3, we get variable $x_1$ (connection to right island) and $x_2$ (connection to down) (see Figure 8). Every $x_i$ can have the

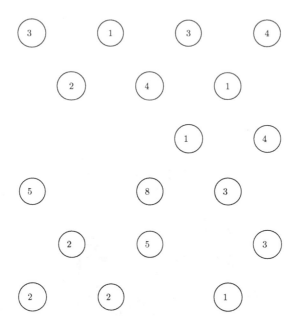

**Fig. 7.** Bridges

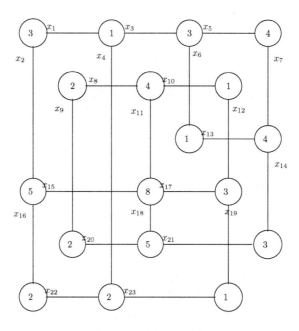

**Fig. 8.** Bridges model

value $0, 1, 2$, so we include the polynomials $x_i(x_i - 1)(x_i - 2)$. The condition that the bridge $x_i$ cannot cross the bridge $x_j$ is equivalent to the equation $x_i x_j = 0$. Last, we have the linear relations given by the sum of bridges that start from an island. In the previous example, we get

$$x_i(x_i - 1)(x_i - 2), i = 1, \ldots, 23,$$

$$x_4 x_8, x_4 x_{15}, x_4 x_{20}, x_6 x_{10}, x_{12} x_{13}, x_9 x_{15}, x_{19} x_{21},$$

$$x_1 + x_2 - 3, x_1 + x_3 + x_4 - 1, x_3 + x_5 + x_6 - 3,$$

$$x_5 + x_7 - 4, x_8 + x_9 - 2, x_8 + x_{10} + x_{11} - 4,$$

$$x_{10} + x_{12} - 1, x_6 + x_{13} - 1, x_7 + x_{13} + x_{14} - 4,$$

$$x_2 + x_{15} + x_{16} - 5, x_{11} + x_{15} + x_{17} + x_{18} - 8,$$

$$x_{12} + x_{17} + x_{19} - 3, x_9 + x_{20} - 2, x_{18} + x_{20} + x_{21} - 5,$$

$$x_{14} + x_{21} - 3, x_{16} + x_{22} - 2, x_4 + x_{22} + x_{23} - 2, x_{19} + x_{23} - 1$$

It is easy to compute the solutions of this system. To extract those that define a connected graph is a different kettle of fish that we do not treat in this little section. The computation in SINGULAR is as follows.

```
ring r0=0,x(1..23),dp; option(redSB);

proc F (int i) { return(x(i)*(x(i)-1)*(x(i)-2)); };

ideal I;

for (i = 1; i<=23; i++) {I[i]=F(i); };

I = I, x(4)*x(8), x(4)*x(15), x(4)*x(20), x(6)*x(10), x(12)*x(13),
x(9)*x(15), x(19)*x(21), x(1)+x(2) -3, x(1)+x(3)+x(4)-1,
x(3)+x(5)+x(6)-3, x(5)+x(7)-4, x(8)+x(9)-2, x(8)+x(10)+x(11)-4,
x(10) +x(12)-1, x(6)+x(13)-1, x(7)+x(13)+x(14)-4,
x(2)+x(15)+x(16)-5, x(11)+x(15)+x(17)+x(18)-8,
x(12)+x(17)+x(19)-3, x(9)+x(20)-2, x(18)+x(20)+x(21)-5,
x(14)+x(21)-3, x(16)+x(22)-2, x(4)+x(22)+x(23)-2, x(19)+x(23)-1;

ideal Isol = std(I);

Isol;

Isol[1]=x(23)-1 Isol[2]=x(22)-1 Isol[3]=x(21)-1 Isol[4]=x(20)-2
Isol[5]=x(19) Isol[6]=x(18)-2 Isol[7]=x(17)-2 Isol[8]=x(16)-1
Isol[9]=x(15)-2 Isol[10]=x(14)-2 Isol[11]=x(13) Isol[12]=x(12)-1
Isol[13]=x(11)-2 Isol[14]=x(10) Isol[15]=x(9) Isol[16]=x(8)-2
Isol[17]=x(7)-2 Isol[18]=x(6)-1 Isol[19]=x(5)-2 Isol[20]=x(4)
Isol[21]=x(3) Isol[22]=x(2)-2 Isol[23]=x(1)-1
```

## 4.4 Minesweeper

The target of this well-known Windows game is to uncover all the tiles that do not have a mine under them. When we click on a tile, if there is a mine under it, the game is over. If there is no mine under it, you will be given a number. The number will tell you how many mines are touching that tile (left, right, above and below). We assign variables to each unknown tile, with values 0 or 1. The relations between them are of the form $\sum_i x_i - k$.

*Remark 9.* There is a classical and interesting problem: given a board of positions with numbers, is it valid? In other words, is there any way in which the mines could be arranged in the hidden squares that would be consistent with those numbers? This problem is known to be NP-complete [8]. With the previous model, we have an algorithm to decide the consistency, through the computation of a Gröbner basis. A theoretical consequence of the polynomial modelling is that we obtain that the consistency of a system of polynomial equations of degree two (almost linear!) is NP-complete.

## 5    Fake Shortcuts and Experimental Facts

Of course we have tried the following (a priori) tricks to speed up our implementation to manage Sudokus based in Gröbner bases:

- Work in a field of 9 elements instead of characteristic 0. No significant improvements.
- Work with the 9-th roots of the unit as colours. No significant improvements.
- Change the numbers of the colours to $-4, -3, -2, -1, 0, 1, 2, 3, 4$ to obtain nicer coefficients. No significant improvements.
- Use symmetric polynomials instead of the $G_i$ of section 2. No significant improvement.
- In the available options to compute Groebner bases with Singular, the option intStrategy has been used: it avoids division of coefficients during standard basis computations. Without this option computations are often much slower.

On the other hand, we have tried to solve our Sudoku systems with some different available methods. Here are the initial experimental results:

- **Numerical methods:** In most examples, usual Newton-Rapshon (cf. [9]) methods —the way in which an engineer would possibly try to solve our systems— have not succeeded. It is a work in progress to show that for systems produced by Sudokus the usual numerical methods diverge for a big enough family of examples. It would mean that Sudoku systems could be regarded as ill-conditioned systems richer by far than the classical Wilkinson's monster (cf. [5]).
- **Numerical homotopy methods:** The numerical homotopy methods implemented in Jan Verschelde's software package PHCpack ([14]) are another way of solving algebraic systems of polynomial equations of great interest. They are known to be well suited to treat the multilinear case. They have been used, for example, to obtain totally mixed Nash equilibria (cf. [13]). Neither have they obtained correct solutions in most examples.

Sudoku systems seems to be somewhat resistant to a non-purely-symbolic approach, and we think that this pathological behavior demands itself a deeper understanding.

# References

1. W. W. Adams and P. Loustaunau. *An introduction to Gröbner bases*, volume 3 of *Graduate Studies in Mathematics*. American Mathematical Society, Providence, RI, 1994.
2. Bayer, Dave *The division algorithm and the Hilbert scheme*. Ph. D. Thesis. Harvard University, June 1982.
3. T. Becker and V. Weispfenning. *Gröbner bases*, volume 141 of *Graduate Texts in Mathematics*. Springer-Verlag, New York, 1993. A computational approach to commutative algebra, In cooperation with Heinz Kredel.
4. Cox, D., J. Little, D. O'Shea. *Ideals , Varieties and Algorithms*. Springer, Berlin, 1997.
5. Cox, D., J. Little, D. O'Shea. *Using Algebraic Geometry*. Springer, Berlin, 1998.
6. B. Felgenhauer and F. Jarvis. Enumerating possible sudoku grids, 2005. http://www.afjarvis.staff.shef.ac.uk/ sudoku/sudoku.pdf. [Online; accessed 30-December-2005].
7. G.-M. Greuel, G. Pfister, and H. Schönemann. SINGULAR 3.0. http://www.singular.uni-kl.de. A Computer Algebra System for Polynomial Computations, Centre for Computer Algebra, University of Kaiserslautern, 2005.
8. R. Kaye. Minesweeper is NP-complete. *Math. Intelligencer*, 22(2):9–15, 2000.
9. C. T. Kelley. *Iterative methods for linear and nonlinear equations*, volume 16 of *Frontiers in Applied Mathematics*. Society for Industrial and Applied Mathematics (SIAM), Philadelphia, PA, 1995. With separately available software.
10. Knuth, D.E. *Dancing links*. Preprint.
11. M. Kreuzer and L. Robbiano. *Computational commutative algebra. 1.* Springer-Verlag, Berlin, 2000.
12. E. Pegg. Sudoku variations, 2005. http://www.maa.org/editorial/mathgames/ mathgames_09_05_05.html. [Online; accessed 12-December-2005].
13. Sturmfels, B. Solving Systems of Polynomial Equations. Amer.Math.Soc., CBMS Regional Conferences Series, No 97, Providence, Rhode Island, 2002.
14. J. Verschelde. PHCpack: A general-purpose solver for polynomial systems by homotopy continuation, 2005. ACM Transactions on Mathematical Software volume 25, number 2: 251–276, 1999.

# Simplicial Perturbation Techniques and Effective Homology*

Rocio Gonzalez-Díaz, Belén Medrano**,
Javier Sánchez-Peláez, and Pedro Real

Departamento de Matemática Aplicada I,
Universidad de Sevilla, Seville (Spain)
{rogodi, belenmg, fjsp, real}@us.es
http://www.us.es/gtocoma

**Abstract.** In this paper, we deal with the problem of the computation of the homology of a finite simplicial complex after an "elementary simplicial perturbation" process such as the inclusion or elimination of a maximal simplex or an edge contraction. To this aim we compute an algebraic topological model that is a special chain homotopy equivalence connecting the simplicial complex with its homology (working with a field as the ground ring).

## 1 Introduction

Simplicial complexes are widely used in geometric modelling. In order to classify them from a topological point of view, a first algebraic invariant that can be used is homology. We can cite two relevant algorithms for computing homology groups of a simplicial complex $K$ in $\mathbf{R}^n$: The classical matrix algorithm [8] based on reducing certain matrices to their Smith normal form. And the incremental algorithm [1] consisting of assembling the complex, simplex by simplex, and at each step the Betti numbers of the current complex are updated. Both methods run in time at most $O(m^3)$, where $m$ is the number of simplices of the complex. Here, we deal with the problem of obtaining the homology $\mathcal{H}$ of a finite simplicial complex $K$ and a chain contraction (a special chain homotopy equivalence [8]) of the chain complex $C(K)$ to $\mathcal{H}$. This notion is a special case of effective homology [10]. We call it an algebraic topological model for $K$ (or AT-model for $K$).

Since the emergence of the Homological Perturbation Theory [5, 6], chain contractions have been widely used [10, 9, 5, 6]. The fundamental tool in this area is the Basic Perturbation Lemma (or BPL) which can be seen as a real algorithm such that the input is a chain contraction between two chain complexes $(C, d)$ and $(C', d')$ and a perturbation $\delta$ of $d$. The output is a chain contraction between the perturbed chain complexes $(C, d+\delta)$ and $(C', d'+d_\delta)$. Here, we are interested

---

* Partially supported by the PAICYT research project FQM–296 "Computational Topology and Applied Math" from Junta de Andalucía.
** Fellow associated to University of Seville under a Junta de Andalucia research grant.

V.G. Ganzha, E.W. Mayr, and E.V. Vorozhtsov (Eds.): CASC 2006, LNCS 4194, pp. 166–177, 2006.

in the following complementary problem: given a chain contraction between two chain complexes $(C, d)$ and $(C', d')$ and a "perturbation" of $C$, is it possible to obtain a chain contraction between the perturbed chain complexes $(\tilde{C}, \tilde{d})$ and $(\tilde{C}', \tilde{d}')$?. Being the ground ring any field, Algorithms 1, 3 and 4 of this paper are positive answers to this question in the particular case of AT-models for simplicial complexes and "elementary simplicial perturbations" such as inclusion or elimination of a maximal simplex or an edge contraction. Moreover, Algorithm 1 is a version for AT-models of the incremental algorithm developed in [1]. It is described in [2, 3, 4] when the ground ring is $\mathbf{Z}/2\mathbf{Z}$ and applied to Digital Images for computing digital cohomology information.

## 2    Algebraic Topological Models

Now, we give a brief summary of concepts and notations. The terminology follows Munkres book [8]. For the sake of clarity and simplicity, we only will define the concepts that are really essential in this paper. We will consider that the ground ring is any field.

**Simplicial Complexes.** Considering an ordering on a vertex set $V$, a $q$–simplex with $q + 1$ affinely independent vertices $v_0 < \cdots < v_q$ of $V$ is the convex hull of these points, denoted by $\langle v_0, \ldots, v_q \rangle$. If $i < q$, an $i$–face of $\sigma$ is an $i$–simplex whose vertices are in the set $\{v_0, \ldots, v_q\}$. A simplex is *maximal* if it does not belong to any higher dimensional simplex. A *simplicial complex* $K$ is a collection of simplices such that every face of a simplex of $K$ is in $K$ and the intersection of any two simplices of $K$ is a face of each of them or empty. The set of all the $q$–simplices of $K$ is denoted by $K^{(q)}$. The *dimension of* $K$ is the dimension of the highest dimensional simplex in $K$.

**Chains and Homology.** Let $K$ be a simplicial complex. A $q$–*chain* $a$ is a formal sum of simplices of $K^{(q)}$. The $q$–chains form the $q$th *chain group* of $K$, denoted by $C_q(K)$. The *boundary* of a $q$–simplex $\sigma = \langle v_0, \ldots, v_q \rangle$ is the $(q - 1)$–chain: $\partial_q(\sigma) = \sum_{i=0}^{q}(-1)^i \langle v_0, \ldots, \hat{v}_i, \ldots, v_q \rangle$, where the hat means that $v_i$ is omitted. By linearity, $\partial_q$ can be extended to $q$–chains. The collection of boundary operators connect the chain groups $C_q(K)$ into the *chain complex* $C(K)$: $\cdots \xrightarrow{\partial_2} C_1(K) \xrightarrow{\partial_1} C_0(K) \xrightarrow{\partial_0} 0$. An essential property is that $\partial_q \partial_{q+1} = 0$. In a more general framework, a *chain complex* $\mathcal{C}$ is a sequence $\cdots \xrightarrow{d_2} C_1 \xrightarrow{d_1} C_0 \xrightarrow{d_0} 0$ of abelian groups $C_q$ and homomorphisms $d_q$, such that for all $q$, $d_q d_{q+1} = 0$. The set of all the homomorphisms $d_q$ is called the *differential* of $\mathcal{C}$. A $q$–chain $a \in C_q$ is called a $q$–*cycle* if $d_q(a) = 0$. If $a = d_{q+1}(a')$ for some $a' \in C_{q+1}$ then $a$ is called a $q$–*boundary*. Denote the groups of $q$–cycles and $q$–boundaries by $Z_q$ and $B_q$ respectively. We say that $a$ is a *representative* $q$–*cycle* of a homology generator $\alpha$ if $\alpha = a + B_q$. Define the $q$th *homology group* to be the quotient group $Z_q/B_q$, denoted by $H_q(\mathcal{C})$. The $q$th *Betti number* $\beta_q$ is the rank of $H_q(\mathcal{C})$. Intuitively, $\beta_0$ is the number of components of connected pieces, $\beta_1$ is the number of independent "holes" and $\beta_2$ is the number of "cavities".

**Chain Contractions** [7]. A *chain contraction* of a chain complex $\mathcal{C}$ to another chain complex $\mathcal{C}'$ is a set of three homomorphisms $(f, g, \phi)$ such that: $f : \mathcal{C} \to \mathcal{C}'$ and $g : \mathcal{C}' \to \mathcal{C}$ are chain maps; $fg$ is the identity map of $\mathcal{C}'$; $\phi : \mathcal{C} \to \mathcal{C}$ is a chain homotopy of the identity map $id$ of $\mathcal{C}$ to $gf$, that is, $\phi\partial + \partial\phi = id - gf$. Moreover, the annihilation properties $f\phi = 0$, $\phi g = 0$ and $\phi\phi = 0$ are required. Important properties of chain contractions are: $\mathcal{C}'$ has fewer or the same number of generators than $\mathcal{C}$; $\mathcal{C}$ and $\mathcal{C}'$ have isomorphic homology groups.

**AT-model.** Given a field as the ground ring, an AT-model for a simplicial complex $K$ is the set $(K, h, f, g, \phi)$ where $h$ is a set of generators of a chain complex $\mathcal{H}$ isomorphic to the homology of $K$ and $(f, g, \phi)$ is a chain contraction of $C(K)$ to $\mathcal{H}$. This implies that $f\partial = 0$ and $\partial g = 0$.

**First Basic Perturbation Lemma** (BPL) [5, 6]. Let $(f, g, \phi)$ be a chain contraction of the chain complex $(C, d)$ to the chain complex $(C', d')$. Let $\delta : C \to C$ be a morphism of degree $-1$, called *perturbation*, such that $\phi\delta$ is pointwise nilpotent and $(d + \delta)^2 = 0$. Then $(f_\delta, g_\delta, \phi_\delta)$ given by $f_\delta = f - f\delta\Delta\phi$, $\phi_\delta = \Delta\phi$, $g_\delta = \Delta g$ is a chain contraction of $(C, d + \delta)$ to $(C', d' + d_\delta)$ where $d'_\delta = f\delta\Delta\phi$ and $\Delta = \sum_{i=0}^{\infty}(-1)^i(\phi\delta)^i$.

# 3   Simplicial Perturbations on AT-Models

Given a simplicial complex $K$ with $m$ simplices, an *elementary simplicial perturbation* of $K$ is one of these operations: an inclusion of a simplex, a deletion of a simplex, an edge contraction. Algorithms 1, 3 and 4 compute AT-models after elementary simplicial perturbation. Moreover, the first one is a particular application of the BPL, showing in this way that there is a relation between the simplicial perturbation (where the ground graded groups are changed) and the algebraic perturbation (the original one, where just the differential changes).

**Theorem 1.** *Given an AT-model for $K$ and an elementary simplicial perturbation of $K$, an AT-model for the perturbed complex $\tilde{K}$ can be computed using Algorithm 1, 3 or 4 in $\mathcal{O}(m^2)$.*

It is necessary to say that if we do not have an AT-model of $K$ as input, the computation of an AT-model of $\tilde{K}$ can be done applying Algorithm 1 $m + 1$ times in the worst case, so the complexity, in this case, is $\mathcal{O}(m^3)$.

## 3.1   AT-Models After Adding a Simplex

An incremental algorithm for computing AT-models with coefficients in $\mathbf{Z}/2\mathbf{Z}$ appears in [2, 3, 4]. Here, we give an extension of the algorithm with coefficients in any field and prove that it is a particular application of BPL.

Given a chain $a$ and a simplex $\sigma$, define $c_\sigma(a)$ as the coefficient of $\sigma$ in $a$.

**Algorithm 1.** *Incremental algorithm for computing AT-models for simplicial complexes in any dimension with coefficients in any field.*

INPUT: An AT-model $(K, h, f, g, \phi)$ for $K$ and a $q$-simplex $\sigma$ not in $K$ such that $K \cup \{\sigma\}$ is a simplicial complex.

If $f\partial(\sigma) = 0$, then
$\quad h := h \cup \{\sigma\}$, $f(\sigma) := \sigma$, $g(\sigma) := \sigma - \phi\partial(\sigma)$ and $\phi(\sigma) := 0$.
Else let $\beta \in h$ such that $\lambda := c_\beta(f\partial(\sigma)) \neq 0$, then
$\quad h := h \backslash \{\beta\}$, $f(\sigma) := 0$, $\phi(\sigma) := 0$.
$\quad$ For every $\mu \in K$
$\quad\quad$ If $\lambda_\mu := c_\beta(f(_\kappa(\mu)) \neq 0$ then
$\quad\quad\quad f(\mu) := f(\mu) - \lambda_\mu \lambda^{-1} f\partial(\sigma)$,
$\quad\quad\quad \phi(\mu) := \phi(\mu) + \lambda_\mu \lambda^{-1}(\sigma - \phi\partial(\sigma))$.
$\quad\quad$ End if.
$\quad$ End for.
End if.

OUTPUT: The set $(K \cup \{\sigma\}, h, f, g, \phi)$.

**Theorem 2.** *The output of Algorithm 1 defines an AT-model for $K \cup \{\sigma\}$.*

**Proof.** Use BPL considering two cases.

– If $f\partial(\sigma) = 0$, let $\mathcal{C}$ be the chain complex generated by the simplices of $K \cup \{\sigma\}$ with differential $d$ given by $d(\mu) := \partial(\mu)$ if $\mu \in K$ and $d(\sigma) := 0$. Let $\mathcal{H}$ be the chain complex with null differential generated by $h \cup \{\sigma\}$. Let $(f', g', \phi')$ be the chain contraction of $\mathcal{C}$ to $\mathcal{H}$ defined by $f'(\mu) := f(\mu)$ and $\phi'(\mu) := \phi(\mu)$ if $\mu \in K$; $g'(\alpha) := g(\alpha)$ if $\alpha \in h$; $f'(\sigma) := \sigma$, $g'(\sigma) := \sigma$ and $\phi'(\sigma) := 0$. Let $\delta : \mathcal{C} \to \mathcal{C}$ be defined by $\delta(\mu) := 0$ if $\mu \in K$ and $\delta(\sigma) := \partial(\sigma)$. It is easy to see that $\phi'\delta$ is pointwise nilpotent (since $\delta\phi'\delta = 0$) and $(d+\delta)^2 = 0$. Apply BPL to the morphism $\delta$ and the chain contraction $(f', g', \phi')$ of $\mathcal{C}$ to $\mathcal{H}$ to obtain the chain contraction $(f_\delta, g_\delta, \phi_\delta)$ of $C(K \cup \{\sigma\})$ to $\mathcal{H}$.
$\quad$ Let us prove now that $(f_\delta, g_\delta, \phi_\delta)$ is the chain contraction obtained in Algorithm 1. We have that $f_\delta(\mu) = (f' - f'\delta\phi')(\mu) = f(\mu)$ and $\phi_\delta(\mu) = (\phi' - \phi'\delta\phi')(\mu) = \phi(\mu)$ if $\mu \in K$; $f_\delta(\sigma) = (f' - f'\delta\phi')(\sigma) = \sigma$ and $\phi_\delta(\sigma) = (\phi' - \phi'\delta\phi')(\sigma) = 0$; Finally, $g_\delta(\alpha) = (g' - \phi'\delta g')(\alpha) = g(\alpha)$ if $\alpha \in h$ and $g_\delta(\sigma) = (g' - \phi'\delta g')(\sigma) = \sigma - \phi\partial(\sigma)$.
– If $f\partial(\sigma) \neq 0$, let $\beta \in h$ such that $\lambda := c_\beta(f\partial(\sigma)) \neq 0$. Let $\mathcal{C}$ be the chain complex generated by the set $K \cup \{\sigma, e\}$ where $e$ is an element of dimension $q - 1$ that will be eliminated at the end. The differential $d$ of $\mathcal{C}$ is given by: $d(\mu) := \partial(\mu)$ if $\mu \in K$, $d(\sigma) := e$ and $d(e) := 0$. Let $\mathcal{H}$ be the chain complex with null differential generated by $h$. Let $(f', g', \phi') : \mathcal{C} \to \mathcal{H}$ be the chain contraction given by $f'(e) := 0$, $f'(\sigma) := 0$, $f'(\mu) := f(\mu)$ if $\mu \in K$; $g'(\beta) := g(\beta) - \lambda^{-1}e$, $g'(\alpha) := g(\alpha)$ if $\alpha \in h$; $\phi'(e) := \sigma$, $\phi'(\sigma) := 0$, $\phi'(\mu) := \phi(\mu) + \lambda_\mu \lambda^{-1}\sigma$ if $\mu \in K$ and $\lambda_\mu := c_\beta(f(\mu)) \neq 0$ and $\phi'(\mu) := \phi(\mu)$ if $\mu \in K$ and $c_\beta(f(\mu)) = 0$. Let $\delta : \mathcal{C} \to \mathcal{C}$ be defined by $\delta(\mu) := 0$ if $\mu \in K$, $\delta(e) := 0$ and $\delta(\sigma) := \partial(\sigma) - e$.

It is easy to see that $\phi'\delta$ is pointwise nilpotent (since $\delta\phi'\delta = 0$) and $(d + \delta)^2 = 0$. Apply the BPL to the morphism $\delta$ and the chain contraction $(f', g', \phi')$ of $\mathcal{C}$ to $\mathcal{H}$ to obtain the chain contraction $(f_\delta, g_\delta, \phi_\delta)$ of $\mathcal{C}'$ to $\mathcal{H}$ where $\mathcal{C}'$ is generated by $K \cup \{\sigma\} \cup \{e\}$, with differential $d + \delta$.

Let us prove now that $(f_\delta, g_\delta, \phi_\delta)$ of $C(K \cup \{\sigma\})$ to $\mathcal{H}'$ (generated by $h \setminus \{\beta\}$ with null differential) is the chain contraction obtained in Algorithm 1. We have that $f_\delta(\sigma) = (f' - f'\delta\phi')(\sigma) = 0$; $f_\delta(\mu) = f(\mu) - \lambda_\mu\lambda^{-1}f\partial(\sigma)$ if $\mu \in K$ and $c_\beta(f\partial(\mu)) \neq 0$ and $f_\delta(\mu) = f(\mu)$ if $\mu \in K$ and $c_\beta(f\partial(\mu)) = 0$. On the other hand, $g_\delta(\alpha) = (g' - \phi'\delta g')(\alpha) = g(\alpha)$ if $\alpha \in h \setminus \{\beta\}$. Finally, $\phi_\delta(\sigma) = (\phi' - \phi'\delta\phi')(\sigma) = 0$; $\phi_\delta(\mu) = \phi(\mu) + \lambda_\mu\lambda^{-1}(\sigma - \phi\partial(\sigma))$ if $\mu \in K$ and $\lambda_\mu = c_\beta(f\partial(\mu)) \neq 0$ and $\phi_\delta(\mu) = \phi(\mu)$ if $\mu \in K$ and $\lambda_\mu = c_\beta(f\partial(\mu)) = 0$.

**Fig. 1.** The simplicial complexes $A$, $B$ and $C$

Using the algorithm above it is possible to design a procedure for computing AT-models for finite simplicial complexes in any dimension with coefficients in any field.

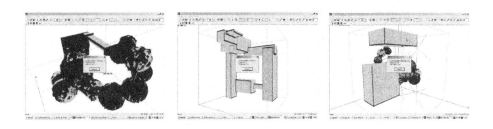

**Fig. 2.** The simplicial complexes $D$, $E$, and $F$

**Algorithm 2.** *Computing an AT-model for a simplicial complex of any dimension with coefficients in any field.*

INPUT: **A simplicial complex $K$ and an ordered-by-increasing-dimension set of all the simplices of $K$: $\{\sigma_0, \ldots, \sigma_n\}$.**

```
Define K₀ := {σ₀},  h := {σ₀},  f(σ₀) := σ₀,  g(σ₀) := σ₀,  φ(σ₀) := 0.
For i = 1 to i = m do
    apply Algorithm 1 to Kᵢ₋₁ and σᵢ.
    Kᵢ := Kᵢ₋₁ ∪ {σᵢ}.
End for.
```

OUTPUT: An AT-model $(K_m, h, f, g, \phi)$ for $K$.

The implementation of the algorithm described before has been made by J. Sánchez-Peláez and P. Real. In Figures 1, 2, 3, 4, and 5, examples of computations (using this implementation) of AT-models for three-dimensional simplicial complexes are shown.

*Example 1.* Consider the simplicial complexes $A$, $B$, $C$, $D$, $E$ and $F$ shown in Figures 1 and 2. In Figures 3, 4, and 5, representative cycles of the homology generators of these complexes are shown. In the following table, we present the Betti numbers obtained and the running time for computing AT-models for these simplicial complexes in a Pentium 4, 3.2 GHz, 1Gb RAM.

**Fig. 3.** The representative 1-cycles (holes) of $A$, $B$, and $C$

| Simplicial complex $K$ | Number of simplices of $K$ | Time | $\beta_0$ | $\beta_1$ | $\beta_2$ |
|---|---|---|---|---|---|
| $A$ | 4586 | 7 seconds | 1 | 14 | 15 |
| $B$ | 13421 | 12 seconds | 3 | 46 | 39 |
| $C$ | 3286 | 4 seconds | 1 | 17 | 10 |
| $D$ | 18842 | 30 seconds | 1 | 27 | 5 |
| $E$ | 26308 | 50 seconds | 2 | 9 | 3 |
| $F$ | 31113 | 38 seconds | 138 | 419 | 13 |

## 3.2  AT-Models After Deleting a Maximal Simplex

Now, an algorithm for computing an AT-model for a simplicial complex $K$ of any dimension with coefficients in any field after deleting a maximal simplex of $K$ is described.

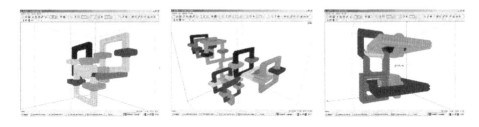

**Fig. 4.** The representative 2-cycles (cavities) of $A$, $B$, and $C$

**Algorithm 3.** *Decremental algorithm for computing AT-models for simplicial complexes of any dimension with coefficients in any field.*

INPUT: An AT-model $(K, h, f, g, \phi)$ for $K$ and a maximal $q$-simplex $\sigma$ in $K$.

If there exists $\beta \in h$ such that $\lambda := c_\sigma(g(\beta)) \neq 0$, then
    $h := h \backslash \{\beta\}$.
    For every $\mu \in K \backslash \{\sigma\}$ and $\alpha \in h$
        If $\lambda_\mu := c_\beta(f(\mu)) \neq 0$ then $f(\mu) := f(\mu) - \lambda_\mu \beta$.
        End if.
        If $\lambda'_\mu := c_\sigma(\phi(\mu)) \neq 0$ then $\phi(\mu) := \phi(\mu) - \lambda'_\mu \lambda^{-1} g(\beta)$.
        End if.
        If $\lambda_\alpha := c_\sigma(g(\alpha)) \neq 0$ then $g(\alpha) := g(\alpha) - \lambda_\alpha \lambda^{-1} g(\beta)$.
        End if.
    End for.
Else let $\gamma \in K$ be a $(q-1)$-simplex not in $h$ then
      $h := h \cup \{\gamma\}$ and $\quad g(\gamma) := \partial(\sigma)$.
    For every $\mu \in K \backslash \{\sigma\}$
        If $\lambda'_\mu := c_\sigma(\phi(\mu)) \neq 0$ then
            $f(\mu) := f(\mu) + \lambda'_\mu \gamma$ and $\phi(\mu) := \phi(\mu) - \lambda'_\mu \phi \partial(\sigma)$.
        End if.
    End for.
End if.

OUTPUT: The set $(K \backslash \{\sigma\}, h, f, g, \phi)$

Observe that the simplex $\gamma$ always exists since the morphism $f$ is onto and $f\partial(\sigma) = 0$.

**Theorem 3.** *The set $(K \backslash \{\sigma\}, h, f, g, \phi)$ defines an AT-model for $K \backslash \{\sigma\}$.*

**Proof.** Let us denote by $(K, h_K, f_K, g_K, \phi_K)$ an AT-model for $K$ and by $(K \backslash \{\sigma\}, h, f, g, \phi)$ the AT-model for $K \backslash \{\sigma\}$ obtained using Algorithm 3. In order to prove that the output of Algorithm 3 is an AT-model for $K \backslash \{\sigma\}$, we check the two most important properties wich are $fg = id$ and $id - gf = \phi\partial + \partial\phi$. The rest of the properties are left to the reader.

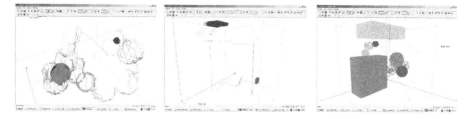

**Fig. 5.** The holes and cavities of $D$, $E$, and $F$

We distinguish two cases.

- If there exists $\beta \in h_K$ such that $\lambda := c_\sigma(g(\beta)) \neq 0$ (recall that $h := h_K \setminus \{\beta\}$).
  - If $\alpha \in h$ such that $\lambda_\alpha := c_\sigma(g_K(\alpha)) \neq 0$ then $fg(\alpha) = f(g_K(\alpha) - \lambda_\alpha \lambda^{-1} g_K(\beta)) = f_K g_K(\alpha) - \lambda_\alpha \lambda^{-1}(f_K g_K(\beta) - \beta) = \alpha$.
    Otherwise, $fg(\alpha) = fg_K(\alpha) = f_K g_K(\alpha) = \alpha$.
  - If $\mu \in K \setminus \{\sigma\}$ and $\lambda'_\mu := c_\sigma(\phi_K(\mu)) \neq 0$ then $(\phi\partial + \partial\phi)(\mu) = \phi_K \partial(\mu) + \partial(\phi_K(\mu) - \lambda'_\mu \lambda_1^{-1} g_K(\alpha_1)) = \mu - g_K f_K(\mu) = \mu - gf_K(\mu) = \mu - gf(\mu)$.
    If $c_\sigma(\phi_K(\mu)) = 0$ and $\lambda_{\partial\mu} := c_\sigma(\phi_K \partial(\mu)) \neq 0$ then $(\phi\partial + \partial\phi)(\mu) = \phi_K \partial(\mu) - \lambda_{\partial\mu} \lambda_1^{-1} g_K(\alpha_1) + \partial\phi_K(\mu) = \mu - g_K f_K(\mu) - \lambda_{\partial\mu} \lambda_1^{-1} g_K(\alpha_1) = \mu - gf(\mu)$.
    In other case, $(\phi\partial + \partial\phi)(\mu) = \phi_K \partial(\mu) + \partial\phi_K(\mu) = \mu - g_K f_K(\mu) = \mu - gf(\mu)$.
- Otherwise, let $\gamma \in K$ be a $(q-1)$-simplex not in $h_K$ (recall that $h := h_K \cup \{\gamma\}$ and $g(\gamma) := \partial(\sigma)$).
  - $fg(\gamma) = f\partial(\sigma) = f_K \partial(\sigma) + \gamma = \gamma$.
    If $\alpha \in h_K$, $fg(\alpha) = fg_K(\alpha) = f_K g_K(\alpha) = \alpha$.
  - If $\mu \in K \setminus \{\sigma\}$ and $\lambda'_\mu := c_\sigma(\phi_K(\mu)) \neq 0$ then $(\phi\partial + \partial\phi)(\mu) = \phi_K \partial(\mu) + \partial\phi_K(\mu) - \lambda_\mu \partial\phi_K \partial(\sigma) = \mu - g_K f_K(\mu) - \lambda_\mu \partial(\sigma) = \mu - gf(\mu)$.
    In other case, $(\phi\partial + \partial\phi)(\mu) = \phi_K \partial(\mu) + \partial\phi_K(\mu) = \mu - g_K f_K(\mu) = \mu - gf(\mu)$.

*Example 2.* Let us show a simple example of the computation of an AT-model after deleting a maximal simplex. Consider a simplicial complex $K$ whose set of maximal simplices is $\{\langle 0,1,2 \rangle, \langle 1,2,3 \rangle\}$. The data of an AT-model for $K$ is showed in the following table.

| $K$ | $h$ | $f$ | $g$ | $\phi$ |
|---|---|---|---|---|
| $\langle 0 \rangle$ | $\langle 0 \rangle$ | $\langle 0 \rangle$ | $\langle 0 \rangle$ | $0$ |
| $\langle 1 \rangle$ | | $\langle 0 \rangle$ | | $\langle 0,1 \rangle$ |
| $\langle 2 \rangle$ | | $\langle 0 \rangle$ | | $\langle 0,2 \rangle$ |
| $\langle 3 \rangle$ | | $\langle 0 \rangle$ | | $\langle 0,1 \rangle + \langle 1,3 \rangle$ |
| $\langle 0,1 \rangle$ | | $0$ | | $0$ |
| $\langle 0,2 \rangle$ | | $0$ | | $0$ |
| $\langle 1,3 \rangle$ | | $0$ | | $0$ |
| $\langle 2,3 \rangle$ | | $0$ | | $\langle 1,2,3 \rangle$ |
| $\langle 1,2 \rangle$ | | $0$ | | $\langle 0,1,2 \rangle$ |
| $\langle 0,1,2 \rangle$ | | $0$ | | $0$ |
| $\langle 1,2,3 \rangle$ | | $0$ | | $0$ |

After applying Algorithm 3 to the AT-model $(K, h, f, g, \phi)$ for $K$ and the maximal simplex $\langle 1, 2, 3 \rangle$ we obtain the AT-model $(K \setminus \{\langle 1, 2, 3 \rangle\}, h, f, g, \phi)$ for $K \setminus \{\langle 1, 2, 3 \rangle\}$ whose data are:

| $K \setminus \{\langle 1, 2, 3 \rangle\}$ | $h$ | $f$ | $g$ | $\phi$ |
|---|---|---|---|---|
| $\langle 0 \rangle$ | $\langle 0 \rangle$ | $\langle 0 \rangle$ | $\langle 0 \rangle$ | $0$ |
| $\langle 1 \rangle$ | | $\langle 0 \rangle$ | | $\langle 0, 1 \rangle$ |
| $\langle 2 \rangle$ | | $\langle 0 \rangle$ | | $\langle 0, 2 \rangle$ |
| $\langle 3 \rangle$ | | $\langle 0 \rangle$ | | $\langle 0, 1 \rangle + \langle 1, 3 \rangle$ |
| $\langle 0, 1 \rangle$ | | $0$ | | $0$ |
| $\langle 0, 2 \rangle$ | | $0$ | | $0$ |
| $\langle 1, 3 \rangle$ | | $0$ | | $0$ |
| $\langle 2, 3 \rangle$ | $\langle 2, 3 \rangle$ | $\langle 2, 3 \rangle$ | $\langle 2, 3 \rangle - \langle 1, 3 \rangle + \langle 1, 2 \rangle$ | $-\langle 0, 1, 2 \rangle$ |
| $\langle 1, 2 \rangle$ | | $0$ | | $\langle 0, 1, 2 \rangle$ |
| $\langle 0, 1, 2 \rangle$ | | $0$ | | $0$ |

### 3.3    AT-Models After Edge Contractions

Finally, we deal with the problem of obtaining an AT-model for a simplicial complex $K$ after an edge contraction.

An *edge contraction* is given by the vertex map

$$f^{(0)} : K^{(0)} \to L^{(0)} = K^{(0)} - \{\langle b \rangle\}$$

where $f^{(0)}(\langle b \rangle) = \langle a \rangle$ and $f^{(0)}(\langle v \rangle) = \langle v \rangle$ for all $v \neq b$.

Let $K$ be a simplicial complex and $B$ a subset of $K$. Define

$$\overline{B} = \{\sigma' \in K : \sigma' \text{ is a face of } \sigma \in B\},$$

$$St\, B = \{\sigma \in K : \sigma' \in B \text{ is a face of } \sigma\} \quad \text{and} \quad Lk\, B = \overline{St\, B} - St\, \overline{B}.$$

The following algorithm computes an AT-model for a simplicial complex $K$ after and edge contraction in three steps. The goal of the first step is to obtain a chain contraction of $C(K)$ to the chain complex associated to a new simplicial complex (that we also denote by $K$) satisfying that $Lk_K\{\langle a \rangle\} \cap Lk_K\{\langle b \rangle\} = Lk_K\{\langle a, b \rangle\}$. In the second step, an AT-model $(K, h, f, g, \phi)$ for the new simplicial complex $K$ obtained at the first step is computed. At the final step, an AT-model for the simplicial complex $K$ after an edge contraction, by composing the chain contractions obtained at the previous steps is obtained.

**Algorithm 4.** *Computing AT-models after edge contractions.*

INPUT: An AT-model $AT := (K, h_K, f_K, g_K, \phi_K)$ for $K$, an edge $\langle a, b \rangle \in K$ and an ordered-by-increasing-dimension set of all the simplices of $Lk_K\{\langle a \rangle\} \cap Lk_K\{\langle b \rangle\} \setminus Lk_K\{\langle a, b \rangle\}$:
$\{\sigma_1 = \langle w_0^1, ..., w_{m_1}^1 \rangle, ..., \sigma_n = \langle w_0^n, ..., w_{m_n}^n \rangle\}$.

STEP 1:

For $i = 1$ to $i = n$:

    apply Algorithm 1 to the AT-model $AT$ and the simplex
    $\langle a, b, w_0^i ..., w_{m_i}^i \rangle$.
    Define $K := K \cup \{\langle a, b, w_0^i, ..., w_{m_i}^i \rangle\}$.

End for.

STEP 2:

For every simplex $\sigma \in K \setminus St\{\langle b \rangle\}$:

    Define $f(\sigma) := \sigma$, $g(\sigma) := \sigma$ and $\phi(\sigma) := 0$.

End for.

Define $f(\langle b \rangle) := \langle a \rangle$, $\phi(\langle b \rangle) := \langle a, b \rangle$, $f(\langle a, b \rangle) := 0$ and $\phi(\langle a, b \rangle) := 0$.

For every simplex $\langle v_0, \ldots, v_n \rangle \in Lk_K^a (b)$:

    $f(\langle b, v_0, \ldots, v_n \rangle) := \langle a, v_0, \ldots, v_n \rangle$ and $\phi(\langle b, v_0, \ldots, v_n \rangle) := 0$.

    If $n = 0$ then $g(\langle a, v_0 \rangle) := \langle b, v_0 \rangle + \langle a, b \rangle$.

    Else $g(\langle a, v_0, \ldots, v_n \rangle) := \langle b, v_0, \ldots, v_n \rangle + \sum_s (-1)^i \langle a, b, v_0, \ldots, \hat{v}_l, \ldots, v_n \rangle$

        where $S = \{i : 0 \le i \le n, \langle v_0, \ldots, \hat{v}_i, \ldots, v_n \rangle \in Lk_K \{\langle a, b \rangle\}\}$,

    End if.

End for.

For every simplex $\langle v_0, \ldots, v_n \rangle \in Lk_K \{\langle a, b \rangle\}$:

    $f(\langle a, b, v_0, \ldots, v_n \rangle) := 0$ and $\phi(\langle a, b, v_0, \ldots, v_n \rangle) := 0$,

    $f(\langle b, v_0, \ldots, v_n \rangle) := \langle a, v_0, \ldots, v_n \rangle$ and $\phi(\langle b, v_0, \ldots, v_n \rangle) := \langle a, b, v_0, \ldots, v_n \rangle$,

    $g(\langle a, v_0, \ldots, v_n \rangle) := \langle b, v_0, \ldots, v_n \rangle + \sum_{i=0}^{n} (-1)^i \langle a, b, v_0, \ldots, \hat{v}_i, \ldots, v_n \rangle$.

End for.

STEP 3:

Define $L := \{f(\sigma) : \sigma \in K\}$.

For every $\mu \in L$ and $\alpha \in h_K$:

    $f_L(\mu) := f_K g(\mu)$, $\phi_L(\mu) := f \phi_K g(\mu)$ and $g_L(\alpha) := f g_K(\alpha)$.

End for.

OUTPUT: The set $(L, h_K, f_L, g_L, \phi_L)$.

Here, we define by $Lk_K^a (b)$ the set of all the simplices in $Lk_K \{\langle b \rangle\}$ without having $a$ as a vertex. Moreover, without loss of generality, we suppose that $a < b < v$ for any vertex $v$ of $K$.

**Theorem 4.** *Given a simplicial complex $K$ and an edge $\langle a, b \rangle \in K$, the set $(L, h_K, f_L, g_L, \phi_L)$ defines an AT-model for the simplicial complex $L$ obtained from $K$ after contracting the edge $\langle a, b \rangle$.*

**Proof.** Let $(K, h_K, f_K, g_K, \phi_K)$ be an AT-model for a simplicial complex $K$ and let $\langle a, b \rangle$ be an edge in $K$. We only prove that the set $(K, h, f, g, \phi)$ (obtained at the second step) is an AT-model for the new simplicial complex $K$ (obtained at the first step). To this aim, we check that $fg = id$ and $id - gf = \phi \partial + \partial \phi$. The rest of the properties are left to the reader.

- If $\sigma \in K \setminus St\{\langle b \rangle\}$ then, by definition, $fg(\sigma) = \sigma$ and $(\phi\partial + \partial\phi)(\sigma) = 0 = \sigma - gf(\sigma)$.
- $(\phi\partial + \partial\phi)(\langle b \rangle) = \partial(\langle a, b \rangle) = \langle b \rangle - \langle a \rangle = \langle b \rangle - gf(\langle b \rangle)$.
  $(\phi\partial + \partial\phi)(\langle a, b \rangle) = \phi(\langle b \rangle - \langle a \rangle) = \langle a, b \rangle - gf(\langle a, b \rangle)$.
- If $\langle v_0, \ldots, v_n \rangle \in Lk_K^a(b)$ then
  - If $n = 0$, $fg(\langle a, v_0 \rangle) = f(\langle b, v_0 \rangle + \langle a, b \rangle) = \langle a, v_0 \rangle$.
    $(\phi\partial + \partial\phi)(\langle b, v_0 \rangle) = \phi(\langle v_0 \rangle - \langle b \rangle) = -\langle a, b \rangle = \langle b, v_0 \rangle - gf(\langle b, v_0 \rangle)$.
  - Otherwise,
    $$fg(\langle a, v_0, \ldots, v_n \rangle) = f(\langle b, v_0, \ldots, v_n \rangle + \sum_s (-1)^i \langle a, b, v_0, \ldots, \hat{v}_i, \ldots, v_n \rangle)$$
    $$= \langle a, v_0, \ldots, v_n \rangle.$$
    $$(\phi\partial + \partial\phi)(\langle b, v_0, \ldots, v_n \rangle) = \phi(\langle v_0, \ldots, v_n \rangle + \sum_s (-1)^{i+1} \langle b, v_0, \ldots, \hat{v}_i, \ldots, v_n \rangle)$$
    $$= \sum_s (-1)^{i+1} \langle a, b, v_0, \ldots, \hat{v}_i, \ldots, v_n \rangle = (id - gf)(\langle b, v_0, \ldots, v_n \rangle).$$
- If $\langle v_0, \ldots, v_n \rangle \in Lk_K\{\langle a, b \rangle\}$ then
  $fg(\langle a, v_0, \ldots, v_n \rangle) = f(\langle b, v_0, \ldots, v_n \rangle) = \langle a, v_0, \ldots, v_n \rangle$.
  $(\phi\partial + \partial\phi)(\langle a, b, v_0, \ldots, v_n \rangle) = \phi(\langle b, v_0, \ldots, v_n \rangle) = \langle a, b, v_0, \ldots, v_n \rangle = (id - gf)(\langle a, b, v_0, \ldots, v_n \rangle)$.

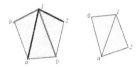

**Fig. 6.** Simplicial complexes $K$ and $L$

*Example 3.* Now we give a simple example of the computation of an AT-model after an edge contraction. Consider the simplicial complexes $K$ and $L$ (obtained from $K$ after contracting the edge $\langle a, b \rangle$) shown in Figure 6. The simplices of $Lk_K\{\langle a \rangle\}$ are in red, and the ones of $Lk_K\{\langle b \rangle\}$ are in blue. In this case, $Lk_K\{\langle a, b \rangle\} = Lk_K\{\langle a \rangle\} \cap Lk_K\{\langle b \rangle\}$. In the following table, the non-trivial results of the contraction $(f, g, \phi)$ from $C(K)$ to $C(L)$ (obtained using Algorithm 4) are given:

| $K$ | $L$ | $f$ | $g$ | $\phi$ |
|---|---|---|---|---|
| $\langle b \rangle$ | | $\langle a \rangle$ | | $\langle a, b \rangle$ |
| $\langle b, 1 \rangle$ | | $\langle a, 1 \rangle$ | | $\langle a, b, 1 \rangle$ |
| $\langle b, 2 \rangle$ | | $\langle a, 2 \rangle$ | | $0$ |
| $\langle a, b \rangle$ | | $0$ | | $0$ |
| $\langle a, b, 1 \rangle$ | | $0$ | | $0$ |
| $\langle b, 1, 2 \rangle$ | | $\langle a, 1, 2 \rangle$ | | $0$ |
| | $\langle a, 1 \rangle$ | | $\langle b, 1 \rangle + \langle a, b \rangle$ | |
| | $\langle a, 2 \rangle$ | | $\langle b, 2 \rangle + \langle a, b \rangle$ | |
| | $\langle a, 1, 2 \rangle$ | | $\langle b, 1, 2 \rangle + \langle a, b, 1 \rangle$ | |

# 4    Conclusions

Homological Perturbation Theory deals with algorithms for manipulating explicit chain homotopy equivalences between differential graded modules under suitable perturbations in the differentials of the modules. Here, we start a study for extending this theory to "module perturbations" produced in the underlying module structures. Working with a field as the ground ring, we reduce our perturbation homological analysis to particular chain contractions: AT-models for finite simplicial complexes. Taking as input an AT-model for a simplicial complex $K$ and perturbing it with changes in the graded module structure of $C(K)$ (addition or deletion of simplices or edge contractions), we have designed algorithms for restructuring the chain contraction information including these changes. More interesting simplicial perturbations such as "parallel" addition or elimination of simplices will be studied in a near future. Moreover, to extend these positive algorithmic results to general differential graded modules seems to be possible establishing an ordering for the generators of the underlying module structure.

# References

1. Delfinado, C., Edelsbrunner, H.: An incremental algorithm for Betti numbers of simplicial complexes on the 3–sphere. Comput. Aided Geom. Design **12** (1995) 771–784
2. González–Díaz, R., Real, P.: Towards digital cohomology. Lecture Notes in Computer Science **2886** (2003) 92–101
3. González–Díaz, R., Real, P.: On the cohomology of $3D$ digital images. Discrete Applied Math **147** (2005) 245–263
4. Gonzalez-Diaz, R., Medrano, B., Real, P., Sánchez-Peláez, J.: Algebraic topological analysis of time-sequence of digital images. Lecture Notes in Computer Science **139** (2005) 208–219
5. Gugenheim, V.K.A.M. , Lambe, L., Stasheff, J.: Perturbation theory in differential homological algebra, II. Illinois J. Math. **35** (3) (1991) 357–373
6. Huebschmann, J., Kadeishvili, T.: Small models for chain algebras. Math. Z. **207** (1991) 245–280
7. MacLane, S.: Homology. Classic in Math., Springer–Verlag, Berlin, 1995
8. Munkres, J.R.: Elements of Algebraic Topology. Addison–Wesley Co. 1984
9. Real, P.: Homological perturbation theory and associativity. Homology, Homotopy and its Applications **2**, No. 5 (2000) 51–88
10. Sergeraert, F.: The computability problem in algebraic topology. Adv. Math. **104**, No. 1 (1994) 1–29

# Numerical Study of Stability Domains of Hamiltonian Equation Solutions

E.A. Grebenicov[1], D. Kozak-Skoworodkin[2], and D.M. Diarova[3]

[1] Computing Center of RAS, Moscow
greben@ccas.ru
[2] University of Podlasie, Poland
kdorota@ap.siedlce.pl
[3] Institute of Oil and Gas, Atyrau, Kazakhstan
ddiarova@mail.ru

The computer algebra methods are effective means for the search of approximate and exact solutions of differential equations of theoretical physics, celestial mechanics, astrodynamics, and other natural sciences. Before appearance of Programming Systems such as *Mathematica, Maple* etc., we knew for classical Newtonian three-body problem only Euler exact collinear and Lagrange triangular solutions, for many-body problem – the rotating regular tetragon solution found by A. Dziobek [1] and the general homographic solution theory developed by A. Winter [2] in the 30es of the 20th century. An amount of similar research [3, 4, 5, 6, 7, 8, 9, 10] has grown recently due to the fact that the existence of central configurations of the many-body problem is eventually reduced to the solution of the systems of nonlinear algebraic-irrational equations, which can be solved only by the computer algebra methods, thanks to exceptional properties of them.

We demonstrated that each of the exact particular solution of the Newtonian $n-$body problem differential equations generates a new dynamic model – the $(n+1)$ restricted problem [11] being a generalization of the noted Poincaré–Jacobi model, the so-called restricted three-body problem [12]. The latter is brought about by the well-known two-body problem.

The successful application of the computer algebra to the study of the general many-body problem solution stability, particularly, of the stationary solutions of restricted problems in Lyapunov sense, has made possible the essential results of qualitative celestial mechanics and astrodynamics, and to demonstrate the existence of non-degenerate domain of the equilibrium positions. At the same time, the dimensions of gravity domain of the stable equilibrium positions have never been studied previously.

The aim of our research is the numerical experiment for determination of the gravity domain dimensions of the stationary solutions, stable in Lyapunov sense, of the restricted many-body problems.

We have studied two variants of the restricted 12-body problem [13]. First, we examined zero mass motion in the gravity field of many bodies, one of them with a mass of $m_0 = 1$, and others - $m \neq 0$, forming a regular decagon, which rotates around its geometric center, with a mass of $m_0 = 1$, at the angular velocity expressly established by dynamic and geometric parameters of the system. We

V.G. Ganzha, E.W. Mayr, and E.V. Vorozhtsov (Eds.): CASC 2006, LNCS 4194, pp. 178–191, 2006.

examined the motion of inactively gravitating mass in the gravity field the above dynamic system generates.

The French mathematicians D.Bang and B.Elmabsout established [14] the existence of the exact solutions to the differential equations of the Newtonian problem of *incompletelysymmetrical* multiple bodies in Euclidian space, when mutually attracted bodies form some regular polygons, but several ones with well-defined positions in the Euclidean configuration space. A model of the restricted 12-body problem, based on the above results, where gravitating bodies form two concentric pentagons, is proposed for study.

It is well known that the problem of stability in the Lyapunov sense of particular solutions of differential equations of the Newtonian restricted many-body models may be solved only with the Kolmogorov – Arnold – Moser theory (the KAM-theory, in abbreviated form [15, 16]) with preliminary normalization of the Hamiltonian equations [17]. Similar results for some specific values of $n$ are studied in [18, 19, 20].

We demonstrated in our articles that the stable and unstable positions of equilibrium of the restricted 12-body problem "alternate" in configuration space, therefore, it is necessary to measure the attraction domain of every stable point. This problem may be solved if we will need to create trajectories with initial points in a small neighborhood of stable position of equilibrium and examine the trajectories behavior within sufficiently long period of time, which may also be implemented by means of computational methods. If we analyzed the dimensions of attraction domain analytically, we would get even less precise results. For example, the application of numerical integration by Runge–Kutta method of the differential equations of the restricted seven-body problem has made it possible to measure a radius of these gravitational fields which is almost eight times as less than the distance to the first unstable point [18], which is not an insignificant value on the given scale. Modelling of gravitational field dimensions for other values of $n$ (where $n$ is a number of gravitational masses) shows that with a growth of values of $n$ dimensions of the fields steadily reduce.

Now we examine the incompletely symmetrical restricted problem of twelve bodies.

Let us have eleven mutually attracted $P_0, P_1, P_2, ..., P_{10}$ bodies with masses of $M_0, M_1 = M_2 = ... = M_5 = m_1, M_6 = M_7 = ... = M_{10} = m_2$. Moreover, the $P_1, P_2, ..., P_{10}$ bodies form two regular pentagons (let us assume $P_0P_1 = 1, P_0P_6 = \alpha$) with a common center $P_0$ (Fig. 1).

The masses in vertices of each of the pentagons are mutually equal, the first pentagon is oriented against the other one by an angle of $\pi/5$. According to the Bang–Elmabsout theory both pentagons revolve around the $P_0$ center at the following angular velocity [20]

$$\omega^2 = M_0 + \sqrt{1 + \tfrac{2}{\sqrt{5}}} m_1 +$$

$$+ m_2 \left( \frac{1}{(1+\alpha)^2} + \frac{\sqrt{2}(4+(-1+\sqrt{5})\alpha)}{(2-(1-\sqrt{5})\alpha+2\alpha^2)^{3/2}} + \frac{\sqrt{2}(4+(-1-\sqrt{5})\alpha)}{(2-(1+\sqrt{5})\alpha+2\alpha^2)^{3/2}} \right). \tag{1}$$

The sufficient condition for the existence of exact homographic solutions of this problem is as follows [20]

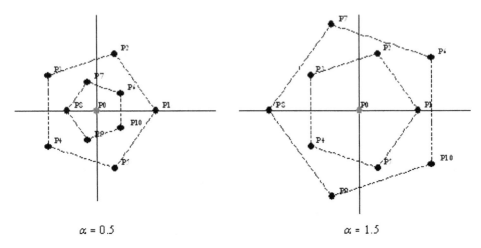

$\alpha = 0.5$          $\alpha = 1.5$

**Fig. 1.**

$$m_2 = \left(-\sqrt{1+\tfrac{2}{\sqrt{5}}}\,m_1 - M_0 + \tfrac{M_0}{\alpha^3} + \right.$$
$$+\tfrac{m_1}{\alpha}\left(\tfrac{1}{(1+\alpha)^2} + \tfrac{\sqrt{2}(-1+\sqrt{5}+4\alpha)}{(2-(1-\sqrt{5})\alpha+2\alpha^2)^{3/2}} + \tfrac{\sqrt{2}(-1-\sqrt{5}+4\alpha)}{(2-(1+\sqrt{5})\alpha+2\alpha^2)^{3/2}}\right)\right)\cdot$$
$$\left(-\tfrac{\sqrt{1+\tfrac{2}{\sqrt{5}}}}{\alpha^3} + \tfrac{1}{(1+\alpha)^2} + \tfrac{\sqrt{2}(4+(-1+\sqrt{5})\alpha)}{(2-(1-\sqrt{5})\alpha+2\alpha^2)^{3/2}} + \tfrac{\sqrt{2}(4+(-1-\sqrt{5})\alpha)}{(2-(1+\sqrt{5})\alpha+2\alpha^2)^{3/2}}\right)^{-1}, \qquad (2)$$

which shows the dependence between masses $M_0, m_1, m_2$ and distance $\alpha$ for the Newtonian problem of eleven bodies. This condition for the actual existence of an exact solution to the 11-body problem, was geometrically shown by the Bang–Elmabsout homographic solution [14].

As we have mentioned above, this model generates the other one – the incompletely symmetrical restricted model of twelve bodies whose differential equations in the steadily revolving Cartesian coordinates $P_0 xy$ [18] are:

$$\begin{cases} \dfrac{d^2x}{dt^2} = 2\omega\dfrac{dy}{dt} + \omega^2 x + \dfrac{\partial U}{\partial x}, \\[2mm] \dfrac{d^2y}{dt^2} = -2\omega\dfrac{dx}{dt} + \omega^2 y + \dfrac{\partial U}{\partial y}, \end{cases} \qquad (3)$$

where

$$\begin{cases} U(x,y) = f\left(\dfrac{M_0}{\Delta_0} + m_1\sum\limits_{i=1}^{5}\dfrac{1}{\Delta_i} + m_2\sum\limits_{j=1}^{5}\dfrac{1}{\Delta_j}\right), \\[2mm] \Delta_0 = \sqrt{x^2+y^2}, \\[1mm] \Delta_i = \sqrt{(x-x_i)^2+(y-y_i)^2}, \\[1mm] x_i = \cos\dfrac{2\pi(i-1)}{5}, \quad y_i = \sin\dfrac{2\pi(i-1)}{5}, \quad i=1,...5, \\[2mm] \Delta_j = \sqrt{(x-x_j)^2+(y-y_j)^2}, \\[1mm] x_j = \alpha\cos\left(\dfrac{2\pi(j-1)}{5}+\dfrac{\pi}{5}\right), \quad y_j = \alpha\sin\left(\dfrac{2\pi(j-1)}{5}+\dfrac{\pi}{5}\right), \quad j=1,...5, \end{cases}$$

and $\omega^2$ is given by equality (1).

We have proved that a number of equilibrium positions in this model depend on values of $m_1$ and $\alpha$ and change between 20 at 50. For example, if $M_0 = 1, \alpha = 0.99958, m_1 = 0.0001$, we have 30 positions of equilibrium (Fig. 2) [20].

It has been proved that for any $m_1 > 0$ and $\alpha > 0$ the stationary solutions $N_1,..., N_{20}$ (Fig. 2) of the restricted 12 incompletely symmetrical body problem are unstable in Lyapunov sense, and there are such values of $m_1$ and $\alpha$ that make the stationary solutions $S_1,..., S_{10}$ (Fig. 2) stable in the Lyapunov sense [13].

It is known also, if the linear system frequency values (3) are bounded by the tertiary resonant correlation $\sigma_1 = 2\sigma_2$, the $S_1, ..., S_{10}$ equilibrium positions (Fig. 2) of the restricted 12 incompletely symmetrical body problem are unstable in the Lyapunov sense, and if the linear system frequency values (3) are bounded by the quartic resonant correlation $\sigma_1 = 3\sigma_2$, the $S_1, ..., S_{10}$ equilibrium positions (Fig. 2) of the restricted 12 incompletely symmetrical body problem are stable in the Lyapunov sense [13].

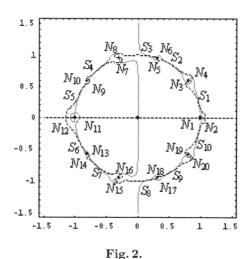

Fig. 2.

Applying Computing System "Mathematica" [10] and its algorithms and techniques, we solve equation (3).

Let $M_0 = 1$.

We study point $S_1$ (Fig. 2) stable in the Lyapunov sense, with the following coordinates:

$$x_1 = 0.9178291468070109..., \quad y_1 = 0.39834574906341325..., \qquad (4)$$

measured given that $m_1 = 0.001$ and $\alpha = 1.000284$ [9].

By means of "Mathematica", we can solve the system of differential equations (3) with certain initial conditions (4), for example, for $0 < t < 10000$, in the form of interpolation functions to ensure that during insignificant changes the integration error within such a range of values does not exceed $10^{-11}$.

Let $x(0) = x_1$,   $y(0) = y_1$,   $x'(0) = 0$,   $y'(0) = 0$. The solution of equations (3) can then be found by means of the following instruction ($f$ and $g$ are the right-hand sides of equations (3)):

**eq1=NDSolve[{x"[t]-2$\omega$y'[t]==f,x[0]==x₁,x'[0]==0,**
**y"[t]+2$\omega$x'[t]==g,y[0]==y₁,y'[0]==0},{x,y},{t,0,10000}]**

**{{x→InterpolatingFunction[{{0.,10000.}},<>],**
**y→InterpolatingFunction[{{0.,10000.}},<>]}}**

Creation of the graphs of the resultant functions is governed by the below procedure [10]:

**ParametricPlot[Evaluate[{x[t],y[t]}/.eq1],{t,0,t₁},**
**AxesLabel→{"x[t]","y[t]"},AxesOrigin→{x1,y1}]**

with solutions being the interpolation functions, which may be graphically demonstrated for different integration intervals by replacing $t_1$ with specific values (coordinate axes go through the $S_1$ point), see Fig. 3.

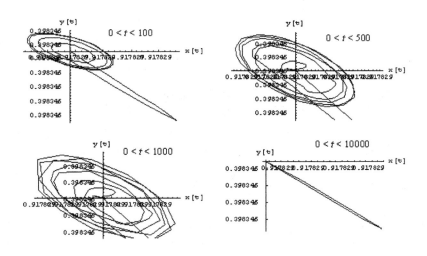

**Fig. 3.**

Considering a scale of the coordinate axes we can clearly see in Fig. 3 that the trajectory does not move far away from the initial point during a sufficiently long period of time. In Fig. 3 the scale is less than $10^{-6}$.

Let $\Delta r(t)$ be the local distance of point on the trajectory from the stationary point $S_1$ for $t$. We may then demonstrate the behavior of the function $\Delta r(t)$ :

a) given $0 < t < 80$

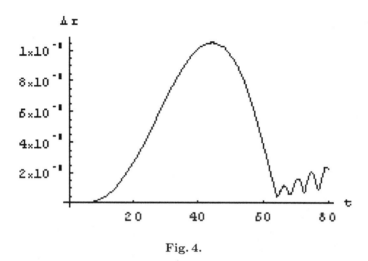

Fig. 4.

b) given $80 < t < 10000$

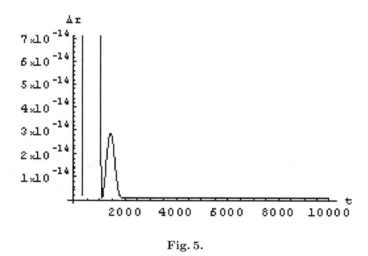

Fig. 5.

Now change the initial conditions by slightly perturbing the initial coordinates according to the below formulae

$$x(0) = x_1 + a\text{Cos}[\varphi], \quad y(0) = y_1 + a\text{Sin}[\varphi], \quad x'(0) = 0, \quad y'(0) = 0. \quad (5)$$

For example, let $\varphi = 0$, $a= 0.000226$. We then find a solution of equations (3) with initial conditions following the below procedure:

**eq2=NDSolve[{x"[t]-2ωy'[t]==f,x[0]==x₁+aCos[φ],x'[0]==0,**
**y"[t]+2ωx'[t]==g,y[0]==y₁+aSin[φ],y'[0]==0},{x,y},{t,0,1000}]**

{{x→InterpolatingFunction[{{0.,1000.}},<>],
y→InterpolatingFunction[{{0.,1000.}},<>]}}

The result of interpolation functions may then be demonstrated graphically for different integration time intervals, see Figs. 4 and 5. The sizes of figures are: $\Delta x < 0.03$, $\Delta y < 0.06$.

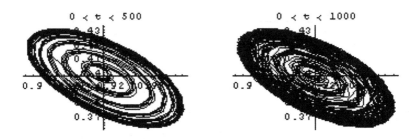

Fig. 6.

Using Fig. 6, we can also demonstrate the behavior of the function $\Delta r(t)$:

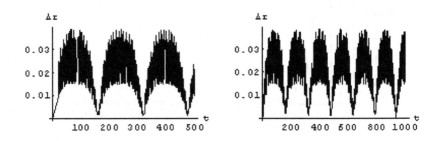

Fig. 7.

Under such initial conditions, the trajectory does not move significantly away from the position of equilibrium $S_1$ being considered. Based on the above computational experiments (Figs. 3 – 7) we may state with certain degree of confidence that the position of equilibrium $S_1$ is stable because the integration interval is quite considerable.

Now, let $\varphi = 0$, $a= 0.000227$. The solution of equations (3) can then be demonstrated as follows

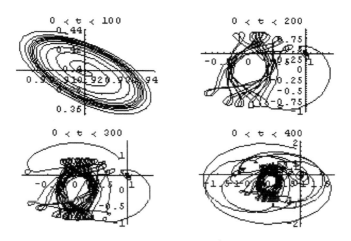

**Fig. 8.**

We are now going to display both diagrams in one coordinate system, combining the trajectory shown in Fig. 6 b) and Fig. 2 on Fig. 9 a), and similarly, combine the last image from Fig. 8 with Fig. 2 on Fig. 9 b).

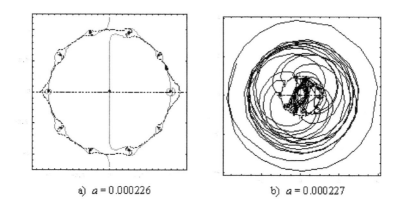

a) $a = 0.000226$          b) $a = 0.000227$

**Fig. 9.**

It is clearly seen in Fig. 9 a) that the trajectory "rotates" around the position of equilibrium $S_1$ and does not move away significantly, whereas in Fig. 9 b) the trajectory moves off.

Based on the computational experiments (Figs. 6–9) we may assume that applying $\varphi = 0$ and modifying initial conditions according to formulae (5) up to $a$, being in the interval of $0.000226 < a < 0.000227$ (let us call it $a_{max}$), we have the trajectory rotating around the position of equilibrium $S_1$ and not moving far away.

In Table 1 below, the intervals of $a_{max}$ for other values of $\varphi$ are specified.

Table 1.

| $\varphi$ | $a_{max}$ |
|---|---|
| $0, \pi$ | $0.00022 < a_{max} < 0.00023$ |
| $\pi/10, 11\pi/10$ | $0.00020 < a_{max} < 0.00021$ |
| $2\pi/10, 12\pi/10$ | $0.00021 < a_{max} < 0.00022$ |
| $3\pi/10, 13\pi/10$ | $0.00024 < a_{max} < 0.00025$ |
| $4\pi/10, 14\pi/10$ | $0.00031 < a_{max} < 0.00032$ |
| $5\pi/10, 15\pi/10$ | $0.00051 < a_{max} < 0.00052$ |
| $6\pi/10, 16\pi/10$ | $0.0017 < a_{max} < 0.0018$ |
| $7\pi/10, 17\pi/10$ | $0.00091 < a_{max} < 0.00092$ |
| $8\pi/10, 18\pi/10$ | $0.00040 < a_{max} < 0.00041$ |
| $9\pi/10, 19\pi/10$ | $0.00027 < a_{max} < 0.00028$ |
| $13\pi/20, 33\pi/20$ | $0.0023 < a_{max} < 0.0024$ |

According to the calculations, the real field of a gravitation of point $S_1$ is similar to an ellipse with a smaller axis about 1795 times smaller than the distance from the $S_1$ point for the first unstable point equal to 0.398608820196195. This distance is equal to 0.398608820196195, and this field is shown in Fig. 10.

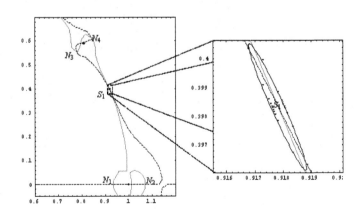

Fig. 10.

Let us now assume that the linear system frequency values are bounded by the quartic resonant correlation $\sigma_1 = 3\sigma_2$. In this case, the $S_1$ point coordinates are as follows

$$x_1 = 0.9737314994387726..., \quad y_1 = 0.22815846219324198..., \tag{6}$$

and they have been calculated with $m_1 = 0.0001$ and $\alpha = 0.999905351291365....$ [13].

We can now solve by means of "Mathematica" the system of differential equations under initial conditions in interpolation function forms:

**eq3=NDSolve[{x"[t]-2ωy'[t]==f,x[0]==x₁,x'[0]==0,**
**y"[t]+2ωx'[t]==g,y[0]==y₁,y'[0]==0},{x,y},{t,0,1000}]**

{{x→InterpolatingFunction[{{0.,1000.}},<>],
y→InterpolatingFunction[{{0.,1000.}},<>]}}

We have for different values of $t$ the following results:

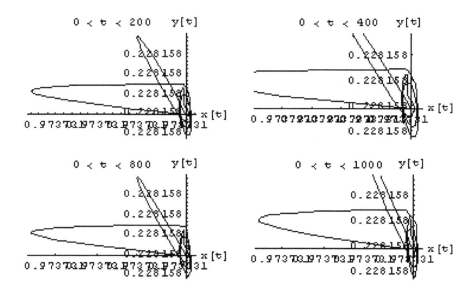

Fig. 11.

We have that $|\Delta x| < 10^{-6}, |\Delta y| < 10^{-6}$ for $0 < t < 1000$.

Let us now modify initial conditions by "perturbing slightly" initial coordinates according to formulae (5). Let, for instance, $\varphi = 0$, $a = 0.0037$. We now have the graphs shown in Fig. 12 for two intervals of time.

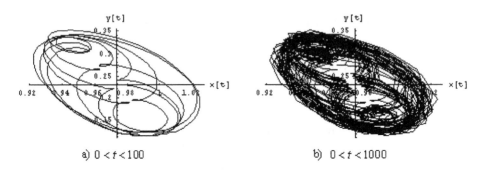

a) $0 < t < 100$    b) $0 < t < 1000$

**Fig. 12.**

For these initial conditions the trajectory does not move away significantly from the position of equilibrium $S_1$. These computational experiments presented in figures 11 – 12 show that, for case of the existence of frequency quartic resonance, the position of equilibrium $S_1$ is quite stable as trajectories remain in the limited part of space for large values of time.

For other set of values $\varphi = 0$, $a = 0.0038$, we have the following:

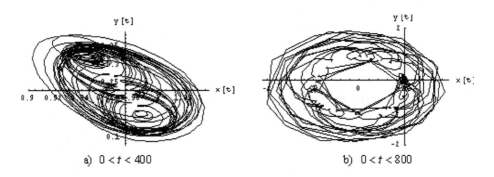

a) $0 < t < 400$    b) $0 < t < 800$

**Fig. 13.**

Under such initial conditions, the trajectory moves away from the position of equilibrium $S_1$ (if one takes into account linear scales on coordinate axes).

We are now going to display both diagrams in the same coordinate system. Let us combine the trajectories shown in Fig. 12 b) and Fig. 2 (the result is presented in Fig. 14 a), and also the trajectories shown in Fig. 13 b) and Fig. 2 (the result is presented in Fig. 14 b).

These computational experiments definitely point to that when the linear system frequency values are bounded by the quartic resonant correlation $\sigma_1 = 3\sigma_2$, then, at $\varphi = 0$ and while modifying initial conditions according to formulae (5) up to $a_{max}$ in the range of $0.0037 < a_{max} < 0.0038$, we have the trajectory

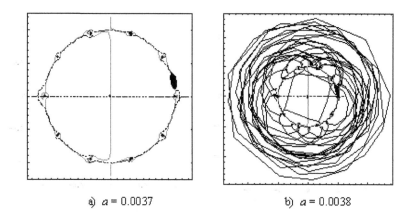

a) $a = 0.0037$                         b) $a = 0.0038$

Fig. 14.

rotating and not moving away far from the position of equilibrium $S_1$, at least on a large enough interval of time.

Table 2 specifies the ranges of $a_{max}$ for other $\varphi$ values considering there is the quartic resonance $\sigma_1 = 3\sigma_2$.

Table 2.

| $\varphi$ | $a_{max}$ |
|---|---|
| $0, \pi$ | $0.0037 < a_{max} < 0.0038$ |
| $\pi/10, 11\pi/10$ | $0.0031 < a_{max} < 0.0032$ |
| $2\pi/10, 12\pi/10$ | $0.0039 < a_{max} < 0.0038$ |
| $3\pi/10, 13\pi/10$ | $0.0046 < a_{max} < 0.0047$ |
| $4\pi/10, 14\pi/10$ | $0.0070 < a_{max} < 0.0071$ |
| $5\pi/10, 15\pi/10$ | $0.015 < a_{max} < 0.016$ |
| $6\pi/10, 16\pi/10$ | $0.058 < a_{max} < 0.059$ |
| $7\pi/10, 17\pi/10$ | $0.0095 < a_{max} < 0.0096$ |
| $8\pi/10, 18\pi/10$ | $0.0049 < a_{max} < 0.0050$ |
| $9\pi/10, 19\pi/10$ | $0.0037 < a_{max} < 0.0038$ |

According to the computations we may state that the actual gravitational field of $S_1$ during the existence of the quartic resonance represents an ellipse too (Fig. 15), although the size of the field is bigger than when there is no resonance.

To examine the stable point gravity fields much more considerable resources must be exploited. We may only calculate by means of the "Mathematica" software the decreasing (smaller and smaller) ranges of the $a_{max}$ value, but the geometric features of the gravitational field will remain the same, similar to Fig. 10 and Fig. 15.

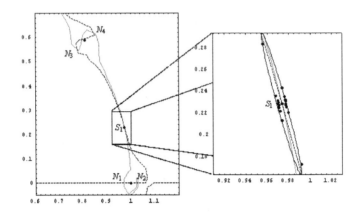

Fig. 15.

# References

1. Dziobek, O.: Die Mathematischen Theorien der Planeten, Bewegung (1888)
2. Wintner, A.: The analytical Foundations of Celestial Mechanics. Princeton Univ. Press, Princeton (1941)
3. Elmabsout, B.: Nouvelles configuration d'eqilibre relatif poure le problème des $N$ corps. C.R. Acad. Sci. Paris (1990)
4. Grebenicov, E.A.: Two new dynamical models in celestial mechanics. Rom. Astron. J. **8** (1) (1998) 13–19
5. Zemtsova, N.I.: Stability of the stationary solutions of the differential equations of restricted Newtonian problem with incomplete symmetry. Nonlinear Dynamics and Systems Theory **3** (1) (2003) 105–116
6. Grebenicov, E.A., Zemtsova, N., Ikhsanov, E.: Linear stability of stationary solutions of the ring-shaped Newton ten-body problem, in: Computer Algebra in Scientific Computing, CASC 2003, Techn. Univ. Munich, Munich (2003) 179–185
7. Bang, D., Elmabsout, B.: Configurations polygonales en equilibre relative. C.R. Acad. Sci. Paris, Serie Iib. **329** (2001) 243–248
8. Ikhsanov, E.V.: Stabilty of equilibrium state in restricted 10-body problem for resonance case of $4^{th}$ order. In: The Questions of Modeling and Analysis in the Problems of Making Decision (in Russian), V.A.Bereznev (Ed.). Computing Center RAS, Moscow (2004) 16–23
9. Palmore, A., Julian I.: Central configurations and relative equilibria in the $n$-body problem. Celestial Mech. **21** (1980) 21–24
10. Wolfram, S.: The Mathematica – Book. Cambridge University Press, Cambridge, 1996
11. Grebenicov, E.A.: The homografic dynamics for Newtonian gravitation (in Russian). Vestnik Brest Univ. **3** (24) (2005) 11–22
12. Szebehely, V.: Theory of Orbits,. Academic Press, New York and London, 1967
13. Kozak-Skoworodkin, D.: The System "Mathematica" for Qualitative Investigations of Many Body Newtonian Problem (in Russian). Published by RUFP, Moscow, 2005

14. Grebenicov, E.A., Prokopenya, A.N.: On the existence of a new class of the exact solutions in the planar Newtonian many-body problem (in Russian). In: The Questions of Modeling and Analysis in the Problems of Making Decision, V.A. Bereznev (Ed.). Computing Center RAS, Moscow (2004) 39–57

15. Arnold, V.I.: About stability of equilibrium positions of Hamiltonian systems in general eliptic case (in Russian). DAN USSR **137** (2) (1961) 255–257

16. Moser, J.K.: Lectures on Hamiltonian Systems. Courant Institute of Mathematical Science, New York, 1968

17. Markeev A.P.: Libration Points in Celestial Mechanics and Cosmodynamics (in Russian). Nauka, Moscow, 1974

18. Ikhsanov, E.: The computer methods of Hamiltonian normalization for restricted problems of celestial mechanics (in Russian). Published by RUFP, Moscow, 2004

19. Grebenikov, E., Kozak-Skoworodkin, D., Jakubiak M.: Investigation of the stability problem for the critical cases of the Newton many-body problem. In: Computer Algebra in Scientific Computing, CASC 2005, Kalamata, Greece (2005)

20. Grebenicov, E.A., Kozak-Skoworodkin, D., Jakubiak, M.: Methodes of Computer Algebra in Many-Body Problem (in Russian). Published by RUFP, Moscow, 2002

# Numeric-Symbolic Computations in the Study of Central Configurations in the Planar Newtonian Four-Body Problem

Evgenii A. Grebenikov[1], Ersain V. Ikhsanov[2], and Alexander N. Prokopenya[3]

[1] Computing Center of RAS,
Vavilova str. 40,
119991 Moscow, Russia
greben@ccas.ru
[2] University of Atyrau,
Kazakhstan
unatatyrau@nursat.kz
[3] Brest State Technical University,
Moskowskaya str. 267,
224017 Brest, Belarus
prokopenya@brest.by

**Abstract.** The planar central configurations in the newtonian problem of four bodies are studied with the computer algebra system *Mathematica*. We have shown that in the case of two equal masses there can exist central configurations in the form of isosceles triangle with three bodies being in its vertices and the fourth body being situated in the axis of symmetry inside or outside the triangle. The number of possible configurations in such cases depends on the masses of the bodies and may be equal to ten, six or two. We have provided evidence numerically that there exist one-parametric family of central configurations in the form of antiparallelogram. We have shown also that central configuration may be deformed continuously by means of changing masses of the bodies and found two-parametric family of central configurations in the neighborhood of the square.

## 1   Introduction

The study of the central configurations in the Newtonian many-body problem is a very old problem in Celestial Mechanics. Many papers have been devoted to its investigation and many interesting results have been obtained (see, for example, [1]). One of the reasons why central configurations are important and interesting is that every such configuration generates an explicit homographic solution of the corresponding $n$-body problem [2]. For example, two bodies form only one central configuration, and general solution of the two-body problem is just a homographic one. Euler and Lagrange found five central configurations in the case of $n = 3$, and one can easily show that there do not exist any other central configurations of three bodies. But for $n \geq 4$, the problem of the existence

V.G. Ganzha, E.W. Mayr, and E.V. Vorozhtsov (Eds.): CASC 2006, LNCS 4194, pp. 192–204, 2006.

of central configurations is still open. Even for $n = 4$ it is not known how many central configurations do exist, and what shapes do they have.

Moulton showed [3] that there exist exactly $n!/2$ collinear central configurations for a given set of positive masses of $n$ bodies when all of them are contained in a straight line. But in non-collinear case similar general statements are not known. Wintner [2] and Smale [4] conjectured that in non-collinear case the number of central configurations is finite for a given set of positive masses of the bodies but this proposition has not been proved yet. And Smale included this problem in his list of the most important mathematical problems for XXI century [5].

Remind that the bodies form a planar central configuration if and only if there exists some positive constant $\sigma$ such that the following system of algebraic equations

$$\sigma x_j = G \sum_{k=0(k \neq j)}^{n-1} m_k \frac{x_j - x_k}{((x_j - x_k)^2 + (y_j - y_k)^2)^{3/2}} ,$$

$$\sigma y_j = G \sum_{k=0(k \neq j)}^{n-1} m_k \frac{y_j - y_k}{((x_j - x_k)^2 + (y_j - y_k)^2)^{3/2}} \quad (j = 0, 1, \ldots, n-1) \quad (1)$$

has a solution. Here $x_j, y_j$ are the cartesian coordinates of the $j$th body, having a mass $m_j$, with respect to the inertial barycentric frame of reference, and $G$ is a gravity constant. Solving system (1), we obtain the constant $\sigma$ and the corresponding positions $(x_j, y_j)$ of the bodies in the $Oxy$ plane when they form central configurations. It should be emphasized that here we do not analyze the dynamics of the system but only its central configurations which correspond to the equilibrium positions of the bodies in the rotating frame [2].

Obviously, algebraic equations (1) are essentially nonlinear, and we can not find all solutions of the system in symbolic form. Nevertheless, in some special cases the system may be solved analytically, and some general results may be obtained. We can also try to find its numerical solutions and investigate their properties combining numeric and symbolic computations. And a modern computer algebra system such as *Mathematica* [6], for example, is just a good tool for doing such calculations.

In the present paper we focus on studying system (1) in the case of four interacting bodies when it reduces to six independent equations with respect to five variables. Symbolic analysis of this system shows that there do not exist non-collinear central configurations of four massive bodies when three of them are situated in the same straight line. Central configuration in the form of rectangle cannot exist in general as well, there can exist only a square with four bodies of equal masses in its vertices. Analyzing system (1) numerically, we have shown that in the case of two equal masses there can exist central configurations in the form of isosceles triangle with three bodies being in its vertices and the fourth body being situated in the axis of symmetry inside or outside the triangle. The number of possible configurations in such cases depends on the masses of the bodies and may be equal to ten, six, or two. If masses of the bodies are equal in pairs there exist central configurations in the form of antiparallelogram and rhombus. In order to demonstrate that central configuration may be deformed continuously by means of changing masses of the bodies we constructed two-parametric family of central configurations in the neighborhood of the square.

## 2    General Analysis of Central Configurations

Geometrical configuration of $n$ bodies $P_0, P_1, \ldots, P_{n-1}$ in the $Oxy$ plane is determined, in general, with $2n$ coordinates $x_j, y_j$ $(j = 0, 1, \ldots, n-1)$ but in barycentric frame only $2(n-1)$ of them are independent because the center of mass of the system is in the origin and, hence, coordinates of the bodies must satisfy the following conditions

$$\sum_{j=0}^{n-1} m_j x_j = 0 \,, \sum_{j=0}^{n-1} m_j y_j = 0 \,. \tag{2}$$

In fact, any non-collinear central configuration is determined with $2(n-2)$ coordinates because it is invariant with respect to homothetic transformations and rotations in the $Oxy$ plane. Actually, without any restrictions in our further analysis, we can introduce relative coordinates $\tilde{x}_j, \tilde{y}_j$ with respect to the body $P_0$, for instance, which are given by

$$\tilde{x}_j = x_j - x_0, \ \tilde{y}_j = y_j - y_0 \ (j = 1, 2, \ldots, n-1) \,. \tag{3}$$

We can also choose such orientation of the axes in the $Oxy$ plane that the body $P_1$, for instance, would be situated at point $(r_1, 0)$, where $r_1$ is a distance between the bodies $P_0$ and $P_1$. Then we can rewrite system (1) in the form

$$\lambda(x_c^* - x_j^*) = -\frac{x_j^*}{((x_j^{*2} + y_j^{*2})^{3/2}} + \sum_{k=1(\neq j)}^{n-1} \frac{\mu_k(x_k^* - x_j^*)}{((x_k^* - x_j^*)^2 + (y_k^* - y_j^*)^2)^{3/2}} \,,$$

$$\lambda(y_c^* - y_j^*) = -\frac{y_j^*}{(x_j^{*2} + y_j^{*2})^{3/2}} + \sum_{k=1(\neq j)}^{n-1} \frac{\mu_k(y_k^* - y_j^*)}{((x_k^* - x_j^*)^2 + (y_k^* - y_j^*)^2)^{3/2}} \,, \tag{4}$$

$$\lambda x_c^* = \sum_{k=1(\neq j)}^{n-1} \frac{\mu_k x_k^*}{(x_k^{*2} + y_k^{*2})^{3/2}}, \ \lambda y_c^* = \sum_{k=1(\neq j)}^{n-1} \frac{\mu_k y_k^*}{(x_k^{*2} + y_k^{*2})^{3/2}} \,, \tag{5}$$

where

$$\lambda = \frac{\sigma r_1^3}{G m_0}, \ x_j^* = \frac{\tilde{x}_j}{r_1}, \ y_j^* = \frac{\tilde{y}_j}{r_1}, \ \mu_j = \frac{m_j}{m_0} \ (j = 1, 2, \ldots, n-1) \,, \tag{6}$$

and coordinates of the center of mass are given by

$$x_c^* = \sum_{k=1}^{n-1} \mu_k x_k^* / (1 + \mu_1 + \ldots + \mu_{n-1}), \ y_c^* = \sum_{k=1}^{n-1} \mu_k y_k^* / (1 + \mu_1 + \ldots + \mu_{n-1}). \tag{7}$$

Note that the system of equations (4)–(5) contains $2(n-1)$ independent equations with respect to only $2(n-2)$ dimensionless variables $x_j^*, y_j^*$ ($j = 2, 3, \ldots, n-1$) because coordinates of the body $P_1$ are now fixed: $x_1^* = 1$, $y_1^* = 0$. Solving the system, we can find positions of the bodies $(x_j^*, y_j^*)$ corresponding to their central configurations, and parameter $\lambda$ which must be a positive constant. Note also that the distance $r_1$ determines only the scale in the $Oxy$ plane and the value of constant $\sigma$ (see (6)) but does not influence the shape and number of central configuration.

It should be noticed that in the case of collinear central configuration when $y_j^* = 0$ ($j = 1, 2, \ldots, n-1$) the second equation in the system (4) is satisfied identically, while the first one determines the system of $(n-1)$ independent equations with respect to $(n-1)$ variables, namely, $x_2^*, x_3^*, \ldots, x_{n-1}^*$ and $\lambda$. Solving this system in the case of three and four interacting bodies numerically we have found that it determines $3!/2 = 3$ and $4!/2 = 12$ collinear central configurations, respectively, what corresponds to the results of Moulton [3].

In the case of non-collinear central configurations the number of independent equations $2(n-1)$ is greater than the number of variables $(2n-3)$, and in some cases this can complicate analysis of the system. Nevertheless, in the case of three interacting bodies, for example, one can solve this system analytically and prove that there exist only two non-collinear central configurations, having a form of equilateral triangle, that was shown by Lagrange yet. Actually, for $n = 3$ the central configuration is determined by two coordinates $x_2^*, y_2^*$ and constant $\lambda$. Then coordinates of the center of mass (7) are given by

$$x_c^* = \frac{\mu_1 + x_2^* \mu_2}{1 + \mu_1 + \mu_2} \ , \quad y_c^* = \frac{y_2^* \mu_2}{1 + \mu_1 + \mu_2} \ , \tag{8}$$

and four independent equations (4) may be written as

$$\lambda \frac{(x_2^* - 1)\mu_2 - 1}{1 + \mu_1 + \mu_2} = -1 + \frac{\mu_2(x_2^* - 1)}{((x_2^* - 1)^2 + y_2^{*2})^{3/2}} \ ,$$

$$\lambda \frac{\mu_2 y_2^*}{1 + \mu_1 + \mu_2} = \frac{\mu_2 y_2^*}{((x_2^* - 1)^2 + y_2^{*2})^{3/2}} \ ,$$

$$\lambda \frac{\mu_1 - x_2^*(1 + \mu_1)}{1 + \mu_1 + \mu_2} = -\frac{x_2^*}{(x_2^{*2} + y_2^{*2})^{3/2}} - \frac{\mu_1(x_2^* - 1)}{((x_2^* - 1)^2 + y_2^{*2})^{3/2}} \ ,$$

$$\lambda \frac{y_2^*(1 + \mu_1)}{1 + \mu_1 + \mu_2} = \frac{y_2^*}{(x_2^{*2} + y_2^{*2})^{3/2}} + \frac{\mu_1 y_2^*}{((x_2^* - 1)^2 + y_2^{*2})^{3/2}} \ . \tag{9}$$

The first two equations of system (9) give

$$(x_2^* - 1)^2 + y_2^{*2} = 1 \ , \quad \lambda = 1 + \mu_1 + \mu_2 \ . \tag{10}$$

On substituting (10) into the second two equations of the system (9), we obtain

$$x_2^{*2} + y_2^{*2} = 1 \ . \tag{11}$$

Then from (10), (11) we easily find

$$x_2^* = \frac{1}{2} \ , \quad y_2^* = \pm \frac{\sqrt{3}}{2} \ . \tag{12}$$

Obviously, the solutions (12) determine two central configurations of three bodies which are just the equilateral triangles, and the system (9) has no other solutions. The constant $\lambda$ is given by (10) and is a positive number.

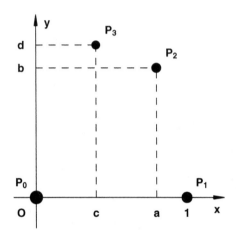

**Fig. 1.** General configuration of four bodies

In order to determine non-collinear central configuration of four bodies in the $Oxy$ plane we need four coordinates (see Fig. 1) which must satisfy the system of eight equations (4)–(5). We consider only six independent equations of this system, for example, equations (5) and equations (4) for $j = 1, 2$ and rewrite them in the form

$$\mu_1 + \frac{a\mu_2}{(a^2 + b^2)^{3/2}} + \frac{c\mu_3}{(c^2 + d^2)^{3/2}} = \lambda \frac{\mu_1 + a\mu_2 + c\mu_3}{1 + \mu_1 + \mu_2 + \mu_3} \quad , \tag{13}$$

$$\frac{b\mu_2}{(a^2 + b^2)^{3/2}} + \frac{d\mu_3}{(c^2 + d^2)^{3/2}} = \lambda \frac{b\mu_2 + d\mu_3}{1 + \mu_1 + \mu_2 + \mu_3} \quad , \tag{14}$$

$$1 + \frac{(1 - a)\mu_2}{((1 - a)^2 + b^2)^{3/2}} + \frac{(1 - c)\mu_3}{((1 - c)^2 + d^2)^{3/2}} =$$
$$= \lambda \frac{1 + (1 - a)\mu_2 + (1 - c)\mu_3}{1 + \mu_1 + \mu_2 + \mu_3} \quad , \tag{15}$$

$$\frac{b\mu_2}{((1 - a)^2 + b^2)^{3/2}} + \frac{d\mu_3}{((1 - c)^2 + d^2)^{3/2}} = \lambda \frac{b\mu_2 + d\mu_3}{1 + \mu_1 + \mu_2 + \mu_3} \quad , \tag{16}$$

$$\frac{a}{(a^2 + b^2)^{3/2}} - \frac{(1 - a)\mu_1}{((1 - a)^2 + b^2)^{3/2}} + \frac{(a - c)\mu_3}{((a - c)^2 + (b - d)^2)^{3/2}} =$$
$$= \lambda \frac{a - (1 - a)\mu_1 + (a - c)\mu_3}{1 + \mu_1 + \mu_2 + \mu_3} \quad , \tag{17}$$

$$\frac{b}{(a^2 + b^2)^{3/2}} + \frac{b\mu_1}{((1-a)^2 + b^2)^{3/2}} + \frac{(b-d)\mu_3}{((a-c)^2 + (b-d)^2)^{3/2}} =$$
$$= \lambda \frac{b + b\mu_1 + (b-d)\mu_3}{1 + \mu_1 + \mu_2 + \mu_3} . \tag{18}$$

System (13)–(18) cannot be solved in general form even with the modern computer algebra systems, and we have to investigate it numerically. Nevertheless, analyzing it symbolically, we can get some general results. For example, we can answer the question whether there exist the non-collinear central configurations with three massive bodies situated in the straight line. Such configuration is realized when $b = 0$ and $d \neq 0$, for example. Then from equations (14), (16) we obtain

$$c = \frac{1}{2} . \tag{19}$$

As we have four different bodies, the conditions $a \neq 0$, $a \neq 1$ must be fulfilled for $b = 0$. Then equation (18) can be satisfied only if $\mu_3 = 0$ when only three bodies have non-zero masses. Similar result is obtained if we consider the case $d = 0$ and $b \neq 0$. Hence, we can conclude that *there does not exist a non-collinear central configuration of four massive bodies when three of them are situated in the same straight line.*

Similarly, we can show that central configuration of four bodies in the form of rectangle is also impossible. Actually, such configuration is realized if $a = 0$, $c = 1$, $d = b$, for example. In this case equations (13), (15), and (17) give

$$\mu_1 = \mu_3 , \quad \mu_2 = 1 . \tag{20}$$

Substituting (20) into equations (14), (16), (18), we immediately obtain

$$\mu_1 = \mu_3 = 1 , \quad b = 1 . \tag{21}$$

It means that *rectangle can exist only as a square with four bodies of equal masses, being situated in its vertices.*

These two examples show that if we choose some values for variables $a$, $b$, $c$, $d$ we can easily check whether the corresponding configuration of four bodies is central because these variables must satisfy equations (13)–(18), and the corresponding constant $\lambda$ must be positive. But we cannot find general solution of system (13)–(18) and so we cannot answer the question how many central configuration of four bodies exist if their masses are given and what are the shapes of these configuration. We can only try to find some numerical solutions of equations (13)–(18) and investigate their properties. Nevertheless, doing such numerical analysis, we can get some new interesting results.

## 3 Symmetric Central Configurations of Four Bodies

In the case of $\mu_2 = 0$, system (13)–(16) is equivalent to equations (9) and determines three collinear configurations of the bodies $P_0, P_1, P_3$ and two equilateral triangles. If two bodies $P_0$ and $P_1$, for example, have equal masses ($\mu_1 = 1$) then three of these configurations are symmetrical with respect of the line $x = \frac{1}{2}$ (see Fig. 1). One can easily check that the corresponding solutions are given by

$$c = \frac{1}{2}, \ d = 0 \ \ or \ \ c = \frac{1}{2}, \ d = \pm\frac{\sqrt{3}}{2} \ . \tag{22}$$

Obviously, symmetry of the system will not be disturbed if the body $P_2$ of non-zero mass is situated on the line $x = 1/2$. As three massive bodies can not be situated in the same straight line if $\mu_2 \neq 0$ then the body $P_3$ must change its position but it can remain in the line $x = 1/2$. Thus, we consider here such configurations of four bodies when $a = c = 1/2$ and positions of the bodies $P_2$ and $P_3$ in the axis of symmetry $x = 1/2$ are determined with two parameters $b$ and $d$, respectively. Then system (13)–(18) reduces to the following three equations

$$\lambda = 2 \left( 1 + \frac{4\mu_2}{(1 + 4b^2)^{3/2}} + \frac{4\mu_3}{(1 + 4d^2)^{3/2}} \right) , \tag{23}$$

$$(2 + \mu_2 + \mu_3) \left( \frac{4b\mu_2}{(1 + 4b^2)^{3/2}} + \frac{4d\mu_3}{(1 + 4d^2)^{3/2}} \right) =$$
$$= (b\mu_2 + d\mu_3) \left( 1 + \frac{4\mu_2}{(1 + 4b^2)^{3/2}} + \frac{4\mu_3}{(1 + 4d^2)^{3/2}} \right) , \tag{24}$$

$$\frac{1}{2}(2 + \mu_2 + \mu_3) \left( \frac{16b}{(1 + 4b^2)^{3/2}} + \frac{(b - d)\mu_3}{|b - d|^3} \right) =$$
$$= (2b + (b - d)\mu_3) \left( 1 + \frac{4\mu_2}{(1 + 4b^2)^{3/2}} + \frac{4\mu_3}{(1 + 4d^2)^{3/2}} \right) . \tag{25}$$

Thus, we obtain two nonlinear algebraic equations (24), (25) with respect to variables $b$, $d$ which contain two parameters $\mu_2$, $\mu_3$, and any their real solution determines symmetric central configuration of four bodies if the corresponding parameter $\lambda$ defined in (23) is a positive number. Using the system *Mathematica*, we can easily find a solution of the system (24)-(25) for any values of $\mu_2$, $\mu_3$ and investigate dependence of the number of symmetrical central configurations and their shape on the masses of the bodies.

It is expedient to start such investigation from the case $\mu_2 = 0$ when equations (24) and (25) become independent and the first one has solutions (22). As the cases $d = \pm\sqrt{3}/2$ are equivalent it is sufficient to consider the case $d = \sqrt{3}/2$ when equation (25) takes the form

$$\mu_3(\sqrt{3} - 2b) \left( \frac{8}{(3 - 4b\sqrt{3} + 4b^2)^{3/2}} - 1 \right) = \frac{32b}{(1 + 4b^2)^{3/2}} - 4b . \tag{26}$$

Numerical analysis of this equation shows that for any $0 < \mu_3 < 1.4689$ it has four roots which are shown in Fig. 2. Note that two of these roots determine positions of the body $P_2$ inside the equilateral triangle what is possible only if mass of the body $P_3$ is only a little bit greater than masses of the bodies $P_0$ and $P_1$. With the parameter $\mu_3$ growth only two positions of the body $P_2$ outside the triangle exist and $b \to \frac{\sqrt{3}}{2} \pm 1$ as $\mu_3 \to \infty$ what means that all three bodies $P_0, P_1, P_2$ tend to be on a circle of radius 1 with the center at point $P_3$.

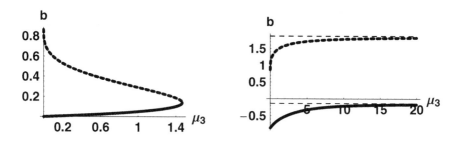

**Fig. 2.** Positions of the body $P_2$: $\mu_2 = 0$, $d = \frac{\sqrt{3}}{2}$

If $d = 0$ and $\mu_2 = 0$ then equation (25) takes the form

$$\mu_3 \left( 8b - \frac{1}{b^2} \right) = 2b \left( 1 - \frac{8}{(1 + 4b^2)^{3/2}} \right) . \tag{27}$$

It has two solutions $b = \pm\sqrt{3}/2$ for $\mu_3 = 0$ which tend to $b = \pm 1/2$ as $\mu_3 \to \infty$. Again with the parameter $\mu_3$ growth all three bodies $P_0, P_1, P_2$ tend to be on a circle of radius $1/2$ with the center in the point $P_3$.

Thus, in the case of $\mu_2 = 0$ the system (24)–(25) determines ten possible central configurations which are symmetric in pairs with respect to the $Ox$ axis. One can expect that for $0 < \mu_2 << 1$ there will exist similar ten central configurations with a little bit disturbed positions of the bodies $P_2$ and $P_3$, in comparison with the case of $\mu_2 = 0$. We have solved system (24)-(25) numerically for $\mu_2 = 1/10$ and different values of $\mu_3$ and found possible positions of the body $P_2$, five of them are shown in Fig. 3, Fig. 4, another five can be easily obtained by means of mirror reflection with respect to the $Ox$ axis. Again there exist two positions of the body $P_2$ inside the triangle $P_0 P_1 P_3$ (see Fig. 3). If parameter $\mu_3$ grows up they move toward each other in the directions shown in Fig. 3 with arrows until their coinciding when parameter $\mu_3$ reaches the value $\mu_3 \approx 1.21048$. Note that configurations shown on Fig. 3 were found earlier by Simo [7] and Albouy [8], who considered the case of equal masses of the bodies. Now we see that these configuration exist in more general case when only two bodies have equal masses.

For $\mu_3 > 1.21048$ there exist only three configurations shown on Fig. 4, where arrows show directions of motion of the body $P_2$ when parameter $\mu_3$ grows. Note that the bodies $P_0, P_1, and P_2$ tend to be on the same circle with the center at point $P_3$ as $\mu_3 \to \infty$. Similar results were obtained in [9], where it is investigated

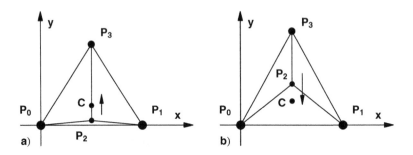

**Fig. 3.** Positions of the body $P_2$ inside the triangle: $\mu_2 = 0.1$, $\mu_3 = 0.7$

the case when only one body has a finite mass while masses of the rest three bodies are negligible.

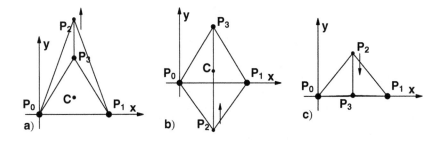

**Fig. 4.** Positions of the body $P_2$ outside the triangle: $\mu_2 = 0.1$, $\mu_3 = 0.7$

Central configurations shown in Fig. 3 and Fig. 4 exist for any values of $0 < \mu_2 \leq 1$, $0 < \mu_3 \leq 1$. But if $\mu_2 = 1$, for example, and $\mu_3 > 1$ then only three configurations shown in Fig. 4 can exist when a more massive body $P_3$ is situated inside or outside the triangle. If mass of the body $P_2$ becomes larger, for example, $\mu_2 = 1.1$ and $\mu_3 > 1$ then only one central configuration shown in Fig. 4b remains. Note that configurations of this kind exist for any values of parameters $\mu_2$, $\mu_3$.

## 4    Central Configuration in the Form of Antiparallelogram

If $\mu_2 = \mu_3$ then one can expect that additional central configuration symmetric with respect to the line $x = 1/2$ exists. It may be determined from the condition

$$a = 1 - c, \; b = d, \; \mu_2 = \mu_3 \; . \tag{28}$$

On substituting (28) into system (13)–(18) we obtain three independent equations

$$\lambda = 2\left(1 + \frac{c\mu_3}{(c^2+d^2)^{3/2}} + \frac{(1-c)\mu_3}{((1-c)^2+d^2)^{3/2}}\right) , \qquad (29)$$

$$(1+\mu_3)\left(\frac{1}{(c^2+d^2)^{3/2}} + \frac{1}{((1-c)^2+d^2)^{3/2}}\right) =$$
$$= 2\left(1 + \frac{c\mu_3}{(c^2+d^2)^{3/2}} + \frac{(1-c)\mu_3}{((1-c)^2+d^2)^{3/2}}\right) , \qquad (30)$$

$$\frac{c}{(c^2+d^2)^{3/2}} - \frac{1-c}{((1-c)^2+d^2)^{3/2}} - \frac{(1-2c)\mu_3}{|1-2c|^3} =$$
$$= -(1-2c)\left(1 + \frac{c\mu_3}{(c^2+d^2)^{3/2}} + \frac{(1-c)\mu_3}{((1-c)^2+d^2)^{3/2}}\right) . \qquad (31)$$

It can be readily seen that in the case of $\mu_3 = 0$ system (30)-(31) has two roots

$$c = \frac{1}{2}, \ d = \pm\frac{\sqrt{3}}{2} , \qquad (32)$$

which correspond to well-known Lagrange triangular solutions in the three-body problem. It means that two massless bodies $P_2$ and $P_3$ are situated at the same point $(1/2, \sqrt{3}/2)$ or $(1/2, -\sqrt{3}/2)$ in the $Oxy$ plane. If $\mu_3 > 0$ then they leave these points and move in the $Oxy$ plane being symmetrical with respect to the line $x = 1/2$. The corresponding configuration of the bodies is just antiparallelogram, and their positions can be found as solutions of system(30)–(31). Graphs of the corresponding functions $c(\mu_3)$, $d(\mu_3)$ are shown in Fig. 5.

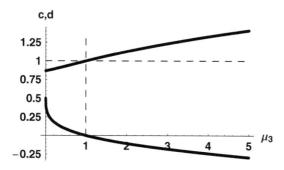

**Fig. 5.** Functions $c(\mu_3)$, $d(\mu_3)$

Thus, with parameter $\mu_3$ varying, antiparallelogram undergoes continuous deformation. Obviously, in the case of $\mu_3 = 1$ it reduces to the square with four equal masses being in its vertices.

If parameters $\mu_2$ and $\mu_3$ vary separately configuration of the bodies will undergo continuous deformation as well. As an example, let us consider a square as unperturbed configuration and add small perturbations to the parameters $\mu_2$, $\mu_3$ according to the rule

$$\mu_2 = 1 + \mu_{21}\varepsilon \ , \quad \mu_3 = 1 + \mu_{31}\varepsilon \ , \tag{33}$$

where $\varepsilon$ is a small parameter. Then we can expect that coordinates of the bodies $a, b, c, d$ will also get some perturbations, and we can represent them as power series of the form

$$a = 1 + a_1\varepsilon + a_2\varepsilon^2 + \dots \ , \quad b = 1 + b_1\varepsilon + b_2\varepsilon^2 + \dots \ ,$$
$$c = c_1\varepsilon + a_2\varepsilon^2 + \dots \ , \quad d = 1 + d_1\varepsilon + d_2\varepsilon^2 + \dots \ . \tag{34}$$

In order to find coefficients in the expansions (34) we substitute (33), (34) into equations (13)-(18), expand them in power series in $\varepsilon$, equate coefficients of $\varepsilon^k$ ($k = 1, 2, \dots$) in the left-hand and right-hand sides of each equation and obtain an infinite sequence of algebraic equations. Solving this system, we obtain in the first order in $\varepsilon$

$$a_1 = \frac{1}{6}(1 - 2\sqrt{2})\mu_{21} + \frac{2}{93}(5 + 11\sqrt{2})\mu_{31} \ ,$$
$$c_1 = -\frac{2}{93}(5 + 11\sqrt{2})\mu_{21} + \frac{1}{6}(-1 + 2\sqrt{2})\mu_{31} \ ,$$
$$b_1 = \frac{1}{62}(17 - 6\sqrt{2})\mu_{21} \ , \quad d_1 = \frac{1}{62}(17 - 6\sqrt{2})\mu_{31} \ . \tag{35}$$

Coefficients of the second order in $\varepsilon$ are given by

$$a_2 = \frac{7}{3844}(17 + 25\sqrt{2})\mu_{21}^2 + \frac{1}{5766}(1130 - 769\sqrt{2})\mu_{21}\mu_{31} +$$
$$+ \frac{7}{11532}(-445 + 137\sqrt{2})\mu_{31}^2 \ ,$$
$$b_2 = \frac{1}{558}(-177 + 88\sqrt{2})\mu_{21}^2 + \frac{1}{2232}(741 - 236\sqrt{2})\mu_{21}\mu_{31} -$$
$$- \frac{2}{8649}(267 + 110\sqrt{2})\mu_{31}^2 \ ,$$
$$c_2 = -\frac{7}{11532}(-445 + 137\sqrt{2})\mu_{21}^2 + \frac{1}{5766}(-1130 + 769\sqrt{2})\mu_{21}\mu_{31} -$$
$$- \frac{7}{3844}(17 + 25\sqrt{2})\mu_{31}^2 \ ,$$
$$d_2 = -\frac{2}{8649}(267 + 110\sqrt{2})\mu_{21}^2 + \frac{1}{2232}(741 - 236\sqrt{2})\mu_{21}\mu_{31} +$$
$$+ \frac{1}{558}(-177 + 88\sqrt{2})\mu_{31}^2 \ . \tag{36}$$

Similarly we can find coefficients of higher order in the expansions (34) and determine coordinates $a, b, c, d$ of the bodies with arbitrary precision. Obviously, parameter $\varepsilon$ appears only in the expressions of the form $\mu_{21}\varepsilon$ and $\mu_{31}\varepsilon$ only.

Hence, after the calculations have been done we can set $\varepsilon = 1$. As a result, we obtain two-parametric family of solutions of the system (13)-(18) in the neighborhood of the square configuration. Now it can be readily seen that changing masses of the bodies we can deform the central configurations continuously. Two such deformed squares are shown in Fig. 6.

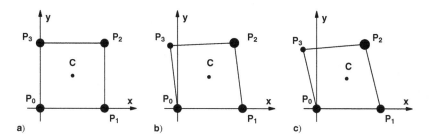

**Fig. 6.** Deformed squares: b) $\mu_{21} = 0.12$, $\mu_{31} = -0.2$; c) $\mu_{21} = 0.24$, $\mu_{31} = -0.4$

## 5   Conclusion

In the present paper we study the central configuration in the newtonian problem of four bodies. We have shown that such configurations are fixed with four coordinates and written the corresponding system of nonlinear algebraic equations (13)–(18) determining these coordinates. Analyzing this system symbolically, we have proved that four bodies having different masses cannot form central configuration being a rectangle, and three of these bodies cannot be situated in the same straight line. If two bodies have equal masses then there can exist central configurations in the form of isosceles triangle with fourth body being situated in the axis of symmetry inside or outside the triangle. The number of possible configurations in such cases depends on the masses of the bodies and may be equal to ten, six, or two.

We have proved that there exists one-parametric family of central configurations in the form of antiparallelogram. We have shown also that central configuration may be deformed continuously by means of changing masses of the bodies. As an example we constructed symbolically two-parametric family of central configurations in the neighborhood of the square.

All numerical and symbolic calculations and visualization of the obtained results have been done with the computer algebra system *Mathematica*.

## References

1. Cors, J.M., Llibre, J., Olle, M.: Central configurations of the planar coorbital satellite problem. Celestial Mechanics and Dynamical Astronomy **89** (2004) 319–342
2. Wintner, A.: The analytical Foundations of Celestial Mechanics. Princeton University Press, 1941

3. Moulton, F.R.: The straight line solutions of the problem of $N$ bodies. Ann. of Math. **12** (1910) 1–17
4. Smale, S.: Problems of the nature of relative equilbria in Celestial Mechanics. Springer Lecture Notes in Math. **197** (1970) 194–198
5. Smale, S.: Mathematical problems for the next century. Math. Intelligencer **20** (1998) 7–15
6. Wolfram, S.: The Mathematica Book. 4th edn. Wolfram Media/Cambridge University Press (1999)
7. Simo, C.: Relative equilibrium solutions in the four body problem. Celest. Mech. **18** (1978) 165–184
8. Albouy, A.: Symetrie des configurations centrales de quatre corps. C. R. Acad. Sci. Paris, Ser. I. **320** (1995) 217–220
9. Casasayas, J., Llibre, J., Nunes, A.: Central configurations of the planar body problem. Celest. Mech. and Dynam. Astron. **60** (1994) 273–288

# A Symbolic-Numerical Algorithm for Solving the Eigenvalue Problem for a Hydrogen Atom in Magnetic Field

Alexander Gusev[1], Vladimir Gerdt[1], Michail Kaschiev[2], Vitaly Rostovtsev[1],
Valentin Samoylov[1], Tatyana Tupikova[1], and Sergue Vinitsky[1]

[1] Joint Institute for Nuclear Research, Dubna, Moscow Region, Russia
vinitsky@thsun1.jinr.ru
[2] Institute of Mathematics and Informatics, BAS, Sofia, Bulgaria

**Abstract.** The boundary-value problem in spherical coordinates for the Shrödinger equation describing a hydrogen like atom in a strong magnetic field is reduced to the problem for a set of radial equations in the framework of the Kantorovich method. The effective potentials of these equations are given by integrals over the angular variable between the oblate angular spheroidal functions depending on the radial variable as a parameter and their derivatives with respect to the parameter. A symbolic-numerical algorithm for evaluating the oblate spheroidal functions and corresponding eigenvalues which depend on the parameter, their derivatives with respect to the parameter and matrix elements is presented. The efficiency and accuracy of the algorithm and of the numerical scheme derived are confirmed by computations of eigenenergies and eigenfunctions for the low-excited states of a hydrogen atom in the uniform magnetic field.

## 1 Introduction

In recent paper [1], a parametric basis of the angular oblate spheroidal functions has been applied to calculate, with help of the Finite Element Method (FEM) codes, the low-lying excited states of a hydrogen atom in a strong magnetic field. However, to build up and examine the various effective approximations of the boundary-value problem in both discrete and continuous spectra for a set of radial equations, we need both symbolic and numerical evaluation of the angular oblate spheroidal functions (AOSF) and eigenvalues depending on a parameter as well as of their derivatives with respect to the radial variable and of matrix elements in a wide range of quantum numbers and values of the radial parameter [2].

With this end in view we consider a symbolic-numerical algorithm for evaluating the AOSF and eigenvalues depending on a parameter, for their derivatives with respect to the radial variable and for the matrix elements. The conventional expansion of these functions by both the unnormalized Legendre polynomials [3, 4] and normalized Legendre polynomials is used. Such expansion leads

V.G. Ganzha, E.W. Mayr, and E.V. Vorozhtsov (Eds.): CASC 2006, LNCS 4194, pp. 205–218, 2006.
© Springer-Verlag Berlin Heidelberg 2006

to an algebraic eigenvalue problem with real symmetric matrices that provides stable calculation with double precision arithmetic by applying the subroutine IMTQL2 from the EISPACK Fortran Library [5]. A symbolic algorithm for evaluating the asymptotic matrix elements with respect to the radial parameter is implemented in MAPLE and is used to continue the calculated numerical values of matrix elements to large values of the radial parameter.

The main goal of this paper is to develop a symbolic algorithm for generation of the inhomogeneous algebraic problems with respect to unknown derivatives of the functions up to arbitrary finite order and for construction of the corresponding matrix elements. The algorithm is explicitly presented and implemented in MAPLE. This provides stable calculation of the derivatives with double precision arithmetic by applying the subroutine F07BRF (ZGBTRS) from the NAG Fortran Library [6]. The developed approach is applied to numerical calculation of effective potentials for Shrödinger equation describing a hydrogen-like atom in a strong magnetic field. The efficiency and accuracy of the proposed algorithm and accompanying numerical schemes is confirmed by computation of eigenenergies and eigenfunctions of a hydrogen atom in the uniform magnetic field.

The paper is organized as follows. In section 2 we briefly describe a reduction of the 2D-eigenvalue problem to the 1D-eigenvalue problem for a set of closed radial equations by means of the Kantorovich method[7]. In section 3 we examine the algorithm for evaluating the AOSF. In section 4 we describe the symbolic-numerical algorithm for generating the nonhomogeneous algebraic problem to calculate derivatives of the AOSF with respect to parameter and corresponding matrix elements. In section 5 the algorithm for asymptotic calculation of matrix elements at large values of radial variables is presented. In section 6 the method is applied to calculating low-lying states of hydrogen atom in strong magnetic field. The rate of convergence is explicitly demonstrated for typical examples. Besides, the obtained results are compared with the best known ones in the literature. In section 7 the conclusions are made, and the possible future applications of the method are discussed.

## 2    Problem Statement

The Schrödinger equation for the hydrogen atom in an axially symmetric magnetic field $\boldsymbol{B} = (0, 0, B)$ and in the spherical coordinates $(r, \theta, \phi)$ can be written as the 2D-equation

$$\left( -\frac{1}{r^2} \frac{\partial}{\partial r} r^2 \frac{\partial}{\partial r} + \frac{A^{(0)}(r, \theta)}{r^2} - \frac{2Z}{r} - \epsilon \right) \Psi(r, \theta) = 0 \tag{1}$$

in the region $\Omega$: $0 < r < \infty$ and $0 < \theta < \pi$. The operator $A^{(0)}(r, \theta)$ is given by

$$A^{(0)}(r, \theta) = -\frac{1}{\sin \theta} \frac{\partial}{\partial \theta} \sin \theta \frac{\partial}{\partial \theta} + \frac{m^2}{\sin^2 \theta} + \gamma m r^2 + \frac{1}{4} \gamma^2 r^4 \sin^2 \theta, \tag{2}$$

where $m = 0, \pm 1, \ldots$ is the magnetic quantum number, $\gamma = B/B_0$, $B_0 \cong 2.35 \times 10^9 G$ is a dimensionless parameter which determines the field strength $B$, and the atomic units (a.u.) $\hbar = m_e = e = 1$ are used under the assumption of infinite mass of the nucleus. In these expressions $\epsilon = 2E$ is the doubled energy (in units of Rydbergs, $1Ry=(1/2)$a.u.) of the bound state $|m\sigma>$ at fixed values of $m$ and z-parity; $\sigma = \pm 1$; $\Psi \equiv \Psi_{m\sigma}(r,\theta) = (\Psi_m(r,\theta) + \sigma\Psi_m(r,\pi-\theta))/2$ is the corresponding wave function. Here the sign of z-parity $\sigma = (-1)^{N_\theta}$ is defined by the (even or odd) number of nodes $N_\theta$ in the solution $\Psi$ with respect to the angular variable $\theta$ in the interval $0 < \theta < \pi$. We will use also units $\hbar = m_e = e = \gamma = 1$ and scaled variables $\hat{r} = r\sqrt{\gamma}$, effective charge $\hat{Z} = Z/\sqrt{\gamma}$ and scaled energy $\hat{\epsilon} = \epsilon/\gamma$. It means that one can put in further consideration the cyclotron frequency $\gamma = 1$ and renormalize the initial charge $Z$ by factor $\sqrt{1/\gamma}$ and initial energy $\epsilon$ by factor $1/\gamma$ only. The wave function satisfies the following boundary conditions in each $m\sigma$ subspace of the full Hilbert space:

$$\lim_{\theta \to 0} \sin\theta \frac{\partial\Psi}{\partial\theta} = 0, \quad \text{if} \quad m = 0, \quad \text{and} \quad \Psi(r,0) = 0, \quad \text{if} \quad m \neq 0, \quad (3)$$

$$\frac{\partial\Psi}{\partial\theta}\left(r, \frac{\pi}{2}\right) = 0, 0 \leq r < \infty, \quad \text{for the even } (\sigma - +1) \text{ parity state}, \quad (4)$$

$$\Psi\left(r, \frac{\pi}{2}\right) = 0, 0 \leq r < \infty, \quad \text{for the odd } (\sigma = -1) \text{ parity state}, \quad (5)$$

$$\lim_{r \to 0} r^2 \frac{\partial\Psi}{\partial r} = 0. \quad (6)$$

The discrete spectrum wave function obeys the asymptotic boundary condition approximated at large $r = r_{max}$ by a boundary condition of the first type

$$\lim_{r \to \infty} r^2\Psi = 0 \quad \to \quad \Psi(r_{max}, \theta) = 0. \quad (7)$$

Here the energy $\epsilon \equiv \epsilon(r_{max})$ plays the role of eigenvalues of the boundary-value problem (1)-(7) on a finite interval $0 \leq r \leq r_{max}$ with additional normalization condition

$$\int_0^{r_{max}} \int_0^{\pi} r^2 \sin\theta |\Psi(r,\theta)|^2 dr d\theta = 1. \quad (8)$$

We consider the Kantorovich expansion of the partial solution $\Psi_i^{m\sigma}(r,\theta)$ using a set of the one-dimensional parametric basis functions $\Phi_j(\theta;r) \equiv \Phi_j^{m\sigma}(\theta;r)$:

$$\Psi_i^{m\sigma}(r,\theta) = \sum_{j=1}^{j_{max}} \Phi_j^{m\sigma}(\theta;r)\chi_j^{(i)}(r). \quad (9)$$

The matrix-valued functions $\chi(r) \equiv \{\chi^{(i)}(r)\}_{i=1}^{j_{max}}$ composed from vector functions $(\chi^{(i)})^T = (\chi_1^{(i)}(r), \ldots, \chi_{j_{max}}^{(i)}(r))$ are unknown. The vector angular functions $(\Phi(\theta;r))^T = (\Phi_1(\theta;r), \ldots, \Phi_{j_{max}}(\theta;r))$ form an orthonormal basis for each value of the radius $r$ which is treated here as a parameter. The angular parametric functions $\Phi_i(\theta;r)$ and potential curves $E_i(r)$ (in Ry) are determined as the solutions of the following one-dimensional parametric eigenvalue problem with operator $A^{(0)} \equiv A^{(0)}(r,\theta)$ from (2) at fixed values $m$ and $\sigma$:

$$A^{(0)}\Phi_j(\theta;r) = E_j^{(0)}(r)\Phi_j(\theta;r), \quad \int_0^{\pi} \Phi_i(\theta;r)\Phi_j(\theta;r)\sin\theta d\theta = \delta_{ij}. \quad (10)$$

The problem (10) with boundary conditions (3)-(5) will be solved for each value of the radial variable $r \in \omega_r$, where $\omega_r = (r_1, r_2, \ldots, r_k, \ldots, r_{max})$ is a set of values of $r$. By substituting the expansion (9) into the variational functional corresponding to the above boundary-value problem (1)–(8), we arrive at an eigenvalue problem for a system of $j_{max}$ ordinary second-order differential equations that determines the energy $\epsilon$ and the coefficients (radial wave functions) $(\boldsymbol{\chi}^{(i)}(r))^T = (\chi_1^{(i)}(r), \chi_2^{(i)}(r), \ldots, \chi_{j_{max}}^{(i)}(r))$ in expansion (9)

$$-\mathbf{I}\frac{1}{r^2}\frac{d}{dr}r^2\frac{d\boldsymbol{\chi}^{(i)}}{dr} + \frac{\mathbf{U}(\mathbf{r})}{r^2}\boldsymbol{\chi}^{(i)} + \mathbf{Q}(r)\frac{d\boldsymbol{\chi}^{(i)}}{dr} + \frac{1}{r^2}\frac{d[r^2\mathbf{Q}(r)\boldsymbol{\chi}^{(i)}]}{dr} = \epsilon_i\,\mathbf{I}\boldsymbol{\chi}^{(i)}. \quad (11)$$

Here $\mathbf{I}$, $\mathbf{U}(r)$, and $\mathbf{Q}(r)$ are finite $j_{max} \times j_{max}$ matrices whose elements are given by the relations, $I_{ij} = \delta_{ij}$, $H_{ij}(r) = H_{ji}(r)$, $Q_{ij}(r) = -Q_{ji}(r)$ (see Figs. 2)

$$U_{ij}(r) = \frac{E_i(r) + E_j(r)}{2}\delta_{ij} + 2Zr + r^2H_{ij}(r), \quad (12)$$

$$H_{ij}(r) = \int_0^\pi \sin\theta \frac{\partial\Phi_i}{\partial r}\frac{\partial\Phi_j}{\partial r}d\theta, \quad Q_{ij}(r) = -\int_0^\pi \sin\theta\Phi_i\frac{\partial\Phi_j}{\partial r}d\theta. \quad (13)$$

The discrete spectrum solutions obey the following asymptotic boundary condition and orthonormal conditions

$$\lim_{r\to 0} r^2\frac{\partial\boldsymbol{\chi}^{(i)}}{\partial r} = 0, \quad \lim_{r\to\infty} r^2\boldsymbol{\chi}^{(i)} = 0 \to \boldsymbol{\chi}^{(i)}(r_{max}) = 0, \quad (14)$$

$$\int_0^{r_{max}} r^2(\boldsymbol{\chi}^{(i)}(r))^T\boldsymbol{\chi}^{(j)}(r)dr = \delta_{ij}. \quad (15)$$

The application of the Kantorovich approach makes the boundary-value problem (1)–(8) equivalent to the following ones:

- Calculation of the potential curves $E_i(r)$ and the angular parametric eigenfunctions $\Phi_i(\theta; r)$ of the spectral problem (10) for a given set of $r \in \omega_r$ at fixed values $|m|$, $\sigma$ and $\gamma = 1$.
- Calculation of the derivatives $\dfrac{\partial\Phi}{\partial r}$ and evaluation of the corresponding integrals (see (12) and(13) ) for constructing the matrix elements of radial coupling $U_{ij}(r)$ and $Q_{ij}(r)$.
- Calculation of the scaled energies $\hat{\epsilon}$ and the radial wave functions $\chi(r)$ as the solutions of one-dimensional eigenvalue problem (11) at fixed values of $m$, $\sigma$, $\gamma = 1$ and effective charge $\hat{Z} = Z/\sqrt{\gamma}$.
- Examination of the convergence of these solutions depending on the number of channels $j_{max}$ and recalculation of the scaled energies $\hat{\epsilon}$ to the initial ones $\epsilon = \hat{\epsilon}\gamma$.

To solve the above problem a numerical algorithm based on FEM has been elaborated [1]. Below we will develop the symbolic-numerical algorithm based on the conventional representation of the angular parametric functions $\Phi_i(\theta; r)$. It is more suitable for our purpose, i.e. for calculating high-order derivatives of the angular parametric functions and matrix elements as well as for studying asymptotics of the effective potentials and radial solutions.

## 3 Calculations of the Angular Oblate Spheroidal Functions

Note that the solutions of the problem (10), (2) with the shifted eigenvalues $\lambda_j(p) = E_j(r) - \gamma m r^2$ correspond to the solutions of the eigenvalue problem for the AOSF [3] with respect to a variable $\eta = \cos\theta$:

$$-\frac{\partial}{\partial\eta}(1-\eta^2)\frac{\partial\Phi_j(\eta;\hat{p})}{\partial\eta} + \left(\frac{m^2}{1-\eta^2} + \hat{p}^2(1-\eta^2)\right)\Phi_j(\eta;\hat{p}) = \hat{\lambda}_j(\hat{p})\Phi_j(\eta;\hat{p}). \quad (16)$$

Here $\hat{\lambda}_j \equiv \hat{\lambda}_j(\hat{p}) = \hat{E}_j(\hat{r},\gamma) - m\hat{r}^2$ is a renormalized eigenvalue because there is only the dependence on $m^2$ in eq.(16), $\hat{p} = \hat{r}^2/2$ is a renormalized parameter, $p = \gamma r^2/2$, and eigenfunctions $\Phi_j(\eta;\hat{p}) \equiv \Phi_j^{m\sigma}(\eta;\hat{p})$ satisfy the orthogonality conditions (10). We find eigenfunctions $\Phi_j(\theta;\hat{r})$ in the form of a series expansion at fixed values $\sigma = \pm 1$ and $m \equiv |m|$,

$$\Phi_j(\eta;\hat{r}) = \sum_{s=(1-\sigma)/2}^{s_{max}} \tilde{c}_{sj}^{m\sigma}(\hat{r})B_{sm}B_{sm}^{-1}\tilde{P}_{m+s}^m(\eta) = \sum_{s=(1-\sigma)/2}^{s_{max}} c_{sj}^{m\sigma}(\hat{r})P_{m+s}^m(\eta). \quad (17)$$

Here $s$ is even (odd) integer at $\sigma = (-1)^s = \pm 1$, $\tilde{P}_{m+s}^m(\eta)$ and $P_{m+s}^m(\eta)$ are the unnormalized and normalized Legendre polynomials defined by the relations [3]

$$-\frac{d}{d\eta}(1-\eta^2)\frac{d}{d\eta}P_{m+s}^m(\eta) + \frac{m^2}{1-\eta^2}P_{m+s}^m(\eta) = \hat{\lambda}_s^{m\sigma}(0)P_{m+s}^m(\eta),$$
$$\hat{\lambda}_s^{m\sigma}(0) = (m+s)(m+s+1), \quad s = 2(j-1) + (1-\sigma)/2, \quad (18)$$

$$\tilde{P}_{m+s}^m(\eta) = B_{sm}P_{m+s}^m(\eta), \quad B_{sm} = \sqrt{\frac{2(s+2m)!}{(2s+2m+1)(s)!}}, \quad (19)$$

$$\int_{-1}^1 P_{m+s}^m(\eta)P_{m+s'}^m(\eta)d\eta = \delta_{ss'}, \quad \int_{-1}^1 \tilde{P}_{m+s}^m(\eta)\tilde{P}_{m+s'}^m(\eta)d\eta = \delta_{ss'}B_{sm}^2. \quad (20)$$

The coefficients $\tilde{c}_{sj}^{m\sigma}(\hat{r}) = c_{sj}^{m\sigma}(\hat{r})B_{sm}^{-1}$ and $c_{sj}^{m\sigma}(\hat{r})$ satisfy the relations

$$\sum_{s=(1-\sigma)/2}^{s_{max}} (c_{sj}^{m\sigma}(\hat{r}))^T c_{sj'}^{m\sigma}(\hat{r}) = \delta_{jj'}. \quad (21)$$

The eigenvalue problems for unnormalized $\tilde{c}_j = \{\tilde{c}_{sj}^{m\sigma}(\hat{r})\}_{(1-\sigma)/2}^{s_{max}}$ and normalized $c_j(\hat{r}) = \{c_{sj}^{m\sigma}(\hat{r})\}_{(1-\sigma)/2}^{s_{max}}$ coefficients, and eigenvalues $\hat{\lambda}_j(\hat{r})$ take the form

$$\tilde{\mathbf{A}}^{(0)}\mathbf{c}_j = \hat{\lambda}_j(\hat{r})\tilde{\mathbf{c}}_j, \quad \mathbf{A}^{(0)}\mathbf{c}_j = \hat{\lambda}_j(\hat{r})\mathbf{c}_j, \quad (22)$$

where matrix $\tilde{\mathbf{A}}^{(0)} = (\mathbf{B})^T\mathbf{A}^{(0)}\mathbf{B}$ and the symmetric matrix $\mathbf{A}^{(0)}$ is given by

$$A_{ss-2}^{(0)} = \frac{\hat{p}^2}{(2|m|+2s-1)}\sqrt{\frac{(s-1)s(2|m|+s-1)(2|m|+s)}{(2|m|+2s-3)(2|m|+2s+1)}},$$

$$A_{ss}^{(0)} = (|m| + s)(|m| + s + 1) + \hat{p}^2 \frac{(2s^2 + 2s + 4|m|s + 2|m| - 1)}{(2|m| + 2s - 1)(2|m| + 2s + 3)},$$

$$A_{ss+2}^{(0)} = \frac{\hat{p}^2}{(2|m|+2s+3)} \sqrt{\frac{(s + 1)(s + 2)(2|m|+s+1)(2|m|+s+2)}{(2|m|+2s+1)(2|m|+2s+5)}}.$$

Note that the coefficients of the matrix $\tilde{\mathbf{A}}^{(0)}$ have no fractional degrees that is a suitable circumstance for the recurrence symbolic calculations [3, 4]. As an example, we show in Fig. 1 the behaviour of coefficients $c_{sj} \equiv c_{sj}^{m=0,\sigma=1}$ and $\tilde{c}_{sj} \equiv \tilde{c}_{sj}^{m=0,\sigma=1}$ in the expansion of states $j = 1(N_\theta = 0)$ and $j = 10(N_\theta = 9)$ with normalized and unnormalized Legendre polynomials calculated at $s_{max} = 200$ and $j_{max} = 10$. Evidently, if the number of state $j$ is growing then the number of oscillations is growing too, but amplitude changes in a more closed region for the normalized expansion. The expansion (17) with the normalized Legendre polynomials was used that provides stability of numerical calculation with double precision arithmetic (relative accuracy is $2 \cdot 10^{-14}$) with help of the subroutine IMTQL2 from the EISPACK Fortran Library [5]. The orthogonality relations (22) were held with an accuracy of the order of $10^{-14}$.

## 4    Algorithm for Evaluating Parametric Derivatives of the Angular Functions and Matrix Elements

A set of the first few integrals under consideration has the form

$$Q_{ij}(r) = -I_{01;ij}, \qquad \frac{\partial Q_{ij}(r)}{\partial r} = -I_{11;ij} - I_{02;ij},$$

$$H_{ij}(r) = \frac{I_{11;ij}}{2}, \qquad \frac{\partial H_{ij}(r)}{\partial r} = \frac{I_{21;ij}}{2} + \frac{I_{12;ij}}{2},$$

where $I_{kn;ij}$ are basic integrals between derivatives $\Phi_j^{(k)}(\theta; r) = \frac{\partial^k \Phi_j(\theta;r)}{\partial r^k}$:

$$I_{kn;ij} = \int \frac{\partial^k \Phi_i(\theta; r)}{\partial r^k} \frac{\partial^n \Phi_j(\theta; r)}{\partial r^n} \sin\theta d\theta, \qquad \frac{\partial^0 \Phi_i(\theta; r)}{\partial r^0} = \Phi_i(\theta; r). \qquad (23)$$

Let us consider the algorithm for calculation of high-order derivatives of eigen-functions $\Phi_j^{(k)}(\theta; r) = \frac{\partial^k \Phi_j(\theta;r)}{\partial r^k}$ and eigenvalues $\lambda_j^{(k)} = \frac{\partial^k \lambda_j^{(0)}}{\partial r^k}$ with respect to the radial variable $r$. The derivatives of functions $\Phi_j(\theta; r)$ at fixed values of $m$ and $\sigma = \pm 1$ can be represented as the following expansion in terms of the normalized Legendre polynomials (17):

$$\frac{\partial^k \Phi_j(\theta; r)}{\partial r^k} = \sum_{s=(1-\sigma)/2}^{s_{max}} c_{sj}^{(k)} P_{m+s}^m(\cos\theta), \qquad c_{sj}^{(k)} \equiv \frac{\partial^k c_{sj}(r)}{\partial r^k}, \qquad (24)$$

where $\mathbf{c}^{(0)} \equiv \{c_{sj}(r)\}$ and $\lambda^{(0)} \equiv \lambda_j(r)$ are solutions of the eigenvalue problem

$$\mathbf{A}^{(0)}\mathbf{c}^{(0)} - \mathbf{c}^{(0)}\lambda^{(0)} = 0, \qquad (25)$$

$$\mathbf{c}^{(0)T}\mathbf{c}^{(0)} = \mathbf{I}, \qquad (26)$$

with symmetric matrix $\mathbf{A}^{(0)} \equiv \{A_{kl}^{(0)}(r)\}$ determined in (22). As follows from (25), we should solve the following linear recurrence system of algebraic equations

$$(\mathbf{A}^{(0)}\mathbf{c}^{(k)} - \mathbf{c}^{(k)}\lambda^{(0)}) + (\mathbf{A}^{(k)}\mathbf{c}^{(0)} - \mathbf{c}^{(0)}\lambda^{(k)}) = \mathbf{b}_{(k)}, \quad \mathbf{A}^{(k)} \equiv \frac{\partial^k \mathbf{A}^{(0)}}{\partial r^k}, \, (27)$$

$$\mathbf{b}_{(k)} \equiv \sum_{n=1}^{k-1} \frac{k!}{n!(k-n)!}(\mathbf{c}^{(k-n)}\lambda^{(n)} - \mathbf{A}^{(n)}\mathbf{c}^{(k-n)}), \qquad \mathbf{b}_{(0)} \equiv 0.$$

If one takes into account that $\lambda^{(0)}$ is an eigenvalue of the operator defined in (25), then the problem (27) has a solution *if and only if the right-hand side term is orthogonal to the eigenfunction* $\mathbf{c}^{(0)}$. Multiplying (27) by $\mathbf{c}^{(0)}{}^T$ and using the normalization condition (26), we obtain the expression for $\lambda^{(k)}$

$$\lambda^{(k)} = \mathbf{c}^{(0)}{}^T \mathbf{A}^{(k)}\mathbf{c}^{(0)} - \mathbf{c}^{(0)}{}^T \mathbf{b}_{(k)} \tag{28}$$

and the system of the inhomogeneous algebraic equations for unknown vector $\mathbf{c}^{(k)}$

$$\mathbf{K}\mathbf{c}^{(k)} \equiv \mathbf{A}^{(0)}\mathbf{c}^{(k)} - \mathbf{c}^{(k)}\lambda^{(0)} = \mathbf{b}^{(k)}, \tag{29}$$

where the right-hand side is defined by $\lambda^{(n)}$ and $\mathbf{c}^{(k-n)}$ determined at the previous steps

$$\mathbf{b}^{(k)} \equiv -\sum_{n=1}^{k} \frac{k!}{n!(k-n)!}(\mathbf{A}^{(n)}\mathbf{c}^{(k-n)} - \mathbf{c}^{(k-n)}\lambda^{(n)}).$$

Now the problem (29) has a solution, but it is not unique. From the normalization condition (26) we obtain the required additional equality

$$\sum_{n=0}^{k} \frac{k!}{n!(k-n)!}\mathbf{c}^{(k-n)}{}^T \mathbf{c}^{(n)} = 0, \tag{30}$$

providing the uniqueness of the solution (29). Since $\lambda^{(0)}$ is an eigenvalue of (25), the matrix $\mathbf{K}$ in (29) is degenerate. The algorithm for solving (29) can be written in three steps as follows:

**Step k1.** Calculate solutions $\mathbf{v}^{(k)}$ and $\mathbf{w}$ of the auxiliary inhomogeneous systems of algebraic equations

$$\bar{\mathbf{K}}\mathbf{v}^{(k)} = \bar{\mathbf{b}}^{(k)}, \quad \bar{\mathbf{K}}\mathbf{w} = \mathbf{d}, \tag{31}$$

with non-degenerate matrix $\bar{\mathbf{K}}$ and right-hand sides $\bar{\mathbf{b}}^{(k)}$ and $\mathbf{d}$

$$\bar{K}_{ss'} = \begin{cases} K_{ij}, \, (s-S)(s'-S) \neq 0, \\ \delta_{ss'}, \, (s-S)(s'-S) = 0, \end{cases} \quad \bar{b}_s^{(k)} = \begin{cases} b_s^{(k)}, \, s \neq S, \\ 0, \quad s = S, \end{cases} \quad d_s = \begin{cases} K_{sS}, \, s \neq S, \\ 0, \quad s = S, \end{cases}$$

where $S$ is element of vector $\mathbf{c}^{(0)}$ with maximum absolute value.

**Step k2.** Evaluate coefficient $\gamma^{(k)}$

$$\gamma^{(k)} = -\frac{\gamma_1^{(k)} - F_N^{(k)}}{(c_s^{(0)} - \gamma_2)}, \quad \gamma_1^{(k)} = \mathbf{v}^{(k)}{}^T \mathbf{c}^{(0)}, \quad \gamma_2 = \mathbf{w}^T \mathbf{c}^{(0)} \tag{32}$$

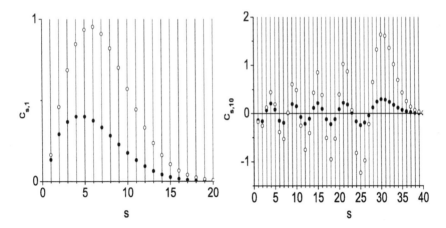

**Fig. 1.** Behaviour of normalized coefficients $c_{sj}$ (filled circles) and non-normalized coefficients $\tilde{c}_{sj}$ (hollow circles) at $r = 15$ and $\gamma = 1$ for the first ($j = 1$, left panel) and the tenth ($j = 10$, right panel) even solutions

from the normalization condition (30)

$$\mathbf{c}^{(0)T}\mathbf{c}^{(k)} \equiv F_N^{(k)} = -\frac{1}{2}\sum_{n=1}^{k-1}\frac{k!}{n!(k-n)!}\mathbf{c}^{(k-n)T}\mathbf{c}^{(n)}. \tag{33}$$

**Step k3.** Evaluate vector $\mathbf{c}^{(k)}$

$$c_s^{(k)} = \begin{cases} v_s^{(k)} - \gamma^{(k)}w_s, & s \neq S, \\ \gamma^{(k)}, & s = S. \end{cases} \tag{34}$$

The algorithm (28)–(30) for calculation of matrix elements and their derivatives was implemented in a MAPLE and FORTRAN that provides stability of numerical calculation with double precision arithmetic (relative accuracy is $2 \cdot 10^{-14}$) with help of the subroutine F07BRF (ZGBTRS) from the NAG Fortran Library Routine Document [6].

## 5 Algorithm for Evaluating the Asymptotics of Matrix Elements

Symbolic algorithms for evaluating asymptotics of the matrix elements at small and large $r$ values were implemented in MAPLE using conventional series expansions by the direct and inverse powers of $p$ for the AOSF [3].

At small $r$ asymptotic values of matrix elements $E_j$, $H_{jj'}$ and $Q_{jj'}$ characterized by $l = |m| + s = 2j - 2 + |m|$ for even states ($\sigma = +1$) and $l = |m| + s = 2j - 1 + |m|$ for odd states ($\sigma = -1$) are the series expansion by the power of $r$ at some finite values $l_l, l_r$

$$E_l(r) = E_l^{(0)} + E_l^{(2)}r^2 + \sum_{k=1}^{k_{max}} r^{4k}E_l^{(4k)}, \quad E_l^{(4k-3)} \equiv E_l^{(4k-1)} \equiv E_l^{(4k+2)} \equiv 0,$$

$$H_{l_i l_r}(r) = \sum_{k=1}^{k_{max}} r^{4k-2} H_{l_i l_r}^{(4k-2)}, \quad H_{l_i l_r}^{(4k-3)} \equiv H_{l_i l_r}^{(4k-1)} \equiv H_{l_i l_r}^{(4k)} \equiv 0, \tag{35}$$

$$Q_{l_i l_r}(r) = \sum_{k=1}^{k_{max}} r^{4k-1} Q_{l_i l_r}^{(4k-1)}, \quad Q_{l_i l_r}^{(4k-3)} \equiv Q_{l_i l_r}^{(4k-2)} \equiv Q_{l_i l_r}^{(4k)} \equiv 0.$$

The calculation was performed by the algorithm implemented in MAPLE up to $k_{max} = 12$. Below we display the first few coefficients of matrix elements:

$$E_l^{(0)} = l(l+1), \quad E_l^{(2)} = \gamma m, \quad E_l^{(4)} = \frac{\gamma^2}{2} \frac{l^2 + l - 1 + |m|^2}{(2l-1)(2l+3)},$$

$$E_l^{(8)} = \frac{\gamma^4}{8} [(20l + 20l^2 + 33)|m|^4 + (-24l^4 - 48l^3 + 2l^2 + 26l - 30)|m|^2$$
$$+4l^6 + 12l^5 - 3l^4 - 26l^3 + 2l^2 + 17l - 3]/[(2l-3)(2l-1)^3(2l+3)^3(2l+5)],$$

$$Q_{ll+2}^{(3)} = \frac{\gamma^2}{2} \frac{\sqrt{(l+1)^2 - |m|^2}\sqrt{(l+2)^2 - |m|^2}}{\sqrt{2l+1}(2l+3)^2\sqrt{2l+5}},$$

$$Q_{ll+2}^{(7)} = \frac{\gamma^4}{2} \frac{\sqrt{(l+1)^2 - |m|^2}\sqrt{(l+2)^2 - |m|^2}(4|m|^2 - 1)}{(2l-1)\sqrt{2l+1}(2l+3)^4\sqrt{2l+5}(2l+7)},$$

$$Q_{ll+4}^{(7)} = \frac{\gamma^4}{8} \frac{\sqrt{(l+1)^2 - |m|^2}\sqrt{(l+2)^2 - |m|^2}\sqrt{(l+3)^2 - |m|^2}\sqrt{(l+4)^2 - |m|^2}}{\sqrt{2l+1}(2l+3)^2(2l+5)^2(2l+7)^2\sqrt{2l+9}},$$

$$H_{ll}^{(6)} = \frac{\gamma^4}{2} \Big( (16l^4 + 32l^3 + 248l^2 + 232l + 201)|m|^4 + (-10l^2 - 224l^4$$
$$-96l^5 + 118l - 288l^3 - 32l^6 - 195)|m|^2 + 16l^8 + 64l^7 + 46l + 40l^6 - 127l^4$$
$$-104l^5 + 71l^2 - 6l^3 - 6\Big)/\Big((2l-3)(2l-1)^4(2l+3)^4(2l+5)\Big),$$

$$H_{ll+4}^{(6)} = \frac{-\gamma^4\sqrt{(l+1)^2 - |m|^2}\sqrt{(l+2)^2 - |m|^2}\sqrt{(l+3)^2 - |m|^2}\sqrt{(l+4)^2 - |m|^2}}{4\sqrt{2l+1}(2l+3)^2(2l+5)(2l+7)^2\sqrt{2l+9}}.$$

This asymptotic behavior of effective potentials allows us to use the above boundary conditions (14) at $r \to 0$ to find regular and bounded solutions.

As an example, we describe briefly the algorithm for evaluating matrix elements (12) and (13) at large $r$ in the vicinity of $\eta = 1 - y/(2p)$ as series expansions by the inverse power of $p$ without taking into account the exponential small terms. Following [10], we define function $\Phi_n(p, \eta)$ in the form

$$\Phi_n(p, y) = \exp(-y/2)\left(\frac{y}{4p}\right)^{|m|/2}\left(1 - \frac{y}{4p}\right)^{|m|/2} F_n(y), \tag{36}$$

Here $F_n(y)$ are solutions of an eigenvalue problem for equation (16) in a region $\eta \in D_1$ with $D_1 = [1 - \eta_1, 1]$, $\eta_1 = o(p^{1/2-\epsilon})$, $0 < \epsilon < 1/2$ with the corresponding spectral parameter $\beta_n(p) = -(|m| + 1)/2 + \lambda_n/(4p)$ at fixed $m = |m|$:

$$y\frac{d^2 F_n(y)}{dy^2} + (|m| + 1 - y)\frac{dF_n(y)}{dy} + \beta_n(p)F_n(y) \tag{37}$$

$$= \frac{1}{4p}\left(y^2\frac{d^2 F_n(y)}{dy^2} + (2|m| + 2 - y)\frac{dF_n(y)}{dy} + (|m| + 1)(|m| - y)F_n(y)\right),$$

with an additional normalization condition

$$\frac{1}{2p}\int_0^\infty \Phi_i(p,y)\Phi_j(p,y)dy = \tag{38}$$

$$\frac{2}{(4p)^{|m|+1}}\int_0^\infty F_i(y)F_j(y)\exp(-y)y^{|m|}\left(1-\frac{y}{4p}\right)^{|m|}dy = \delta_{ij},$$

$$\left(1-\frac{y}{4p}\right)^{|m|} = \sum_{k=0}^{|m|}\left(\frac{y}{4p}\right)^k\prod_{l=1}^{k}\frac{l-m}{l}, \quad \left(\frac{y}{4p}\right)<1.$$

We find $F_n(y)$ as a series expansion in the inverse powers of $p$ at $k_{max}>|m|$:

$$F_n(y)=(4p)^{\frac{|m|+1}{2}}\sum_{k=0}^{k_{max}}\left(\frac{1}{4p}\right)^k\sum_{s=-k}^{k}c_{s,n}^{(k)}L_{n+s}^{(|m|)}(y), \quad \beta_n(p)=\beta_n^{(0)}+\sum_{k=1}^{k_{max}}\left(\frac{1}{4p}\right)^k\beta_n^{(k)}.$$

Here $L_n^{(|m|)}(y)$ are Laguerre polynomials that satisfy equations [3]

$$y\frac{d^2L_n^{(|m|)}(y)}{dy^2}+(|m|+1-y)\frac{dL_n^{(|m|)}(y)}{dy}+\beta_n^{(0)}L_n^{(|m|)}(y)=0, \tag{39}$$

where $\beta_n^{(0)}=n$, and the orthonormalization conditions

$$\langle n_l|n_r\rangle \equiv \int_0^\infty dy y^{|m|}e^{-y}L_{n_l}^{(|m|)}(y)L_{n_r}^{(|m|)}(y) = \frac{(n_l+|m|)!}{n_l!}\delta_{n_l n_r}, \tag{40}$$

$$\langle n|y^a|n\rangle = \frac{(n+|m|)!}{n!}(2n+|m|+1)^a, \quad a>-|m|-1. \tag{41}$$

The substitution of (39) into (37) leads to recurrence relations with respect to unknown coefficients $c_{s,n}^{(k)}$ and $\beta_n^{(k)}$ of the above series expansion for $s\geq 1$

$$sc_{s,n}^{(k)} = ((n+s+|m|+1)(2n+2s+|m|+1)-(n+s+|m|)(|m|+1))c_{s,n}^{(k-1)}$$
$$-(n+s)(n+s+|m|)c_{s-1,n}^{(k-1)}-(n+s+|m|+1)(n+s+1)c_{s+1,n}^{(k-1)}$$
$$+\sum_{k'=1}^{k-|s|}\beta_n^{(t)}c_{s-1,n}^{(k-k')} \tag{42}$$

with initial data $\beta_n^{(0)}=n$ and $c_{0,n}^{(0)}=2^{-1/2}(n!/(n+|m|)!)^{1/2}$. Note, the unknown coefficients $c_{0,n}^{(k)}$, $k\geq 1$ are not specified from recurrence relation (42). From orthonormalization conditions (38) we have the following equation for $c_{0,n}^{(k)}$:

$$c_{0,n}^{(k)} = -\frac{1}{2p}\sum_{s=1}^{k}\sum_{t'=s-k}^{k-s}\sum_{t=-s}^{s}c_{t,n}^{(s)}c_{t',n}^{(k-s)}\langle n+t|\left(1-\frac{y}{4p}\right)^{|m|}|n+t'\rangle$$

$$= -\frac{2}{(4p)^{|m|+1}}\sum_{s=1}^{k}\sum_{t'=s-k}^{k-s}\sum_{t=-s}^{s}\sum_{k'=0}^{\min(k,m)}c_{t,n}^{(s)}c_{t',n}^{(k-k'-s)}\langle n+t|\left(\frac{y}{4p}\right)^{k'}\left(\prod_{l=1}^{k'}\frac{l-m}{l}\right)|n+t'\rangle.$$

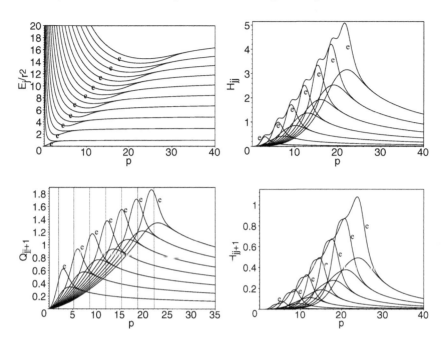

**Fig. 2.** Some potentials $H_{ij}(r)$, $Q_{ij}(r)$ and potential curves $E_j(r)/r^2$ versus the parameter $p$ for even (marked by symbol "e") and odd parity at $|m| = 0$ and $\gamma = 1$

In terms of (36) the asymptotics of matrix elements (12) and (13) at large $r$

$$Q_{n_l,n_r}(r) = -\frac{1}{2p} \int_0^\infty dy \Phi_{n_l}(y,r) \left( \frac{\partial \Phi_{n_r}(y,r)}{\partial r} + \frac{2y}{r} \frac{\partial \Phi_{n_r}(y,r)}{\partial y} \right), \tag{43}$$

$$H_{n_l,n_r}(r) = \frac{1}{2p} \int_0^\infty dy \left( \frac{\partial \Phi_{n_l}(y,r)}{\partial r} + \frac{2y}{r} \frac{\partial \Phi_{n_l}(y,r)}{\partial y} \right)\left( \frac{\partial \Phi_{n_r}(y,r)}{\partial r} + \frac{2y}{r} \frac{\partial \Phi_{n_r}(y,r)}{\partial y} \right)$$

are the series expansion by the inverse power of $r$ without the exponential terms

$$r^{-2}E_n(r) = E_n^{(0)} + \sum_{k=1}^{k_{max}} r^{-2k} E_n^{(2k)}, \quad H_{n_l,n_r}(r) = \sum_{k=1}^{k_{max}} r^{-2k} H_{n_l,n_r}^{(2k)}, \tag{44}$$

$$Q_{n_l,n_r}(r) = \sum_{k=1}^{k_{max}} r^{-2k+1} Q_{n_l,n_r}^{(2k-1)}, \quad E_n^{(2k-1)} \equiv 0, \quad H_{n_l,n_r}^{(2k-1)} \equiv 0, \quad Q_{n_l,n_r}^{(2k)} \equiv 0.$$

In these formulas asymptotic quantum numbers $n_l$, $n_r$ denote transversal quantum numbers that are connected with the unified numbers $j$, $j'$ by the formulas $n_l = j - 1$, $n_r = j' - 1$ for both even and odd parity. Evaluating the exponential small corrections can be done using additional series expansion of the solution in the region $D_2 = [0, 1 - \eta_2]$, $\eta_2 < \eta$, $\eta_2 = o(p^{-1/2-\varepsilon})$ in accordance with [10].

The calculation was performed by the algorithm implemented in MAPLE up to $k_{max} = 12$. Below we display the first few coefficients of the diagonal matrix of potential curves $E$ at fixed $m$

$$E_n^{(0)} = \gamma(2n + |m| + m + 1), \quad E_n^{(2)} = -2n^2 - 2n|m| - 2n - |m| - 11, \tag{45}$$
$$E_n^{(4)} = \gamma^{-1}(-4n - 1 - 6n^2 - 2|m| - 4n^3 - |m|^2 - 6n^2|m| - 2n|m|^2 - 6n|m|),$$
$$E_n^{(6)} = \gamma^{-2}(-136n - 96n|m|^3 - 23 - 336n^2 - 330n^4|m| - 68|m| - 426n^2|m|^2$$
$$-330n^4 - 444n^3 - 72|m|^2 - 132n^5 - 32|m|^3 - 10n|m|^4 - 5|m|^4$$
$$-284n^3|m|^2 - 666n^2|m| - 286n|m|^2 - 660n^3|m| - 96n^2|m|^3 - 336n|m|).$$

We display also matrix elements with $m = 0$:
$Q$ is antisymmetric matrix with matrix elements

$$Q^{(1)}_{n_l,n_r} = (n_l + 1)\delta_{n_l+1,n_r} - (n_r + 1)\delta_{n_l,n_r+1},$$
$$Q^{(3)}_{n_l,n_r} = (2\gamma)^{-1}[2(n_l + 1)^2\delta_{n_l+1,n_r} - 2(n_r + 1)^2\delta_{n_l,n_r+1}$$
$$+(n_l + 1)(n_l + 2)\delta_{n_l+2,n_r} - (n_r + 1)(n_r + 2)\delta_{n_l,n_r+2}],$$

$H$ is symmetric matrix with matrix elements

$$H^{(2)}_{n_l,n_r} = (2n_r^2 + 2n_r + 1)\delta_{n_l,n_r} - (n_l+1)(n_l+2)\delta_{n_l+2,n_r} - (n_r+1)(n_r+2)\delta_{n_l,n_r+2},$$
$$H^{(4)}_{n_l,n_r} = \gamma^{-1}[2(2n_r + 1)(n_r^2 + n_r + 1)\delta_{n_l,n_r}$$
$$+(n_l + 1)(n_l^2 + 2n_l + 2)\delta_{n_l+1,n_r} + (n_r + 1)(n_r^2 + 2n_r + 2)\delta_{n_l,n_r+1}$$
$$-(n_l + 1)(n_l + 2)(2n_l + 3)\delta_{n_l+2,n_r} - (n_r + 1)(n_r + 2)(2n_r + 3)\delta_{n_l,n_r+2}$$
$$-(n_l + 1)(n_l + 2)(n_l + 3)\delta_{n_l+3,n_r} - (n_r + 1)(n_r + 2)(n_r + 3)\delta_{n_l,n_r+3}].$$

Note that $E_j^{(2)} + H_{jj}^{(2)} = 0$, i.e., at large $r$ the centrifugal terms are eliminated in Eq. (11). It means that the leading terms of radial solutions $\chi_{ji_o}(r)$ have the same asymptotics as the Coulomb functions with a zero angular momentum and the effective charge $\hat{Z}$ in terms of the scaled radial variable $\hat{r}$.

## 6   Numerical Results

In this section we present our numerical results for the energy spectrum of a hydrogen atom in a magnetic field. Ten angular basis functions of the Kantorovich expansion (9) at $j_{max} = 10$ that are eigensolutions of the problem (10) were calculated together with the needed matrix elements (12) and (13) to solve eigenvalue problem (11)-(15) for the first ten equations ($j_{max} = 10$) of the system (11). The problem (10) was solved by the conventional expansion (17) at $s_{max} = 200$ of the regular and bound solutions of Eq. (16). The ten eigensolutions ($j_{max} = 10$) of the algebraic problem (22) and the corresponded matrix elements (12) and (13) were calculated by the algorithms described in sections 3 and 4. The obtained results coincide with those obtained by FEM [1] with ten digits accuracy. The numerical values of effective potentials are shown in

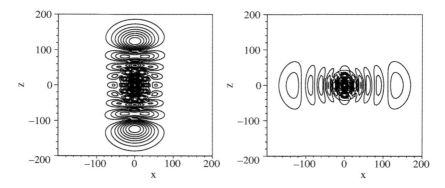

**Fig. 3.** The probability density isolines of the Zeeman wave states $|N, N_r, m\rangle$ with even parity $\sigma = +1$ and $m = 0$ in a homogeneous magnetic field: (lhs) of minimal energy correction $|9, 0, 0\rangle$ and (rhs) of maximal energy correction $|9, 8, 0\rangle$

**Table 1.** Convergence of the method for energy $E(N = 9, N_r = 0, 2, 4, 6, 8, m = 0, \sigma = +1)$ (in a.u.) of even wavefunctions with the number of coupled channels $n_{max} = 8$, $\gamma = 1.472 \times 10^{-5}$ and $shift = 0.0112$

| $N_r$ | $j_{max} = 2$ | $j_{max} = 4$ | $j_{max} = 6$ | $j_{max} = 8$ |
|---|---|---|---|---|
| 0 | -0.00781242971347 | -0.00617279526323 | -0.00617279808777 | -0.00617279808777 |
| 2 | -0.00781225974455 | -0.00617270287945 | -0.00617274784933 | -0.00617274784933 |
| 4 | -0.00617272642538 | -0.00617258450255 | -0.00617268955914 | -0.00617268955914 |
| 6 | -0.00617245301145 | -0.00617243588598 | -0.00617258283911 | -0.00617258283911 |
| 8 | -0.00499982705326 | -0.00499993540325 | -0.00617243586258 | -0.00617243586258 |

Figs. 2 that are in a good agreement with the asymptotic values from Eq. (35) and (44).

Note that the power series expansion of eigenvalues and matrix elements have a finite radius of convergence because there are branch points of eigenvalues $\lambda_j$ in the complex plane of the parameter $p$ [8, 9]. As a consequence, one can see that the maximums of the matrix elements correspond to real parts of such branch points in a complex plane of the parameter $p$. Their positions were estimated by solving an equation, $p^2 - \lambda_j(p) - (1 - m^2) = 0$, using asymptotic values (45) in $\lambda_j(p) = E_j(r) - \gamma m r^2$ at large $p$ that are pointed out in Fig. 2 by vertical lines.

The finite element grid $\omega_r$ of $r$ has been chosen as follows : 0 (100) 3 (70) 20 (80) 100 (the number in parentheses denotes the number of finite elements of order $k = 4$ in each interval). This grid is composed of 999 nodes. The maximum number of unknowns in the system (11) ($n_{max} = 10$) is 9990. The results of calculating the energy values and the rate of convergence of the method versus number $j_{max}$ of basis functions for $N = 9$, $m = 0$ and $\gamma = 1.472 \times 10^{-5}$ are shown in Table 1. The probability density isolines of the Zeeman wave states are shown in Fig. 3.

# 7   Conclusion

A new effective method of calculating wave functions of a hydrogen atom in a strong magnetic field is developed. The method is based on the Kantorovich approach to parametric eigenvalue problems in spherical coordinates. The rate of convergence is examined numerically and illustrated by a set of typical examples. The results are in a good agreement with calculations done by other authors. The elaborated symbolic-numerical algorithm for calculating high derivatives in matrix elements allows us to generate effective approximations for the finite set of radial equations describing an open channel. The developed approach yields a good tool for calculation of threshold phenomena in formation and ionization of (anti)hydrogen like atoms and ions in magnetic traps. One can calculate the dynamics of the Zeeman states in time dependent external electric fields with help of the algorithm of an unitary decomposition of the time evolution operator [11]. The latter has a certain perspective for computer simulation of the ion motion in the trapped models of quantum computers [12].

**Acknowledgement.** This work was partly supported by the Russian Foundation for Basic Research (grant No. 04-01-00784) and by Grant I-1402/2004-2006 of the Bulgarian Foundation for Scientific Investigations.

# References

1. Dimova, M.G., Kaschiev, M.S., Vinitsky, S.I.: The Kantorovich method for high-accuracy calculations of a hydrogen atom in a strong magnetic field: low-lying excited states. Journal of Physics B: At. Mol. Phys. **38** (2005) 2337–2352
2. Chuluunbaatar, O. et al.: On an effective approximation of the Kantorovich method for calculations of a hydrogen atom in a strong magnetic field. In: Proc. SPIE **6165** (2006) 67–83
3. Abramovits, M., Stegun, I.A.: Handbook of Mathematical Functions, Dover, New York, 1965
4. http://mathworld.wolfram.com/SpheroidalWaveFunction.html
5. http://www.netlib.org/eispack/
6. http://www.nag.co.uk/numeric/numerical_libraries.asp
7. Kantorovich, L.V., Krylov, V.I.: Approximate Methods of Higher Analysis, Wiley, New York, 1964
8. Oguchi, T.: Eigenvalues of spheroidal wave functions and their branch points for complex values of propagation constants. Radio Sci. **5** (1970) 1207–1214
9. Skorokhodov, S. L. and Khristoforov, D. V.: Calculation of the branch points of the eigenfunctions corresponding to wave spheroidal functions. Journal of Computational Mathematics and Mathematical Physics **46** (2006) 1132-1146
10. Damburg, R.J., Propin, R.Kh.: On asymptotic expansions of electronic terms of the molecular ion $H_2^+$. J. Phys. B: At. Mol. Phys. **1** (1968) 681–691
11. Gusev A.A. et al.: Symbolic-numerical algorithm for solving the time-dependent Shroedinger equation by split-operator method. In: Proc. 8th Int. Workshop, CASC2005 Kalamata, Greece, Sept. 12–16, 2005, V.G. Ganzha, E.W. Mayr and E.V. Vorozhtsov (eds.), Springer-Verlag, Berlin, Heidelberg (2005) 245–258
12. The Physics of Quantum Information, D. Bouwmeester, A. Ekert and A. Zeilinger (eds.), Springer-Verlag, Berlin, Heidelberg, 2000

# On Decomposition of Tame Polynomials and Rational Functions

Jaime Gutierrez[1] and David Sevilla[2]

[1] Faculty of Sciences,
University of Cantabria,
Santander E–39071, Spain
jaime.gutierrez@unican.es

[2] Dpt. of Computer Science and Software Engineering,
Concordia University,
Montréal, QC, H3G 1M8 Canada
dsevilla@cs.concordia.ca

**Abstract.** In this paper we present algorithmic considerations and theoretical results about the relation between the orders of certain groups associated to the components of a polynomial and the order of the group that corresponds to the polynomial, proving it for arbitrary tame polynomials, and considering the case of rational functions.

## 1 Introduction

The general functional decomposition problem can be stated as follows: given $f$ in a class of functions, we want to represent $f$ as a composition of two "simpler" functions $g$ and $h$ in the same class, i.e. $f = g \circ h = g(h)$. Although not every function can be decomposed in this manner, when such a decomposition does exist many problems become significantly simpler.

Univariate polynomial decomposition has applications in computer science, computational algebra, and robotics. In fact, computer algebra systems such as AXIOM, MAPLE, MATHEMATICA, and REDUCE support polynomial decomposition for univariate polynomials. For some time, this problem was considered computationally hard: the security of a cryptographic protocol was based on its hardness, see [3]. A polynomial time algorithm is given in [13], requiring $O(ns \log r)$ or $O(n^2)$ field operations, where $n = \deg f$, $r = \deg g$, and $s = \deg h$. It works over any commutative ring in the *tame case*, that is, when the ring contains a multiplicative inverse of $r$, and assumes that the polynomials involved are monic. Independently, [8] presented a similar algorithm, running in time $O(n^2)$ sequentially and $O(n \log^2 n)$ in parallel. Several papers have been published on different extensions and variations of this problem; see for instance [5],[6], [4], [11] and [7].

In [18] a polynomial time algorithm to decompose a univariate rational function over any field is presented with efficient polynomial factorization. The paper [1] presented two exponential-time algorithms to decompose rational functions, which are quite efficient in practice. They have been implemented in the MAPLE

V.G. Ganzha, E.W. Mayr, and E.V. Vorozhtsov (Eds.): CASC 2006, LNCS 4194, pp. 219–226, 2006.

package CADECOM, which is designed for performing computations in rational function fields; see [9].

In this paper we will focus on certain structural properties of decomposition of polynomials and rational functions in one variable. Namely, for each polynomial or rational function $f$ in one variable, we can consider the group of transformations of the form

$$z \; \mapsto \; \frac{az + b}{cz + d} \quad \text{such that} \quad f(z) = f\left(\frac{az + b}{cz + d}\right).$$

The relation between the degree of a rational function and the order of its corresponding group can provide valuable information about the structure of the different decompositions of the function. In particular, the following result appears in [2]:

**Theorem 1 ([2]).** *Let $p_1, \ldots, p_m \in \mathbb{C}[x]$ be non-constant and $k_1, \ldots, k_m, k$ be the orders of the groups $G(p_1), \ldots, G(p_m), G(p_1 \circ \cdots \circ p_m)$. Then $k$ divides $k_1 \cdots k_m$.*

One of our goals is to generalize this result to a wide class of polynomials, namely the *tame polynomials*, and also consider other generalizations, like the case of rational functions. We think that it can be used to obtain better algorithms for decomposing non tame polynomials, see [6].

## 2   Polynomial and Rational Decomposition

Our starting point is the decomposition of polynomials and rational functions in one variable. First we will define the basic concepts of this topic in full generality.

**Definition 1.** *Let $\mathbb{K}$ be any field, $x$ a transcendental over $\mathbb{K}$ and $\mathbb{K}(x)$ the field of rational functions in the variable $x$ with coefficients in $\mathbb{K}$. In the set $T = \mathbb{K}(x) \backslash \mathbb{K}$ we define the binary operation of composition as*

$$g(x) \circ h(x) = g(h(x)) = g(h).$$

*We have that $(T, \circ)$ is a semigroup, the element $x$ being its neutral element.*

*If $f = g \circ h$, we call this a decomposition of $f$ and say that $g$ is a* component *on the left of $f$ and $h$ is a* component *on the right of $f$. We call a decomposition* trivial *if any of the components is a unit with respect to decomposition; the units in $(T, \circ)$ are precisely the elements of the form*

$$\frac{ax + b}{cx + d}, \quad a, b, c, d \in \mathbb{K}, \quad ad - bc \neq 0.$$

*Given two decompositions $f = g_1 \circ h_1 = g_2 \circ h_2$ of a rational function, we call them* equivalent *if there exists a unit $u$ such that*

$$h_1 = u \circ h_2, \; g_1 = g_2 \circ u^{-1},$$

*where the inverse is taken with respect to composition.*

*Given $f \in T$, we say that it is* indecomposable *if it is not a unit and all its decompositions are trivial.*

*We define a* complete decomposition *of $f \in \mathbb{K}(x)$ to be $f = g_1 \circ \cdots \circ g_r$ where $g_i$ is indecomposable. The notion of equivalent complete decompositions is straightforward from the previous concepts.*

**Definition 2.** *Given a non–constant rational function $f(x) \in \mathbb{K}(x)$ where $f(x) = f_N(x)/f_D(x)$ with $f_N, f_D \in \mathbb{K}[x]$ and $(f_N, f_D) = 1$, we define the* degree *of $f$ as*

$$\deg f = \max\{\deg f_N,\ \deg f_D\}.$$

*We also define $\deg a = 0$ when $a \in \mathbb{K}$.*

**Remark.** From now on, we will use the previous notation when we refer to the numerator and denominator of a rational function. Unless explicitly stated, we will take the numerator to be monic, even though multiplication by constants will not be relevant.

Now we introduce some basic results about univariate decomposition, see [1] for more details.

**Lemma 1.**

(i) *For every $f \in T$, $\deg f = [\mathbb{K}(x) : \mathbb{K}(f)]$.*
(ii) $\deg (g \circ h) = \deg g \cdot \deg h$.
(iii) *$f(x)$ is a unit with respect to composition if and only if $\deg f = 1$, that is,*
$$f(x) = \frac{ax + b}{cx + d} \text{ with } a, b, c, d \in \mathbb{K} \text{ and } ad - bc \neq 0.$$
(iv) *Every non–constant element of $\mathbb{K}(x)$ is cancellable on the right with respect to composition. In other words, if $f(x), h(x) \in T$ are such that $f(x) = g(h(x))$ then $g(x)$ is uniquely determined by $f(x)$ and $h(x)$.*

Now we relate decomposition and Field Theory by means of the following extended version of Lüroth's theorem.

**Theorem 2.** *Let $\mathbb{K}(\mathbf{x}) = \mathbb{K}(x_1, \ldots, x_n)$ be the field of rational functions in the variables $\mathbf{x} = (x_1, \ldots, x_n)$ over an arbitrary field $\mathbb{K}$. If $\mathbb{F}$ is a field of transcendence degree 1 over $\mathbb{K}$ with $\mathbb{K} \subset \mathbb{F} \subset \mathbb{K}(\mathbf{x})$, then there exists $f \in \mathbb{K}(\mathbf{x})$ such that $\mathbb{F} = \mathbb{K}(f)$. Moreover, if $\mathbb{F}$ contains a non–constant polynomial over $\mathbb{K}$, then there exists a polynomial $f \in \mathbb{K}[\mathbf{x}]$ such that $\mathbb{F} = \mathbb{K}(f)$.*

*Proof.* For a proof we refer to [16], Theorems 3 and 4, and [14]. Constructive proofs can be found in [15] for $n = 1$, and in [10] for arbitrary $n$.

Let $f = g \circ h$. Then $f \in \mathbb{K}(h)$, thus $\mathbb{K}(f) \subset \mathbb{K}(h)$. Also, $\mathbb{K}(f) = \mathbb{K}(h)$ if and only if $f = u \circ h$ for some unit $u$. This provides the following bijection between the decompositions of a rational function $f$ and the intermediate fields in the extension $\mathbb{K}(f) \subset \mathbb{K}(x)$.

**Theorem 3.** *Let $f \in \mathbb{K}(x)$. In the set of decompositions of $f$ we have an equivalence relation given by the definition of equivalent decompositions, and we denote as $[(g, h)]$ the class of the decomposition $f = g \circ h$. Then we have the bijection:*

$$\{ [(g, h)] : f = g(h) \} \longleftrightarrow \{ \mathbb{F} : \mathbb{K}(f) \subset \mathbb{F} \subset \mathbb{K}(x) \}$$
$$[(g, h)] \longleftrightarrow \mathbb{F} = \mathbb{K}(h).$$

Of special interest is the case of $f$ being a polynomial. The following corollary to the second part of Theorem 2 shows that, without loss of generality, we can consider only polynomial components.

**Corollary 1.** *Let $f$ be a polynomial with $f = g \circ h$, where $g, h$ are rational functions. Then there exists a unit $u$ such that*

$$g \circ u, \quad u^{-1} \circ h$$

*are polynomials.*

Because of this, we only need to consider polynomial decomposition when our original function is a polynomial. In the next section we will define and analyze the notion that will allow us to obtain information about the decompositions of a polynomial.

## 3   The Fixing Group of a Polynomial

In order to obtain information about the decompositions of a polynomial, we will introduce a concept that comes directly from Galois Theory.

**Definition 3.** *Let $f \in \mathbb{K}(x)$. The* fixing group *for $f$ is*

$$\Gamma_{\mathbb{K}}(f) = \left\{ \frac{ax + b}{cx + d} : f \circ u = f \right\} < PSL(2, \mathbb{K}).$$

*We will drop the subindex when there is no possibility of confusion about the ground field.*

This definition corresponds to one of the classical Galois applications between the intermediate fields of an extension and the subgroups of its automorphism group, as the following diagram shows:

$$\mathbb{K}(x) \longleftrightarrow \quad \{id\}$$

$$| \qquad \qquad |$$

$$\mathbb{K}(f) \longrightarrow \quad \Gamma_{\mathbb{K}}(f)$$

$$| \qquad \qquad |$$

$$\mathbb{K} \quad \longleftrightarrow PSL(2, \mathbb{K})$$

**Remark.** As $\mathbb{K}(f) = \mathbb{K}(f')$ if and only if $f = u \circ f'$ for some unit $u$, we have that the application $\mathbb{K}(f) \mapsto \Gamma_{\mathbb{K}}(f)$ is well–defined.

Next, we state several interesting properties of the fixing group, see [12] for details.

**Theorem 4.**

(i) *Given* $f \in \mathbb{K}(x) \setminus \mathbb{K}$, $|\Gamma_{\mathbb{K}}(f)|$ *divides* deg $f$. *Moreover, for any field* $\mathbb{K}$ *there is always a function* $f \in \mathbb{K}(x)$ *such that* $1 < |\Gamma_{\mathbb{K}}(f)| < $ deg $f$, *for example for* $f = x^2 \, (x-1)^2$ *we have* $\Gamma_{\mathbb{K}}(f) = \{x, 1\text{-}x\}$ *for any* $\mathbb{K}$.

(ii) $|\Gamma_{\mathbb{K}}(f)| = $ deg $f \Rightarrow \mathbb{K}(f) \subseteq \mathbb{K}(x)$ *is normal. Moreover, if the extension* $\mathbb{K}(f) \subseteq \mathbb{K}(x)$ *is separable, then*

$$\mathbb{K}(f) \subseteq \mathbb{K}(x) \text{ is normal} \quad \Rightarrow \quad |\Gamma_{\mathbb{K}}(f)| = \deg f.$$

# 4   Uniqueness of Intermediate Fields of the Same Degree

First, we will define the class of polynomials on which we will work.

**Definition 4.** *A polynomial* $f \in \mathbb{K}[x]$ *is* tame *when* char $\mathbb{K}$ *does not divide* deg $f$.

The following result shows a nice property of tame polynomials.

**Theorem 5.** *Let* $f \in \mathbb{K}[x]$ *be tame and* $f = g_1 \circ h_1 = g_2 \circ h_2$ *such that* deg $h_1 = $ deg $h_2$. *Then there exists a polynomial unit* $u$ *such that* $h_1 = u \circ h_2$.

*Proof.* See [7].

Due to the equivalence given by Theorem 3, the previous theorem is equivalent to the uniqueness of intermediate fields of the same degree; that is, if $\mathbb{K}(h_1)$, $\mathbb{K}(h_2)$ are fields between $K(f)$ and $\mathbb{K}(x)$ and deg $h_1 = $ deg $h_2$, then $\mathbb{K}(h_1) = \mathbb{K}(h_2)$.

This is not true if we drop the tameness hypothesis.

**Example 1 ([17]).** *Let* $\mathbb{K} = \mathbb{F}_2$, $\alpha^2 - \alpha + 1 = 0$ *with* $\alpha \in \mathbb{F}_4$. *We have that*

$$x^4 + x^2 + x = (x^2 + \alpha x)^2 + \alpha^{-1}(x^2 + \alpha x).$$

In the case of rational functions, the result is also false.

**Example 2 ([1]).** *Let*

$$f = \frac{\omega^3 x^4 - \omega^3 x^3 - 8x - 1}{2\omega^3 x^4 + \omega^3 x^3 - 16x + 1}$$

*where* $\omega$ *is a non-real cubic root of unity in* $\mathbb{Q}$. $f$ *is indecomposable in* $\mathbb{Q}(x)$. *However,* $f = f_1 \circ f_2$ *where*

$$f_1 = \frac{x^2 + (4 - \omega)x - \omega}{2x^2 + (8 + \omega)x + \omega} \; , \; f_2 = \frac{x\omega(x\omega - 2)}{x\omega + 1}.$$

**Example 3.** *Let*

$$f = x^2 + \frac{1}{x^2}.$$

*This function has two different decompositions of the same degree that are not equivalent:*

$$f = \frac{1}{x} \circ x^2 = (x^2 - 2) \circ \frac{1}{x}.$$

## 5 Main Result

In relation to the existence of these fields we will discuss the generalization of the following result:

**Theorem 6 ([2]).** *Let $p_1, \ldots, p_m \in \mathbb{C}[x]$ be non-constant and $k_1, \ldots, k_m, k$ be the orders of the groups $\Gamma(p_1), \ldots, \Gamma(p_m), \Gamma(p_1 \circ \cdots \circ p_m)$. Then $k$ divides $k_1 \cdots k_m$.*

We try to generalize this to polynomials with coefficients in any field. First we study the fixing groups of these polynomials.

**Theorem 7.** *Let $\mathbb{K}$ be a field and $f \in \mathbb{K}[x]$ a tame polynomial. Then $\Gamma(f)$ is cyclic.*

*Proof.* First we prove that there are no elements of the form $x + b$ in $\Gamma(f)$ with $b \neq 0$. Let $H = \{x + b : b \in \mathbb{K}, f(x + b) = f(x)\} < \Gamma(f)$.

If char $\mathbb{K} = p > 0$, any element $x + b \in H$ with $b \neq 0$ has order $p$, so the order of $H$ is divisible by $p$. But the order of $H$ divides deg $f$, therefore $H$ is a trivial group. If char $\mathbb{K} = 0$, no elements of the form $x + b$ with $b \neq 0$ have finite order.

Let $a, b, c \in \mathbb{K}$ be such that $ax + b$, $ax + c \in \Gamma(f)$. Then $(ax + b) \circ (ax + c)^{-1} = x + c - b$, thus $b = c$. Therefore, $B = \{a \in \mathbb{K}^* : \exists b \,|\, ax + b \in \Gamma(f)\}$, a subgroup of the multiplicative group $\mathbb{K}^*$, has the same order as $\Gamma(f)$. But $B$ is cyclic, thus there exists $a_0 \in \mathbb{K}^*$ such that $B = \langle a_0 \rangle$. Given the corresponding $a_0 x + b_0 \in \Gamma(f)$, it is clear that every element of $\Gamma(f)$ is a power of it, therefore $\Gamma(f)$ is cyclic.

We can now generalize Theorem 6 to the case of tame polynomials:

**Theorem 8.** *Let $\mathbb{K}$ be any field and $p_1, \ldots, p_m \in \mathbb{K}[x]$ be tame. Let $k_1, \ldots, k_m, k$ be the orders of $\Gamma(p_1), \ldots, \Gamma(p_m), \Gamma(p_1 \circ \cdots \circ p_m)$. Then $k$ divides $k_1 \cdots k_m$.*

*Proof.* It suffices to take $m = 2$ and then use induction. Let $\gamma$ be a generator of the cyclic group $\Gamma(p_1 \circ p_2)$. As $p_1 \circ p_2 = (p_1 \circ p_2) \circ \gamma = p_1 \circ (p_2 \circ \gamma)$, by Theorem 5 there exists a unit $\eta$ such that $p_2 \circ \gamma = \eta \circ p_2$. Then $p_1 \circ p_2 = p_1 \circ p_2 \circ \gamma = p_1 \circ \eta \circ p_2$, therefore $p_1 \circ \eta = p_1$. That is, $\eta \in \Gamma(p_1)$ and its order $l_1$ divides $k_1$.

Also, $p_2 \circ \gamma = \eta \circ p_2$ implies $p_2 \circ \gamma^r = \eta^r \circ p_2$ for each integer $r$. On one hand, taking $r = k$, we have $\eta^k = x$, thus $l_1$ divides $k$. On the other hand, taking $r = l_1$ we have $\gamma^{l_1} \in \Gamma(p_2)$, that has order $l_2 = k/l_1$. Therefore, as $l_1|k_1, l_2|k_2$ y $l_1 l_2 = k$, we have $k|k_1 k_2$.

## 6   Generalizations and Future Work

In the rational case, as the uniqueness of fields of the same degree is not true in general (see Examples 2 and 3), we can think that this theorem cannot be fully generalized. This is indeed the case, as the next example shows.

**Example 4.** *Let*
$$f = \frac{-1 + 33x^4 + 33x^8 - x^{12}}{x^2 - 2x^6 + x^{10}}.$$
*We have that*
$$\Gamma_{\mathbb{C}}(f) = \left\{ \pm x, \pm \frac{1}{x}, \pm \frac{i(x+1)}{x-1}, \pm \frac{i(x-1)}{x+1}, \pm \frac{x+i}{x-i}, \pm \frac{x-i}{x+i} \right\}.$$

*The element $i(x+1)/(x-1)$ has order 3, and a function that is fixed by it is*
$$h = \frac{x^3 + (x \quad 1)x + 1 - i}{(x-1)(x-i)}.$$

*The field $\mathbb{C}(h)$ is not left invariant by every element of $\Gamma_{\mathbb{C}}(f)$, only by the three elements in the subgroup (as they leave the generator fixed). For example it is easy to check that*
$$h \circ (-x) \notin \mathbb{C}(h).$$

Still, the following conjecture can be posed even if the proof is not valid in this case.

**Conjecture 1.** *Theorem 8 is true for every rational function whose degree is not a multiple of the characteristic of the field.*

A different direction that may allow for some generalization is given by the relation between the degrees of the components for tame polynomials:

**Theorem 9 ([17]).** *If $\mathbb{K}(f) \cap \mathbb{K}(g)$ contains a polynomial $h$ such that $\deg h \not\equiv 0 \bmod \operatorname{char} \mathbb{K}$, then*

$$[\mathbb{K}(f) : \mathbb{K}(f) \cap \mathbb{K}(g)] = \frac{\operatorname{lcm}(\deg f, \deg g)}{\deg f},$$

$$[\mathbb{K}(f,g) : \mathbb{K}(f)] = \frac{\deg f}{\gcd(\deg f, \deg g)}.$$

Because of this, it is possible to consider that, as in Theorem 8, not only $k$ divides $k_1 k_2$ but also $\gcd(k_1, k_2)$. The following trivial example shows that this is not true in general:

**Example 5.** *The function $x^4 = x^2 \circ x^2$ does not satisfy the above statement, since $4 \nmid 2$.*

In any case, we consider that it is of interest to study the classes of polynomials and rational functions for which these statements hold.

## Acknowledgement

This work is partially supported by Spanish Ministry of Science grant MTM2004-07086.

## References

1. C. Alonso, J. Gutierrez, T. Recio, *A rational function decomposition algorithm by near-separated polynomials.* J. Symbolic Comput. 19 (1995), no. 6, 527–544.
2. A. F. Beardon, T. W. Ng, *On Ritt's factorization of polynomials.* J. London Math. Soc. (2) 62 (2000), no. 1, 127–138.
3. J. Cade, *A new public-key cipher which allows signatures.* Proc. 2nd SIAM Conf on Appl. Linear Algebra, Raleigh NC (1985).
4. D. Casperson, D. Ford, J. McKay, *An ideal decomposition Algorithm.* J. Symbolic Comput. 21 (1996), no. 2, 133–137.
5. J. von zur Gathen, *Functional decomposition of polynomials: the tame case.* J. Symbolic Comput. 9 (1990), no. 3, 281–299.
6. J. von zur Gathen, *Functional decomposition of polynomials: the wild case.* J. Symbolic Comput. 10 (1990), no. 5, 437–452.
7. J. Gutierrez, *A polynomial decomposition algorithm over factorial domains.* C. R. Math. Rep. Acad. Sci. Canada 13 (1991), no. 2-3, 81–86.
8. J. Gutierrez, T. Recio, C. Ruiz de Velasco, *A polynomial decomposition algorithm of almost quadratic complexity.* Proc. AAECC-6/88; L. N. Computer Science 357 (1989), 471–476.
9. J. Gutierrez, R. Rubio, *CADECOM: Computer Algebra software for functional DE-COMposition.* Proceedings of the Second Workshop on Computer Algebra in Scientific Computing, V. G. Ganzha, E. W. Mayr, E. V. Vorozhtsov, editors, Samarkand, Uzbekistan, Springer–Verlag (2000), 233–248.
10. J. Gutierrez, R. Rubio, D. Sevilla, *On Multivariate Rational Function Decomposition.* J. Symbolic Comput. 33 (2002), no. 5, 545–562.
11. J. Gutierrez, R. Rubio, J. von zur Gathen, *Multivariate Polynomial decomposition.* Applicable Algebra in Engineering, Communication and Computing, 14 (2003), no. 1, 11–31.
12. J. Gutierrez, D. Sevilla, *On Ritt's decomposition Theorem in the case of finite fields.* Finite Fields and Their Applications 12 (2006), no. 3, 403–412.
13. D. Kozen, S. Landau, *Polynomial decomposition algorithms.* J. Symbolic Comput. 7 (1989), no. 5, 445–456.
14. M. Nagata, *Theory of Commutative Fields.* Translations of Mathematical Monographs, Amer. Math. Soc. 125 (1993).
15. E. Netto, *Ünber einen Lüroth-Gordaschen Satz.* Math. Ann. 9 (1895), 310–318.
16. A. Schinzel: *Selected Topics on Polynomials.* Ann Arbor, University Michigan Press, 1982.
17. A. Schinzel, *Polynomials with special regard to reducibility.* Cambridge University Press, New York, 2000.
18. R. Zippel, *Rational Function Decomposition.* Proc. ISSAC-91 (1991), ACM press, 1–6.

# Newton Polyhedra and an Oscillation Index of Oscillatory Integrals with Convex Phases[*]

Isroil A. Ikromov and Akhmadjon Soleev

Department of Mathematics, Samarkand State University,
Samarkand, 703004, Uzbekistan
asoleev@yandex.ru, ikromov1@rambler.ru

**Abstract.** In this paper we obtain an analog of Schultz decomposition for arbitrary convex smooth functions. We prove existence of adapted coordinate systems for analytic convex functions. We show that the oscillation index of oscillatory integrals with analytic phases is defined by the distance between Newton polyhedron constructed in adapted coordinate systems and the origin.

## 1 Introduction

Let $f : (\mathbb{R}^n, 0) \mapsto (\mathbb{R}, 0)$ be a smooth function in a neighborhood of the origin $f(0) = 0$ and $\nabla f(0) = 0$. An oscillatory integral with the phase $f$ is an integral of the form:

$$I(t) = \int_{\mathbb{R}^n} e^{itf(x)} a(x) dx, \qquad (1)$$

where $t$ is a large real parameter and $a$ is a smooth function with compact support.

Following [1] we define an oscillation index: The oscillation index $\beta$ of the germ $s$ of the function $f$ at critical point 0 is infimum of the set of all real numbers $\alpha$ for which for oscillatory integral (1) holds:

$$I(t) = O(|t|^\alpha) \quad \text{as} \quad |t| \to +\infty$$

for every representative $s$ of the germ $f$ and for all amplitude functions $a$ with sufficiently small support (chosen in accordance with the representative $s$ of $f$).

The following well known result is due to Bernstein I.N. [2], [4]: If the phase $f$ is a real analytic function at zero with $\nabla f(0) = 0$ and $a$ is a smooth function with sufficiently small support concentrated at the origin then for the oscillatory integral (1) the following asymptotic expansion

$$I(t) \approx e^{itf(0)} \sum_{\alpha} \sum_{k=0}^{n-1} c_{k\alpha}(a) t^\alpha (\log t)^k \qquad (2)$$

---

[*] This work was supported by State Committee for Science and Technology of the Republic of Uzbekistan, grant No. 1.1.13 and Foundation of Academy of Sciences of Uzbekistan grant no. 76-06.

V.G. Ganzha, E.W. Mayr, and E.V. Vorozhtsov (Eds.): CASC 2006, LNCS 4194, pp. 227–239, 2006.

holds. Moreover, the range of indices $\alpha$ is a finite number of negative arithmetic progressions, $c_{k\alpha}$ is a distribution concentrated at the critical set $\{x : df(x) = 0\}$ .

A.N. Varchenko investigated connections between the principal part of expansion (2) and Newton polyhedron of the phase function at the critical point constructed in special so-called adapted coordinate systems.

A.N. Varchenko proved that if $n = 2$ and the phase function is analytic at the critical point then there exist coordinate systems adapted to them. But, if $n \geq 3$ for the general case such a coordinate system does not exist. More precisely, there exists the polynomial phase function with property: Discrete characteristics of Newton polyhedron of the phase function constructed in any coordinate system do not define the principal part of the asymptotic expansion of the oscillatory integral with this phase function [6].

On the other hand, as has been shown by H. Schultz [8], such coordinate systems exist in the case when $f$ is a smooth finite type convex function. In a general convex (not finite type) function the behavior of $I(t)$ may be quite different. We consider the connection between oscillation index of the oscillatory integral for convex smooth phase functions. Moreover, we obtain adapted coordinate systems for arbitrary convex analytic functions.

## 2   Newton Polyhedra: Some Terminology

Following [3] and [6] we introduce some notations. Let $Z_+ \subset \mathbb{R}_+ \subset \mathbb{R}$ be the set of all nonnegative integers, all nonnegative real numbers, and all real numbers, respectively. Let $K \subset Z_+^n$. Newton polyhedron of a set $K$ is defined by the convex hull in $\mathbb{R}_+^n$ of the set $\cup_{k \in K}(k + \mathbb{R}_+^n)$.

Let $f$ be a smooth function in a neighborhood of zero. Consider the Maclaurin series of this function:

$$f_x \approx \sum_{k \in Z_+^n} c_k x^k, \quad c_n \in \mathbb{R}.$$

Let us write $supp(f_x) = \{k \in Z_+^n \setminus \{0\} : c_k \neq 0\}$.

The Newton's polyhedron of a Maclaurin series of $f$ is defined by Newton's polyhedron of the set $supp(f_x)$. For practical construction of the Newton's polyhedra see [7].

Let us specify a coordinate system in $\mathbb{R}^n$ and denote by $f_x$ the Maclaurin series of the function $f$ in this coordinate system. Let us denote by $d$ a coordinate of intersection of the straight line $x_1 = \ldots = x_n = d$, $d \in \mathbb{R}$, and the boundary of the Newton's polyhedron. This number will be called the distance between Newton's polyhedron and the origin. The distance is denoted by $d(x)$. A principal face is the face of minimal dimension containing the point $(d(x), \ldots, d(x))$.

Let $f$ be as above and $x = (x_1, \ldots, x_n)$ be a local coordinate system at zero in $\mathbb{R}^n$. Let $f_x$ be Maclaurin series of $f$, and $d(x)$ be the distance between the origin and Newton's polyhedron $N(f_x)$. Let us write $h(f) = sup\{d(x)\}$, where supremum is taken over the set of all local smooth coordinate systems $x$ at the origin. The number $h(f)$ is called the height of the function $f$ [6]. The local coordinate system is called to be adapted to the phase function $f$ if $h(f) = d(x)$.

The polynomial function constructed by A.N. Varchenko shows that in general for $n \geq 3$ such a coordinate system does not exist. On the other hand, for the smooth finite type convex function there exist coordinate systems adapted to phase function.

## 3  On the Schultz Decomposition

We obtain an analog of H. Schultz [8] decomposition for the arbitrary smooth convex function. Moreover, we obtain adapted coordinate systems for arbitrary analytic convex functions.

Let $f$ be a smooth convex function in a neighborhood of the origin with $f(0) = 0$, $\nabla f(0) = 0$. Let $e_1, \ldots, e_n$ denote the standard orthonormal basis in $\mathbb{R}^n$ and $\alpha \in Z_+^n$ a multi-index; in particular, $\alpha_k e_k$ is the multi-index with $k$th entry $\alpha_k \in Z_+$ and otherwise 0.

The following result belongs to H. Schultz for the case of smooth convex finite type functions.

**Theorem 1.** *Let $f : (\mathbb{R}^n, 0) \mapsto (\mathbb{R}, 0)$ be a smooth convex function with $\nabla f(0) = 0$. Then by a rotation of the coordinate system the following hold:*
*(a) There exist an integer number $0 \leq m \leq n$ and even positive integers $(\kappa_1, \ldots, \kappa_m)$ such that the function $f$ has the form:*

$$f(x', x'') = \sum_{\rho(\alpha)=1} c_\alpha x'^\alpha + R(x', x''),$$

*where $x' = (x_1, \ldots, x_m)$, $x'' = (x_{m+1}, \ldots x_n)$, $\alpha \in Z_+^m$ is multi-index and $\rho(\alpha) = \alpha_1/\kappa_1 + \ldots + \alpha_m/\kappa_m$. Moreover, the remainder term $R(x', x'')$ can be written in the form*

$$R(x', x'') = \sum_{\rho(\alpha)=1} x'^\alpha R_\alpha(x', x'') + \sum_{\rho(\alpha)<1} x'^\alpha R_\alpha(x''),$$

*here $R_\alpha(0,0) = 0$ for $\rho(\alpha) = 1$, $\alpha \in Z_+^m$ and for any $\alpha$ with $\rho(\alpha) < 1$ the $R_\alpha(x'')$ is a flat function.*
*(b) The polynomial $p(x') = \sum_{\rho(\alpha)=1} a_\alpha x'^\alpha$ is convex, and $p(x') > 0$ for all $x' \neq 0$.*

**Proof.** Following H. Schultz [8] we define the sets:

$$S_k := \{x \in \mathbb{R}^n : \sum_{j=2}^{k} |D_x^j f(0)| = 0\}$$

for integers $k \geq 2$, where $D_x^j f(0) = (\partial/\partial t)^j f(tx)|_{t=0}$; is the $j$th derivative of $f$ at the origin in the direction $x$ for the unit vector $x$. Since $x \in S_k$ is equivalent to $f(tx) \leq C|t|^{k+1}$ in a neighborhood of $t = 0$, we get from the convexity of $f$ for $x, y \in S_k$ and for $\lambda \in \mathbb{R}$:

$$0 \leq f(t(\lambda x)) \leq C|\lambda t|^{k+1} = C_1 |t|^{k+1},$$

$$0 \leq f(\frac{1}{2}t(x+y)) \leq \frac{1}{2}f(tx) + \frac{1}{2}f(ty) \leq C|t|^{k+1},$$

i.e. $\lambda x, x + y \in S_k$. Therefore, the sets $S_k$ are linear subspaces of $\mathbb{R}^n$. By the definition, if $k_1 > k_2$ then $S_{k_1} \subset S_{k_2}$. Thus, we obtain the sequence of linear subspaces of $\mathbb{R}^n$:

$$S_2 \supset S_3 \supset S_4 \supset \ldots \supset S_n \supset \ldots.$$

Therefore, there exists $N$ such that for any $\nu > N$ we have $S_N = S_\nu$.

Now the following assertion holds:

**Proposition 2.** There are at most $n$ even integers $k_1, \ldots, k_l$ with $2 \leq k_1 < \ldots < k_l$ such that
(1) $S_{k_l} \subset \ldots \subset S_{k_1} \subset S_{k_0} := \mathbb{R}^n$;
(2) For any $x \in S_{k_{j-1}} \setminus S_{k_j}$, $D_x^\nu f(0) = 0$ for any $\nu < k_j$ and $D_x^{k_j} f(0) > 0$.

**Proof.** If $S_k = \mathbb{R}^n$ for any $k \geq 2$ then $f$ is a flat function at the origin. In this case there is nothing to prove. Otherwise, we define $k_1$ to be the smallest integer with $S_{k_1} \neq \mathbb{R}^n$. For $x \in \mathbb{R}^n \setminus S_{k_1}$ we have $D_x^\nu f(0) = 0$ for any $\nu < k_1$ and $D_x^{k_1} f(0) \neq 0$ from the the definition of $k_1$. Moreover, since $f(tx)$ has minimum at $t = 0$, then $k_1$ is an even number and $D_x^{k_1} f(0) > 0$.

If $S_{k_1} = S_N$ for any $N > k_1$ then proposition 2 is proved. Otherwise, we continue in the same way having fixed $k_1, \ldots, k_j$, we define $k_{j+1}$ with $S_{k_{j+1}} \neq S_{k_j}$, and we get the desired properties of $k_{j+1}$ as for $k_1$. We repeat this until $S_{k_l} = S_N$ for any $N > k_l$, and proposition 2 is proved.

Assume that $dim S_N = n - m$. If $S_N = \{0\}$ or equivalently $n = m$ then we get the result proved by H. Schultz [8].

First, we define an orthonormal basis in $\mathbb{R}^n$ such that $S_N$ is generated by $\{e_{m+1}, \ldots, e_n\}$.

Now, we rotate the coordinate system such that for each $j$ the space $S_{k_j}/S_N$ (where $S_{k_j}/S_N$ is the factor space) is spanned from $e_i, \ldots, e_m$ where $i = m + 1 - dim(S_{k_j}/S_N)$. We define the numbers $\kappa_j$ to be $k_\nu$ if $e_j \in S_{k_{\nu-1}} \setminus S_{k_\nu}$. Then since $\mu! a_{\mu e_j} = D_{e_j}^\mu f(0)$ it follows that $a_{\mu e_j} = 0$ for $\mu \leq \kappa_j - 1$, and $a_{\kappa_j e_j} > 0$.

Let $\rho(\alpha) = \alpha_1/\kappa_1 + \ldots + \alpha_m/\kappa_m$, $\delta = \min\{\rho(\alpha) : a_\alpha \neq 0\}$ and $p_\delta(x) = \sum_{\rho(\alpha)=\delta} a_\alpha x^\alpha$. If we can show $p_\delta$ is convex, then we get easily $\delta = 1$. If $\delta < 1$, we get $p_\delta(e_j) = 0$ for $j = 1, \ldots, m$ then the convexity of $p_\delta$ gives $p_\delta \equiv 0$, i.e. $a_\alpha = 0$ for $\rho(\alpha) = \delta$ in contradiction to the definition of $\delta$.

To prove the convexity of $p_\delta$ we assume the existence of $x', v' \in \mathbb{R}^m$ with $v'^T p_\delta''(x')v' < 0$. If $Q_\alpha = a_\alpha x^\alpha$ and $A_t$ the diagonal matrix $diag(t^{1/\kappa_1}, \ldots, t^{1/\kappa_m})$, then an elementary computation gives:

$$(A_t v'^T)Q_\alpha''(A_t x')(A_t v') = t^{\rho(\alpha)} v'^T Q_\alpha''(x')v'.$$

Splitting

$$f(x) = \sum_{\alpha \in M} a_\alpha x^\alpha + R(x)$$

with $M = \{\alpha \in Z^n_+ : a_\alpha \neq 0, |\alpha| \leq k_m\}$, we get for $t > 0$

$$(A_t v'^T) f''(A_t x')(A_t v') = t^\delta v'^T p''_\delta(x') v' + \sum_{\alpha \in M, \rho(\alpha) > \delta} t^{\rho(\alpha)} v'^T Q''_\alpha(x') v' +$$

$$(A_t v'^T) R''(A_t x')(A_t v'),$$

here

$$|(A_t v'^T) R''(A_t x')(A_t v')| \leq c t^{1+1/\kappa_m}, \quad 0 < t < 1.$$

Hence

$$(A_t v'^T) f''(A_t x')(A_t v') < 0$$

for sufficiently small $t > 0$ in contradiction to the convexity of $f$.

It remains to prove that $p(x') > 0$ for $x' \neq 0$. Consider $D - \{x' \subset \mathbb{R}^m : p(x') = 0\}$, then the convexity of $p$ (observe that $p \geq 0$) gives that $B$ is a linear space. Indeed, if $p(x') = 0$ for some $x' \neq 0$ then for $0 < t < 1$ we have

$$0 \leq p(tx') = p(tx' + (1-t)0) \leq tp(x') + (1-t)p(0) = tp(x') = 0.$$

Hence $p(tx') = 0$ for any $t \in [0,1]$. Since $p(tx')$ is a polynomial function then $p(tx') \equiv 0$. Thus, if $x' \in B$ then for any $t$ we have $tx' \in B$.

If $p(x') = 0$ and $p(y') = 0$ then

$$0 \leq p\left(\frac{x'}{2} + \frac{y'}{2}\right) \leq \frac{1}{2}p(x') + \frac{1}{2}p(y') = 0$$

Consequently, $\frac{x'}{2} + \frac{y'}{2} \in B$. Therefore, $x' + y' \in B$.

If $B = \{0\}$ we are done, so assume that there is a nonzero $x \in B \subset \mathbb{R}^m$. By the homogeneity of $B$ with respect to $A_t$, $A_t x \in B$ also, and, in particular, $x(t) = |A_t x|^{-1} A_t x \in B$, for $t > 0$. Let $\nu$ be the greatest integer for which $x_\nu$ is nonzero. Then $x(t)$ converges as $t \to +0$ to some $y$, $|y| = 1$, with $y_j = 0$ for all $j$ with $\kappa_j \neq \kappa_\nu$. Since $p$ is continuous, then also $y \in B$, and from the definition of $k_j$ we get $y \in S_{\kappa_\nu - 1} \setminus S_{\kappa_\nu}$. This implies:

$$c_1 t^{\kappa_\nu} \leq f(ty) = \sum_{\alpha \in M} a_\alpha t^{|\alpha|} y^\alpha + R(ty) \leq c_2 t^{\kappa_\nu},$$

for sufficiently small $t > 0$, where $|R(ty)| \leq c t^{\kappa_n + 1}$. Therefore, $\sum_{|\alpha| = \kappa_\nu} a_\alpha y^\alpha > 0$.

Let $M_0$ denote the set $\{\alpha \in M : \alpha_j = 0$ for all $j$ with $\kappa_j \neq \kappa_\nu\}$. Then $y^\alpha = 0$ if $\alpha \notin M_0$, and $\rho(\alpha) = |\alpha|/\kappa_\nu$ if $\alpha \in M_0$.

It follows that

$$p(y) = \sum_{\rho(\alpha)=1} a_\alpha y^\alpha = \sum_{\rho(\alpha)=1, \, \alpha \in M_0} a_\alpha y^\alpha = \sum_{|\alpha|=\kappa_\nu, \, \alpha \in M_0} a_\alpha y^\alpha = \sum_{|\alpha|=\kappa_\nu} a_\alpha y^\alpha > 0$$

in contradiction to $p(y) = 0$. Hence $B = \{0\}$.

Now we show that the remainder term $R(x', x'')$ can be written in the form:

$$R(x', x'') = \sum_{\rho(\alpha)=1} x'^\alpha R_\alpha(x', x'') + \sum_{\rho(\alpha)<1} x'^\alpha R_\alpha(x''),$$

where for any $\alpha$ with $\rho(\alpha) < 1$ the $R_\alpha(x'')$ is a flat function. For the finite type convex functions the terms with $\rho(\alpha) < 1$ do not occur. Thus, the next part of the proof is different from the case considered by H. H. Schultz.

First, we consider the case $m = 1$ and $n = 2$.

**Lemma 3.** If $f$ is a convex smooth function with $m = 1$ and $n = 2$ then there exists a coordinate system such that it can be written in the form:

$$f(x_1, x_2) = x_1^{2k} a(x_1, x_2) + \sum_{l=0}^{2k-1} x_1^l a_l(x_2),$$

where $a_l(x_2)$ are flat functions for $l = 0, \ldots, 2l - 1$ and also $a(0, 0) > 0$.

**Proof of Lemma 3.** It is easy to show that $a_0$ is a flat function because $a_0(x_2) = f(0, x_2)$ is a flat function by the definition.

Lemma 3 will be proved by a contradiction method. Denote by $l_0$ the minimal number of the set $\{1, \ldots, 2k-1\}$ such that $a_{l_0}$ is not a flat function. We introduce the new function by

$$f_1(x_1, x_2) = \sum_{l=l_0}^{2k-1} x_1^l a_l(x_2).$$

We can write the function $f_1(x_1, x_2)$ in the form:

$$f_1(x_1, x_2) = p_1(x_1, x_2) + R(x_1, x_2),$$

where $p_1(x_1, x_2)$ is a non-zero weighted homogeneous polynomial function with weight $(1/2k, \kappa)$ and $R(x_1, x_2)$ is the remainder term. Further we take the straight line $\kappa_1 x_1 + \kappa_2 x_2 = 1$ with $\kappa_1 > 0$, $\kappa_2 > 0$ supported to the Newton polyhedron of $f_1$. We assume that it has a unique point of intersection with Newton's polyhedron of the function $f$. If $f_1$ is not a flat function then such a straight line exists. Then the function $f$ can be written as

$$f(x_1, x_2) = c x_1^l x_2^k + f_{1>}(x_1, x_2),$$

where $c > 0$ is a fixed constant, $x_1^l x_2^k$ is a weighted homogeneous polynomial function of degree one with weight $(\kappa_1, \kappa_2)$, and the Newton polyhedron of the function $f_{1>}(x_1, x_2)$ is contained in the half-plane $\kappa_1 x_1 + \kappa_2 x_2 > 1$. Note that both $l, k$ are even positive integers. Simple computations show that

$$Hess f(x_1, x_2) = -lk(k+l)c^2 x_1^{2l-2} x_2^{2k-2} + h(x_1, x_2),$$

where the Newton polyhedron of the function $h$ is contained in the half-plane $\kappa_1/(1 - 2(\kappa_1 + \kappa_2))x_1 + \kappa_1/(1 - 2(\kappa_1 + \kappa_2))x_2 > 1$.

It is easy to see that $Hess f(x_1, x_2) < 0$ for some sufficiently small values $x_1, x_2$, which contradicts the convexity condition of the function. Lemma 3 is proved.

We write the function $f$ in the form:

$$f(x) = p(x') + R(x', x''),$$

and write the remainder term in the form:

$$R(x', x'') = \sum_{\rho(\alpha)=1} x'^\alpha R_\alpha(x', x'') + \sum_{\rho(\alpha)<1} x'^\alpha R_\alpha(x'').$$

**Lemma 4.** For any $\alpha$ with $\rho(\alpha) < 1$ the function $R_\alpha(x'')$ is a flat function.

**Proof of Lemma 4.** Note that we can consider the restriction of the function in the line $x'' = \tau v''$ (where $v'' \in \mathbb{R}^{n-m}$ is a fixed vector), then the new function $f(x', \tau v'')$ is a convex function of the variables $(x', \tau)$. Thus, we can assume without loss of generality that $m = n - 1$ or $x'' = x_n$. We now define the operator $A_\mu : \mathbb{R}^{n-1} \mapsto \mathbb{R}^{n-1}$ depending on the parameter $\mu > 0$ and defined by:

$$A_\mu(v_1, \ldots, v_{n-1}) = (\mu^{1/\kappa_1} v_1, \ldots, \mu^{1/\kappa_{n-1}} v_{n-1}).$$

Note that for any fixed positive real number $\mu$ the function $f(A_\mu(v), x_n)$ is a convex function of the variables $(v, x_n)$. We now introduce a new function of two variables $(t, x_n)$ defined by $g(t, x_n) := f(A_\mu(tv), x_n)$ Thus, we have

$$g(t, x_n) = \mu p(tv) + \mu \sum_{\rho(\alpha)=1} t^{|\alpha|} v^\alpha R_\alpha(A_\mu(tv), x_n) + \sum_{\rho(\alpha)<1} \mu^{\rho(\alpha)} t^{|\alpha|} v^\alpha R_\alpha(x_n).$$

Since $g(t, x_n)$ is a convex function of the variables $(t, x_n)$ the $Hess_{tx_n} g(t, x_n)$ is a positive function. Denote by $l$ a minimal number such that $v_l \neq 0$. Then $p(tv) = t^{k_l} p_1(t, v)$. Note that $p_1(0, v) \neq 0$. As above (as in the proof of Lemma 2) we prove Lemma 3 by a contradiction method. Denote by

$$\beta = min\{\rho(\alpha) : g_1(t, x_n) := \sum_{\rho(\alpha)=\beta} t^{|\alpha|} v^\alpha R_\alpha(x_n) \text{ is not a flat function}\}.$$

Note that for any $\alpha$ with $\rho(\alpha) = \beta$ we have $\alpha \neq 0$.

Note that if there exists such $\beta < 1$ then the function $g_1(t, x_n)$ can be written in the form:

$$g_1(t, x_n) = ct^l x_n^k + g_{1>}(t, x_n),$$

where $g_{1>}$ is a smooth function having Newton polyhedron contained in the half-plane $\kappa_1 t + \kappa_2 x_n > 1$ and $\kappa_1 l + \kappa_2 k = 1$. As above (see Lemma 3) we have

$$Hess\, g_1(t, x_n) = -c^2 lk(k+l) t^{2l-2} x_n^{2k-2} + h(t, x_n) < 0$$

for sufficiently small values $(t, x_n)$. Moreover, for the function $g(t, x_n)$ we have

$$Hess\, g(t, x_n) = \mu^\beta (Hess\, g_1(t, x_n) + \mu^\delta \psi(\mu, t, x_n)),$$

where $\delta > 0$ and $\psi(\mu, t, x_n)$ is a smooth function. Note that $Hess\ g_1(t, x_n)$ is a finite type function. Therefore, we can choose the parameters $\mu, t, x_n$ such that the inequality $Hess\ g(t, x_n) < 0$ holds, which contradicts the convexity condition of the function $f$. Lemma 3 is proved. Theorem 1 is proved.

**Corollary 5.** Let $f : (\mathbb{R}^n, 0) \mapsto (\mathbb{R}, 0)$ be a convex analytic function $\nabla f(0) = 0$. Then by a rotation of the coordinate system the following holds:
(a) Let $f(x) = \sum a_\alpha x^\alpha$ be the Taylor expansion of $f$ near 0 (in the rotated coordinates). Then there are even integers $k_j \geq 2 (j = 1, \ldots, m \leq n)$ such that for all $\alpha$ with $\rho(\alpha) := \sum_{j=1}^m \alpha_j / k_j < 1$ we have $a_\alpha = 0$; further, if $\alpha = k_j e_j\ (j \leq m)$, then $a_\alpha$ is positive.
(b) The polynomial $p(x') = \sum_{\rho(\alpha)=1} a_\alpha x'^\alpha$ with $x' = (x_1, \ldots, x_m)$ (called the principal part of the Taylor expansion of $f$) is convex, and $p(x') > 0$ for all $x' \neq 0$. The function $f$ has the form:

$$f(x_1, \ldots, x_m, x_{m+1}, \ldots x_n) = \sum_{\rho(\alpha)=1} a_\alpha x'^\alpha + R(x', x''),$$

where the remainder term $R(x', x'')$ can be written in the form

$$R(x', x'') = \sum_{\rho(\alpha)=1} a_\alpha x'^\alpha R_\alpha(x', x''),$$

with $R_\alpha(0, 0) = 0$.

**The proof of Corollary 5 is follows from Theorem 1.**

**Theorem 6.** *Let $f : (\mathbb{R}^n, 0) \mapsto (\mathbb{R}, 0)$ be an analytic convex function in a neighborhood of the origin and $\nabla f(0) = 0$. Then by a rotation of the coordinate system the following hold:*
*1) The coordinate system is adapted to $f$;*
*2) The oscillation index $\beta(f)$ of the function $f$ at zero is equal to $-(h(f))^{-1}$.*

If $f$ is a finite type smooth convex function then the Theorem 5 follows from the H. H. Schultz result [8]. The result follows from Corollary 5 for general convex functions.

## 4    An Algorithm of Finding a Shultz Base

First, we consider some interesting result following from the Theorem 6.

**Proposition 7.** Let $P(x)$ be a homogeneous convex polynomial function with $P(0) = 0$ and $\nabla P(0) = 0$. Then there exists an orthogonal matrix $A$ such that

$$P(Ax) = Q(x_1, \ldots, x_k),$$

where $1 \leq k \leq n$ is a natural number and $Q(x_1, \ldots, x_k)$ is a convex polynomial function satisfying $Q(0) = 0$ and $Q(x_1, \ldots, x_k) \neq 0$ for any $(x_1, \ldots, x_k) \neq 0$.

**Proof.** Indeed we can use Corollary 5 and obtain an orthogonal matrix $A$ such that

$$P(Ax) = Q(x_1, \ldots, x_k) + R(x_1, \ldots, x_n),$$

where $Q$ is a convex polynomial function with $Q(0) = 0$, $\nabla Q(0) = 0$, $R(x)$ is the remainder term. By uniqueness Theorem for analytic functions $R(x) \equiv 0$. Proposition 7 is proved.

Finally, we give a practical computation of the Shultz base. Our algorithm is based on the Newton polyhedron of the function $f$.

**Step 1.** We construct the Newton polyhedron $N(f)$ by using program [7].
**Step 2.** We find a support hyperplane given by $x_1 + x_2 + \ldots + x_n = N_1$ to the Newton polyhedron $N(f)$.
**Step 3.** By using [7] we find a face $p_1$ defined by

$$p_1 := N(f) \cap \{x : x_1 + x_2 + \ldots + x_n = N_1\}.$$

**Step 4.** We take the truncation of the function $f$ corresponding to the face $p_1$ denoting it by $f_{p_1}$.
**Step 5.** We write the function $f$ by using the truncation $f_{p_1}$:

$$f(x) = f_{p_1}(x) + R(x),$$

where $R(x)$ is the remainder term. Note that $f_{p_1}(x)$ is a convex homogeneous polynomial function.
**Step 6.** We find the set

$$S_0 = \{x \in \mathbb{R}^n : f_{p_1}(x) = 0\}.$$

We know that $S_0$ is a linear subspace of the space $\mathbb{R}^n$. Denote by $S_1$ the orthogonal complement of $S_0$ and $k_1 = dim S_1 \geq 1$. Let $\{e_1, \ldots e_{k_1}\}$ be an orthogonal basis of $S_1$. We complete the basis of $\mathbb{R}^n$ by using the basis of $S_1$. Denote by $A_1$ the corresponding orthogonal matrix. Then we obtain

$$f(A_1 x) = Q_1(x_1, \ldots, x_{k_1}) + R_1(x),$$

where $Q_1$ is a convex polynomial function satisfying $Q_1(0) = 0$ and $Q_1(x_1, \ldots, x_{k_1}) \neq 0$ whenever $(x_1, \ldots, x_{k_1}) \neq 0$.
**Step 7.** Now, we consider the function

$$f_1(x'') := R_1(0, \ldots, 0, x''),$$

where $x'' = (x_{k_1+1}, \ldots, x_n)$.

Note that the new function $f_1(x'')$ satisfies all our assumptions proposed for the original function $f$. Now, we repeat all our arguments from step 1 until step 6.

After finitely many steps, we obtain subspaces $S_1, S_2, \ldots, S_m$ satisfying the conditions: $S_i$ and $S_j$ are mutually orthogonal whenever $i \neq j$ and $\mathbb{R}^n = \sum_{j=1}^m S_j$.

The corresponding basis is the Shultz basis for the function $f$. Thus, the problem of finding the Shultz basis is reduced to the classical problem of finding the set of roots for polynomial functions.

1. The POLYHEDRA [7] program is used at steps 1–4 of the algorithm.

2. We use the following MAPLE command
>RootOf $(f_p(x) = 0, x)$;
>evalf(%); to find the plane given by the equation $f_p(x) = 0$;

3. We use the following MAPLE command:
>GramSchmidt $([u1, u2, \ldots, un])$;
>GramSchmidt $([u1, u2, \ldots, un]$, normalized);
to find the orthogonal basis and normalized orthogonal basis of the plane $S_0$ given by the equation $f_p(x) = 0$ and its orthogonal complement.

4. We use the following MAPLE commands:
$subs(x = a, expr)$
$subs(s1, \ldots, sn, expr)$,
where $x, a, expr$ are expressions, $s1, \ldots, sn$ are equations or sets or lists of equations, to find expression of the function $f$ in the new coordinate system and expression of the restriction of the function on orthogonal complement of $S_1$.

**Example.** We consider the polynomial function of the form:

$$f(x, x_2) = p_4(x, x_2) + p_6(x, x_2),$$

where $p_4$ and $p_6$ are homogeneous polynomials of degree 4 and degree 6, respectively.

Let us construct a Shultz basis for the polynomial function $f$:
1. We construct the Newton polygon $N(f)$ for the function $f$;
2. We find a support hyperplane to the Newton polygon given by $\{k_1 + k_2 = N\}$. We have $\{k_1 + k_2 = 4\}$.
3. We find the edge $\gamma$ defined by

$$\gamma := N(f) \cap \{k : k_1 + k_2 = 4\}.$$

4. We take the truncation of the function $f$ corresponding to the edge $\gamma$ denoting it by $f_\gamma$. We obtain $f_\gamma(x, x_2) = p_4(x, x_2)$.
5. We find the set

$$S_0 = \{x \in \mathbb{R}^2 : p_4(x, x_2) = 0\}.$$

We know that $S_0$ is a linear subspace of the space $\mathbb{R}^2$. If $dim(S_0) = 0$ then any orthonormal coordinate system is the Shultz basis. Suppose $dim(S_0) = 1$ and $S_0$ is the subspace generated by a unit vector $g_1$. Denote by $S_1 :=< g_2 >$ the orthogonal complement of $S_0$ and by $A$ the corresponding orthogonal matrix of the transformation defined by $e_1 = Ag_1$, $e_2 = Ag_2$. Denote by $y$ the coordinates of the point $x$ in the coordinate system $\{g_1, g_2\}$. Then we obtain

$$\tilde{p}_4(y) := p_4(Ay) = c_2 y_2^4,$$

where $c_2$ is a positive constant.

Then the coordinate system $\{g_1, g_2\}$ is the Shultz basis for the function $f$. Let us write the Shultz decomposition for the function. We have

$$\tilde{f}(y) := f(Ay) = c_1 y_1^6 + c_2 y_2^4 + c_0 y_1^2 y_2^3 + R(y_1, y_2),$$

where $R$ is the remainder term, and $c_1$, $c_2$ are positive constants.

Note that if $f$ is a convex finite type function then $c_1 y_1^6 + c_2 y_2^4 + c_0 y_1^2 y_2^3$ is the convex polynomial function. A simple computation of Hessian of that polynomial shows that $c_0 = 0$. Thus, we obtain:

$$\tilde{f}(y) := f(Ay) = c_1 y_1^6 + c_2 y_2^4 + R(y_1, y_2).$$

Note that the algorithm corresponds to A.N. Varchenko's algorithm for finding the coordinate systems adapted to $f$ for function of two variables.

Let us consider a simple example.

$$p_4(x, y) = x^4 + 4x^3 y + 6x^2 y^2 + 4xy^3 + y^4$$

and

$$p_6(x, y) = x^6 + 12x^5 y + 60x^4 y^2 + 160x^3 y^3 + 240x^2 y^4 + 192xy^5 + 64y^6.$$

We use the following Maple commands for the example.

```
>   eq := x^4+4*x^3+6*x^2+4*x+1;
```
$$eq := x^4 + 4x^3 + 6x^2 + 4x + 1$$
```
>   solve(eq,x);
```
$$-1, -1, -1, -1$$
```
>   sols := [solve(eq,x)];
```
$$sols := [-1, -1, -1, -1]$$
```
>   with(LinearAlgebra): w1 := <1,-1>: w2 := <1,0>:
    ord := GramSchmidt([w1,w2],normalized);
```
$$ord := \left[\begin{bmatrix} \frac{1}{2}\sqrt{2} \\ -\frac{1}{2}\sqrt{2} \end{bmatrix}, \begin{bmatrix} \frac{1}{2}\sqrt{2} \\ \frac{1}{2}\sqrt{2} \end{bmatrix}\right]$$
```
>   with(LinearAlgebra):
>   V := <1,-1>;
```
$$V := \begin{bmatrix} 1 \\ -1 \end{bmatrix}$$
```
>   GivensRotationMatrix(V, 1, 2);
```
$$\begin{bmatrix} \frac{1}{2}\sqrt{2} & -\frac{1}{2}\sqrt{2} \\ \frac{1}{2}\sqrt{2} & \frac{1}{2}\sqrt{2} \end{bmatrix}$$

> `B:=<<(1/2)*sqrt(2),(1/2)*sqrt(2)>|<-(1/2)*sqrt(2),`
  `(1/2)*sqrt(2)>>;`

$$B := \begin{bmatrix} \dfrac{1}{2}\sqrt{2} & -\dfrac{1}{2}\sqrt{2} \\ \dfrac{1}{2}\sqrt{2} & \dfrac{1}{2}\sqrt{2} \end{bmatrix}$$

> `V:=<t,z>;`

$$V := \begin{bmatrix} t \\ z \end{bmatrix}$$

> `C:=B . V;`

$$C := \begin{bmatrix} \dfrac{1}{2}\sqrt{2}\,t - \dfrac{1}{2}\sqrt{2}\,z \\ \dfrac{1}{2}\sqrt{2}\,t + \dfrac{1}{2}\sqrt{2}\,z \end{bmatrix}$$

> `subs(x=(1/2)*sqrt(2)*t-(1/2)*sqrt(2)*z,}{%`
  `y=(1/2)*sqrt(2)*t+(1/2)*sqrt(2)*z,`
  `x^4+4*x^3*y+6*x^2*y^2+4*x*y^3+y^4+x^6+12*x^5*y+60*x^4*y^2`
  `+160*x^3*y^3+240*x^2*y^4+192*x*y^5+64*y^6);`

$$(1/2*\sqrt{(2)}*t - 1/2*\sqrt{(2)}*z)^4 + 4*(1/2*\sqrt{(2)}*t - 1/2*\sqrt{(2)}*z)^3$$
$$(1/2*\sqrt{(2)}*t + 1/2*\sqrt{(2)}*z) + 6*(1/2*\sqrt{(2)}*t - 1/2*\sqrt{(2)}*z)^2*$$
$$(1/2*\sqrt{(2)}*t + 1/2*\sqrt{(2)}*z)^2 + 4*(1/2*\sqrt{(2)}*t - 1/2*\sqrt{(2)}*z)$$
$$(1/2*\sqrt{(2)}*t + 1/2*\sqrt{(2)}*z)^3 + (1/2*\sqrt{(2)}*t + 1/2*\sqrt{(2)}*z)^4$$
$$+(1/2*\sqrt{(2)}*t - 1/2*\sqrt{(2)}*z)^6 + 12*(1/2*\sqrt{(2)}*t - 1/2*\sqrt{(2)}*z)^5$$
$$(1/2*\sqrt{(2)}*t + 1/2*\sqrt{(2)}*z) + 60*(1/2*\sqrt{(2)}*t - 1/2*\sqrt{(2)}*z)^4$$
$$(1/2*\sqrt{(2)}*t + 1/2*\sqrt{(2)}*z)^2 + 160*(1/2*\sqrt{(2)}*t - 1/2*\sqrt{(2)}*z)^3$$
$$(1/2*\sqrt{(2)}*t + 1/2*\sqrt{(2)}*z)^3 + 240*(1/2*\sqrt{(2)}*t - 1/2*\sqrt{(2)}*z)^2$$
$$(1/2*\sqrt{(2)}*t + 1/2*\sqrt{(2)}*z)^4 + 192*(1/2*\sqrt{(2)}*t - 1/2*\sqrt{(2)}*z)$$
$$(1/2*\sqrt{(2)}*t + 1/2*\sqrt{(2)}*z)^5 + 64*(1/2*\sqrt{(2)}*t + 1/2*\sqrt{(2)}*z)^6;$$

$$\%2^4 + 4\,\%2^3\,\%1 + 6\,\%2^2\,\%1^2 + 4\,\%2\,\%1^3 + \%1^4 + \%2^6 + 12\,\%2^5\,\%1$$
$$+60\,\%2^4\,\%1^2$$
$$+160\,\%2^3\,\%1^3 + 240\,\%2^2\,\%1^4 + 192\,\%2\,\%1^5 + 64\,\%1^6$$
$$\%1 := \frac{1}{2}\sqrt{2}\,t + \frac{1}{2}\sqrt{2}\,z$$
$$\%2 := \frac{1}{2}\sqrt{2}\,t - \frac{1}{2}\sqrt{2}\,z$$

> `simplify(%);`

$$\frac{1}{8}z^6 + \frac{135}{8}t^2 z^4 + \frac{9}{4}t z^5 + \frac{729}{4}t^5 z + \frac{729}{8}t^6 + \frac{1215}{8}t^4 z^2 + \frac{135}{2}t^3 z^3 + 4\,t^4$$

> `sort(%,[t,z]);`

$$\frac{729}{8} t^6 + \frac{729}{4} t^5 z + \frac{1215}{8} t^4 z^2 + \frac{135}{2} t^3 z^3 + \frac{135}{8} t^2 z^4 + \frac{9}{4} t z^5 + \frac{1}{8} z^6 + 4 t^4$$

The last expression is the H. Schultz decomposition for the original function.

## References

1. Arnol'd, V.I.: Remarks on the method of stationary phase and Coxeter numbers. Uspechi Mat. Nauk (in Russian) **28**(5) (1973) 17–44
   English. transl. Russ. Math. Surv. **28**(5) (1973) 19–58
2. Atiyah, M.F.: Resolution of singularities and division of distributions. Comm. Pure Appl. Math. **23**(2) (1970) 145–150
3. Arnol'd, V.I., Gusein-zade, S.M., Varchenko, A.N.: Singularities of Differentiable Maps, Vol. II. Monodromy and Asymptotic of Integrals (in Russian). Nauka, Moscow, 1984. English. transl. Birhäuser, 1988
4. Bernstein, I.N., Gel'fand, I.M.: Meromorphic property of the functions $p^\lambda$//Funct. Anal. Appl. **3**(1) (1969) 68–69
5. Karpushkin, V.N.: Theorems on uniform estimates of oscillatory integrals with phases depending on two variables. In: Proc. Sem. I.G. Petrovskii **10** (1984) 50–68
6. Varchenko, A.N.: Newton polyhedrons and estimates of oscillating integrals. Funct. anal. and appl. **10**(3) (1976) 175–196
7. Soleev, A., Aranson, A.: Calculation of a polyhedron and normal cones of its faces. Preprint Inst. Appl. Math. Russian Acad. Sci., Moscow, 1994, No. 36
8. Schultz, H.: Convex hypersurface of finite type and the asymptotics of their Fourier transforms. Indiana University Math. Journal **40**(3)(1991) 1267–1275

# Cellular Automata with Symmetric Local Rules

Vladimir V. Kornyak

Laboratory of Information Technologies
Joint Institute for Nuclear Research
141980 Dubna, Russia
kornyak@jinr.ru

**Abstract.** The cellular automata with local permutation invariance are considered. We show that in the two-state case the set of such automata coincides with the generalized Game of Life family. We count the number of equivalence classes of the rules under consideration with respect to permutations of states. This reduced number of rules can be efficiently generated in many practical cases by our C program. Since a cellular automaton is a combination of a local rule and a lattice, we consider also maximally symmetric two-dimensional lattices. In addition, we present the results of compatibility analysis of several rules from the Life family.

## 1 Introduction

The number of possible local rules for $q$-state cellular automaton defined on a $(k+1)$-cell neighborhood is double exponential of $k$, namely, $q^{q^{k+1}}$. It is natural to restrict our attention to special classes of local rules. S. Wolfram showed [1] that even simplest 2-state 3-cell automata, which he terms *elementary cellular automata* (there are $2^{2^3} = 256$ possible rules for these automata), demonstrate the unexpectedly complex behavior.

We consider here a class of *symmetric* local rules defined on a $(k + 1)$-cell neighborhood. Here by a 'symmetry' we mean a symmetry with respect to all permutations of $k$ cells (*points* or *vertices*) surrounding $(k+1)$th cell, which time evolution the local rule determines. The reasons to distinguish such rules are

- The number of possible symmetric rules is the single exponential of polynomial of $k$ (see formula (2) below). For example, for $k = 8$, $q = 2$ (the case of the Conway's game of Life) the number of symmetric rules is $262144 \approx 2.6 \times 10^5$, whereas the number of all possible rules $\approx 1.3 \times 10^{154}$.
- The symmetry of the neighborhood under permutations is in a certain sense a discrete analog of general local diffeomorphism invariance which is believed must hold for any fundamental physical theory based on continuum spacetime.[1]

---

[1] The symmetric group of any finite or infinite set $M$ is often denoted by Sym($M$). If $M$ is a manifold, then a diffeomorphism of $M$ is nothing but a special — continuous and differentiable — permutation from Sym($M$).

V.G. Ganzha, E.W. Mayr, and E.V. Vorozhtsov (Eds.): CASC 2006, LNCS 4194, pp. 240–250, 2006.

- This class of rules contains such widely known automata as **Conway's Life**. In fact, as we show below, any symmetric rule is a natural generalization of the **Life** rule.

## 2   Symmetric Local Rules and Generalized Life

### 2.1   Symmetric Rules

We interpret a $(k + 1)$-cell neighborhood of a cellular automaton as a $k$-*star* graph, i.e., rooted tree of height 1 with $k$ leaves. We call this the $k$-*valent* neigh-borhood. We adopt the convention that the leaves are indexed by the numbers $1, 2, \ldots, k$ and the root is numbered by $k + 1$. For example, the trivalent neigh-borhood looks like this

A *local rule* is a function specifying one time step evolution of the state of root

$$x'_{k+1} = f(x_1, \ldots, x_k, x_{k+1}).  \tag{1}$$

We consider the set $R_{\mathbf{S}_k}$ of local rules symmetric with respect to the group $\mathbf{S}_k$ of all permutations of leaves, i.e., variables $x_1, \ldots, x_k$. We will consider also the subset $R_{\mathbf{S}_{k+1}} \subset R_{\mathbf{S}_k}$ of rules symmetric with respect to permutations of all $k+1$ points of the neigborhood. For brevity we shall use the terms $k$-*symmetry* and $(k + 1)$-*symmetry*, respectively.

Obviously the total numbers of $k$- and $(k+1)$-symmetric rules are, respectively,

$$N_{\mathbf{S}_k}^q \;=\; q^{\binom{k+q-1}{q-1}q},  \tag{2}$$

$$N_{\mathbf{S}_{k+1}}^q \;=\; q^{\binom{k+q}{q-1}}.  \tag{3}$$

### 2.2   Life Family

The "Life family" is a set of 2-dimensional, binary cellular automata similar to **Conway's Life** [2], which rule is defined on 9-cell ($3{\times}3$) Moore neighborhood and is described as follows. A cell is "born" if it has exactly 3 alive neighbors, "survives" if it has 2 or 3 such neighbors, and dies otherwise. This rule is sym-bolized in terms of the "birth"/"survival" lists as B3/S23. Another examples of automata from this family are **HighLife** (the rule B36/S23), and **Day&Night** (the rule B3678/S34678). The site [3] contains collection of more than twenty rules from the Life family with Java applet to run these rules and descriptions of their behavior.

Generalizing this type of local rules, we define a *k-valent Life rule* as a *binary* rule on a *k*-valent neighborhood, described by two *arbitrary* subsets of the set $\{0, 1, \ldots, k\}$. These subsets $B, S \subseteq \{0, 1, \ldots, k\}$ contain conditions for the $x_k \rightarrow x'_k$ transitions of the forms $0 \rightarrow 1$ and $1 \rightarrow 1$, respectively. Since the number of subsets of any finite set $A$ is $2^{|A|}$, the number of rules defined by two sets $B$ and $S$ is equal to $2^{k+1} \times 2^{k+1} = 2^{2k+2}$, which in turn is equal to (2) evaluated at $q = 2$. On the other hand, *different* pairs $B/S$ define *different* rules.

Thus, we have the obvious

**Proposition.** *For any $k$ the set of $k$-symmetric binary rules coincides with the set of $k$-valent Life rules.*

This proposition implies, in particular, that one can always express any symmetric binary rule in terms of "birth"/"survival" lists.

### 2.3    Equivalence with Respect to Permutations of States

Exploiting the symmetry with respect to renaming of $q$ states of cellular automata allows us to reduce the number of rules to consider. Namely, it suffices to consider only orbits (equivalence classes) of the rules under $q!$ permutations forming the group $\mathbf{S}_q$. For counting orbits of a finite group $G$ acting on a set $R$ (*rules*, in our context) there is the formula called *Burnside's lemma*. This lemma states (see., e.g., [4]) that the number of orbits, denoted $|R/G|$, is equal to the average number of points $R^g \subset R$ fixed by elements $g \in G$:

$$|R/G| = \frac{1}{|G|} \sum_{g \in G} |R^g|. \tag{4}$$

Thus, the problem is reduced to finding the sets of fixed points.

Since we are mainly interested here in the binary automata, we shall consider further the case $q = 2$ only. For this case the combinatorics is rather simple.

Specializing (2) and (3) for $q = 2$ we have the numbers of binary $k$- and $(k+1)$−symmetric rules, respectively,

$$N_{\mathbf{S}_k} = 2^{2k+2}, \tag{5}$$
$$N_{\mathbf{S}_{k+1}} = 2^{k+2}. \tag{6}$$

The group $\mathbf{S}_2$ contains two elements $e = (0)(1)$ and $c = (01)$ (in cyclic notation). After S.Wolfram, we shall call the permutation $c \in \mathbf{S}_2$ "*black-white*" (shortly $BW$) transformation.

Since the number of fixed points for $e$ is either (5) or (6), all we need is to count the number of fixed points for the BW transformation in both $k$- and $(k+1)$-symmetry cases. The rules in both these cases can be represented, respectively, by the bit strings

$$\alpha_1 \alpha_2 \cdots \alpha_{2k+2} \tag{7}$$

and

$$\alpha_1 \alpha_2 \cdots \alpha_{k+2} \tag{8}$$

in accordance with the tables

| $x_1$ | $x_2$ | $\cdots$ | $x_k$ | $x_{k+1}$ | $x'_{k+1}$ |
|---|---|---|---|---|---|
| 0 | 0 | $\cdots$ | 0 | 0 | $\alpha_1$ |
| 0 | 0 | $\cdots$ | 0 | 1 | $\alpha_2$ |
| 1 | 0 | $\cdots$ | 0 | 0 | $\alpha_3$ |
| 1 | 0 | $\cdots$ | 0 | 1 | $\alpha_4$ |
| $\vdots$ | $\vdots$ | $\vdots$ | $\vdots$ | $\vdots$ | $\vdots$ |
| 1 | 1 | $\cdots$ | 1 | 0 | $\alpha_{2k+1}$ |
| 1 | 1 | $\cdots$ | 1 | 1 | $\alpha_{2k+2}$ |

| $x_1$ | $x_2$ | $\cdots$ | $x_k$ | $x_{k+1}$ | $x'_{k+1}$ |
|---|---|---|---|---|---|
| 0 | 0 | $\cdots$ | 0 | 0 | $\alpha_1$ |
| 1 | 0 | $\cdots$ | 0 | 0 | $\alpha_2$ |
| 1 | 1 | $\cdots$ | 0 | 0 | $\alpha_3$ |
| $\vdots$ | $\vdots$ | $\vdots$ | $\vdots$ | $\vdots$ | $\vdots$ |
| 1 | 1 | $\cdots$ | 1 | 0 | $\alpha_{k+1}$ |
| 1 | 1 | $\cdots$ | 1 | 1 | $\alpha_{k+2}$ |

One can see from these tables that the BW transformation acts similarly on both strings (7) and (8), namely,

$$\alpha_1\alpha_2\cdots\alpha_{n-1}\alpha_n \xrightarrow{BW} \bar{\alpha}_n\bar{\alpha}_{n-1}\cdots\bar{\alpha}_2\bar{\alpha}_1,$$

where bar means complementary transformation of bit, i.e., $\bar{\alpha} = \alpha + 1 \mod 2$, and $n = 2k + 2$ or $n = k + 2$.

The fixed point condition

$$\alpha_1\alpha_2\cdots\alpha_{n-1}\alpha_n = \bar{\alpha}_n\bar{\alpha}_{n-1}\cdots\bar{\alpha}_2\bar{\alpha}_1 \tag{9}$$

implies that the string is defined by a half of bits, i.e., by $k + 1$ or by $(k + 2)/2$ bits. In other words, the numbers of different bit strings satisfying condition (9) are, respectively, $2^{k+1}$ and $2^{(k+2)/2}$ (for $k$ even). For the $(k + 1)$-symmetry case with odd $k = 2m + 1$ condition (9) leads to the contradiction

$$\bar{\alpha}_{m+1} = \alpha_{m+1},$$

which means zero number of bit strings in this case.

Summarizing our calculations (recollecting that $G = \mathbf{S}_2$ in formula (4), i.e., $|G| = 2$), we have:

- in the case of $k$-symmetry
  - number of BW-symmetric rules

$$N_{\mathbf{S}_k BW} = 2^{k+1}, \tag{10}$$

  - number of non-equivalent rules

$$N_{\mathbf{S}_k/BW} = 2^{2k+1} + 2^k, \tag{11}$$

- in the case of $(k + 1)$-symmetry
  - number of BW-symmetric rules

$$N_{\mathbf{S}_{k+1}BW} = \begin{cases} 2^{k/2+1} & \text{if } k = 2m, \\ 0 & \text{if } k = 2m + 1, \end{cases} \tag{12}$$

  - number of non-equivalent rules

$$N_{\mathbf{S}_{k+1}/BW} = \begin{cases} 2^{k+1} + 2^{k/2} & \text{if } k = 2m, \\ 2^{k+1} & \text{if } k = 2m + 1. \end{cases} \tag{13}$$

For example, for trivalent rules the numbers are

$$N_{\mathbf{S}_3 BW} = 16, \quad N_{\mathbf{S}_3/BW} = 136, \quad N_{\mathbf{S}_{3+1}BW} = 0, \quad N_{\mathbf{S}_{3+1}/BW} = 16.$$

# 3   Assembling Neighborhoods into Regular Lattices

$k$-valent neighborhoods can be gathered into a lattice in many ways. In fact, any $k$-regular graph may serve as a space for a cellular automaton. In applications a cellular automaton acts, as a rule, on a lattice embedded in a metric space, usually 2- or 3-dimensional Euclidean space. Any graph, being 1-dimensional simplicial complex, can be embedded into $\mathbb{E}^3$. When dealing with symmetric local rules it is natural to consider equidistant regular systems of cells. We consider here only two-dimensional case as more simple and suitable for visualization of the automaton behavior. Most symmetric 2D lattices correspond to the tilings by congruent regular polygons. The number of different types of such tilings is rather restricted (see, e.g., [5]). To denote a $k$-valent lattice composed of regular $p$-gons we use the *Schläfli symbol* $\{p, k\}$.

## 3.1   2D Euclidean Metric

There are only three regular lattices in $\mathbb{E}^2$ (see Fig. 1).

*3-valent*, $\{6,3\}$      *4-valent*, $\{4,4\}$      *6-valent*, $\{3,6\}$

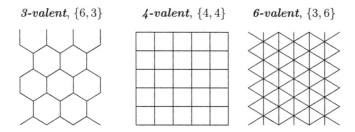

**Fig. 1.** All regular lattices in $\mathbb{E}^2$

Since real computers have finite memory, cellular automata can be simulated only on a finite lattice. Usually the universe of a cellular automaton is a rectangle instead of an infinite plane. There are different ways to handle the edges of the rectangle. One possible method is to fix states of the border cells. This breaks the symmetry of the lattice and is thus not interesting for us. Another way is to glue together the opposite edges of the rectangle.

The toroidal arrangement is a standard practice, but it would be interesting to study cellular automata on nonorientable surfaces also. There are 3 different identifications of opposite sides of a rectangle: the *torus* $\mathbb{T}^2$, the *Klein bottle* $\mathbb{K}^2$, and the *projective plane* $\mathbb{P}^2$. All these spaces (the gluing does not affect their Euclidean metric) are shown in the figure below.

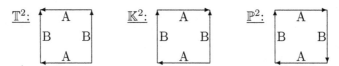

We need to check whether the regular lattices like in Fig. 1, i.e., with the Schläfli symbols $\{p, k\} = \{6, 3\}$, $\{4, 4\}$, $\{3, 6\}$ can be embedded in $\mathbb{T}^2$, $\mathbb{K}^2$, or $\mathbb{P}^2$. To do this we must solve the system of equations

$$V - E + F = \chi(M), \quad pF = kV = 2E, \tag{14}$$

where $V, E, F$ are numbers of vertices, edges, and faces, respectively, $\chi(M)$ is the Euler characteristic of a manifold $M$. Since $\chi(\mathbb{T}^2) = \chi(\mathbb{K}^2) = 0$ and $\chi(\mathbb{P}^2) = 1$, we see that the regular lattices are possible only in the torus and the Klein bottle and impossible in the projective plane, as well as in any other closed surfaces.

Summarizing, for Euclidean metric there are only 3-valent hexagonal, 4-valent square and 6-valent triangular regular lattices in $\mathbb{E}^2$, $\mathbb{T}^2$, and $\mathbb{K}^2$.

## 3.2 Hyperbolic Plane $\mathbb{H}^2$

The hyperbolic (Lobachevsky) plane $\mathbb{H}^2$ allows infinitely many regular lattices. Poincaré proved that regular tilings $\{p, k\}$ of $\mathbb{H}^2$ exist for any $p, k \geq 3$ satisfying $\frac{1}{p} + \frac{1}{k} < \frac{1}{2}$.

For example, let us consider the octivalent Moore neighborhood used in the Life family and shown in the figure below.

The Moore neighborhood can not form regular lattice in the Euclidean plane since the distances of surrounding cells from the center are different. But in the hyperbolic plane regular 8-valent lattices exist and there are infinitely many of them. The simplest one is shown (using the Poincaré disc model projection) in Fig. 2.

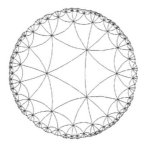

**Fig. 2.** Octivalent regular lattice $\{3, 8\}$ in $\mathbb{H}^2$

### 3.3  Sphere $\mathbb{S}^2$

All regular lattices in the two-dimensional sphere correspond to the Platonic solids which are shown in Fig. 3.

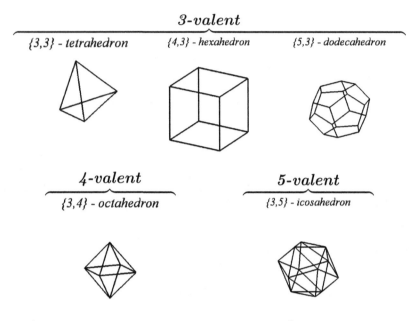

## *3-valent*

| *{3,3} - tetrahedron* | *{4,3} - hexahedron* | *{5,3} - dodecahedron* |

## *4-valent*              *5-valent*

| *{3,4} - octahedron* | *{3,5} - icosahedron* |

**Fig. 3.** All regular lattices in $\mathbb{S}^2$

In the sphere there are infinitely many other lattices, which are close to regular and may serve as spaces for 3-valent symmetric automata. They are called *fullerenes*.

The fullerenes were first discovered in carbon chemistry in 1985, and this discovery was rewarded with the 1996 Nobel Prize in Chemistry. A model of the first revealed fullerene — the carbon molecule $C_{60}$ — is displayed in Fig. 4 (the figure is borrowed from [6]). Later there were discovered other forms of large carbon molecules with structural properties of fullerenes (larger spherical fullerenes, carbon nanotubes, graphenes). Their unique properties promise they will have an important role in future technology, in particular, in nanotechnology engineering.

From a mathematical viewpoint, the structure of fullerene is a 3-valent convex polyhedron with pentagonal and hexagonal faces. In terms of graphs, fullerene can be defined as a 3-regular (3-valent) *planar*, or equivalently, embeddable in $M = \mathbb{S}^2$, graph with all faces of size 5 or 6.

Let us generalize slightly this definition assuming that $M$ is not necessarily $\mathbb{S}^2$, but may be closed surface of other type, orientable or nonorientable. Then

**Fig. 4.** A model of $C_{60}$ carbon molecule (buckyball)

the Euler–Poincaré equation together with the edge balance relations gives the system of equations

$$V - E + f_5 + f_6 = \chi(M), \quad 3V = 5f_5 + 6f_6 = 2E, \tag{15}$$

where $f_5$ and $f_6$ are numbers of pentagons and hexagons, respectively. The general solution of this system is

$$f_5 = 6\chi(M), \tag{16}$$
$$V = 2f_6 + 10\chi(M), \tag{17}$$
$$E = 3f_6 + 15\chi(M). \tag{18}$$

We see that generalized fullerenes are possible only in the sphere $\mathbb{S}^2$, in the projective plane $\mathbb{P}^2$, and in the torus $\mathbb{T}^2$ and Klein bottle $\mathbb{K}^2$. Attempts to consider surfaces with greater genus lead due to (16) to senseless negative numbers of pentagons. Thus, we have for all generalized fullerenes:

$$V = 2f_6 + 20, \quad E = 3f_6 + 30, \quad f_5 = 12, \quad \text{sphere } \mathbb{S}^2;$$
$$V = 2f_6 + 10, \quad E = 3f_6 + 15, \quad f_5 = 6, \quad \text{projective plane } \mathbb{P}^2;$$
$$V = 2f_6, \qquad E = 3f_6, \qquad f_5 = 0, \quad \text{torus } \mathbb{T}^2, \text{ Klein bottle } \mathbb{K}^2.$$

We see that any fullerene in $\mathbb{S}^2$ or in $\mathbb{P}^2$ contains exactly 12 or 6 pentagons, respectively, and arbitrary number of hexagons. In the case of torus or Klein bottle the fullerene structure degenerates into purely hexagonal lattice without pentagons considered already in subsection 3.1. Note that in carbon chemistry one-layer graphite sheets (and similar structures with few pentagons or heptagons added) are called *graphenes*.

## 4   Canonical Decomposition of Some Rules from Life Family

In this section we present the canonical decompositions of relations for three rules from the Life family: the standard *Conway's Life*, *HighLife* and *Day&Night*.

Recall that canonical decomposition [7] of relation on a set of points is the representation of the relation as a combination of its projections onto subsets of points. This decomposition is discrete analog of compatibility analysis of algebraic and differential equations. To apply our approach we interpret the local rule $x_9' = f(x_1, x_2, \ldots, x_9)$ as a relation on 10 points $x_1, x_2, \ldots, x_9, x_9'$.

The **HighLife** is interesting since for it the *replicator* — a self-reproducing pattern — is known explicitly. For **Conway's Life**, the existence of replicators is proved, but no example is known.

The name **Day&Night** reflects the BW-symmetry of this rule. Perhaps, it is the first (conceivably found "by hand") automaton from the Life family with this symmetry property. Note that there are exactly 512 such automata in the Life family, and they all are generated easily (for time < 1 sec) by the C program mentioned in [7].

It might be easier to grasp the relations, if they are written in the form of polynomials over the field $\mathbb{F}_2$. In the below formulas we use *elementary symmetric polynomials* in variables $x_1, \ldots, x_8$. Recall that the $k$th (of degree $k$) elementary symmetric polynomial of $n$ variables $x_1, \ldots, x_n$ is defined by the formula:

$$\sigma_k(x_1, \ldots, x_n) = \sum_{1 \leq i_1 < i_2 < \cdots < i_k \leq n} x_{i_1} x_{i_2} \cdots x_{i_k}.$$

Hereafter we use the notations $\sigma_k \equiv \sigma_k(x_1, \ldots, x_8)$, $\sigma_k^i \equiv \sigma_k(x_1, \ldots, \widehat{x}_i, \ldots, x_8)$, $\sigma_k^{ij} \equiv \sigma_k(x_1, \ldots, \widehat{x}_i, \ldots, \widehat{x}_j, \ldots, x_8)$. The indices $i, j, k, l$ satisfy $1 \leq i < j < k < l \leq 8$.

The polynomials representing **Conway's Life**, **HighLife** and **Day&Night** take the forms, respectively

$$P_{\textbf{Conway's Life}} = x_9' + x_9 \{\sigma_7 + \sigma_6 + \sigma_3 + \sigma_2\} + \sigma_7 + \sigma_3, \tag{19}$$

$$P_{\textbf{HighLife}} = x_9' + x_9 \{\sigma_3 + \sigma_2\} + \sigma_6 + \sigma_3, \tag{20}$$

$$P_{\textbf{Day\&Night}} = x_9' + x_9 \{\sigma_7 + \sigma_6 + \sigma_5 + \sigma_4\} + \sigma_8 + \sigma_7 + \sigma_6 + \sigma_3. \tag{21}$$

Note that these polynomials have degrees 8, 6, 8 and numbers of terms 185, 169, 256, respectively, so application of the Gröbner basis technique to their analysis may take some time (about 1 hour on 1.8GHz AMD Athlon notebook with the Maple 9 Gröbner procedure). Our program computes the decompositions for time < 1 sec.

The decompositions are:

– **Conway's Life**

$$x_9' \{\sigma_3 + \sigma_2 + 1\} + \sigma_7 + \sigma_3 = 0, \tag{22}$$

$$x_9' x_9 \{\sigma_2^i + \sigma_1^i\} + x_9' \{\sigma_2^i + 1\} + x_9 \{\sigma_7^i + \sigma_6^i + \sigma_3^i + \sigma_2^i\} = 0, \tag{23}$$

$$x_9' \{\sigma_3^i + \sigma_2^i + \sigma_1^i + 1\} = 0, \tag{24}$$

$$x_9' (x_9 + 1) \{\sigma_3^{ij} + \sigma_2^{ij} + \sigma_1^{ij} + 1\} = 0, \tag{25}$$

$$x_9' x_i x_j x_k x_l = 0. \tag{26}$$

– **HighLife**

$$x'_9 \{\sigma_3 + \sigma_2 + 1\} + \sigma_7 + \sigma_3 = 0, \quad (27)$$
$$x'_9 x_9 \{\sigma_2^i + \sigma_1^i\} + x'_9 \{\sigma_5^i + \sigma_2^i + 1\} + x_9 \{\sigma_7^i + \sigma_6^i + \sigma_3^i + \sigma_2^i\} = 0, \quad (28)$$
$$x'_9 \{\sigma_7^i + \sigma_3^i + \sigma_2^i + \sigma_1^i + 1\} = 0, \quad (29)$$
$$x'_9 x_9 \left\{\sigma_3^{ij} + \sigma_2^{ij} + \sigma_1^{ij} + 1\right\}$$
$$+ x'_9 \left\{\sigma_6^{ij} + \sigma_5^{ij} + \sigma_4^{ij} + \sigma_3^{ij} + \sigma_2^{ij} + \sigma_1^{ij} + 1\right\} = 0, \quad (30)$$
$$x'_9 x_9 x_i x_j x_k x_l = 0. \quad (31)$$

– **Day&Night**

$$x'_9 \{\sigma_7 + \sigma_6 + \sigma_5 + \sigma_4 + 1\} + \sigma_8 + \sigma_7 + \sigma_6 + \sigma_3 = 0, \quad (32)$$
$$x'_9 x_9 \{\sigma_6^i + \sigma_5^i + \sigma_4^i + \sigma_3^i\} + x'_9 \{\sigma_7^i + \sigma_6^i + \sigma_5^i + \sigma_2^i + 1\}$$
$$+ x_9 \{\sigma_7^i + \upsilon_3^i\} + \sigma_6^i = 0, \quad (33)$$
$$x'_9 \left\{\sigma_5^{ij} + \sigma_4^{ij} + \sigma_3^{ij} + \sigma_2^{ij} + \sigma_1^{ij} + 1\right\} + \sigma_6^{ij} = 0. \quad (34)$$

Note that system (34) of prime relations can be combined into the reducible relation (for terminology like *prime* and *reducible* see [7])

$$x'_9 \{\sigma_4^i + \sigma_2^i + 1\} + \sigma_6^i = 0,$$

which looks nicer (the polynomial representation of relations is somewhat artificial), but depends on larger set of variables.

We see that the decomposition for **Day&Night** differs essentially from the decompositions for **Conway's Life** and **HighLife** having resembling structures.

## 5    Conclusions

We proved that the cellular automata from the Life family are nothing but binary automata with the local rules symmetric with respect to all permutations of the outer cells in the neighborhood.

Then we showed that the number of non-equivalent with respect to renaming of states $k$-valent symmetric binary local rules is equal to $2^{2k+1} + 2^k$. Considering the 3-valent case — the next step up after the 2-valent Wolfram's elementary automata — we see that the total number of non-equivalent symmetric rules is only 136.

All interesting 3-valent 2-dimensional regular (or almost regular) lattices are:

– hexagonal lattice $\{6, 3\}$ in the plane $\mathbb{E}^2$, in the torus $\mathbb{T}^2$ and in the Klein bottle $\mathbb{K}^2$
– tetrahedron $\{3, 3\}$, hexahedron (cube) $\{4, 3\}$ and dodecahedron $\{5, 3\}$ in the sphere $\mathbb{S}^2$
– fullerenes in $\mathbb{S}^2$ and in the projective plane $\mathbb{P}^2$.

Combining these lattices with 136 symmetric rules we obtain a class of cellular automata which looks like quite available for systematic study. Since these automata have more interesting geometry than the elementary ones we may expect more interesting behavior of them.

## Acknowledgments

I am very grateful to Vladimir Gerdt for detailed discussions about the present paper. This work was supported in part by the grants 04-01-00784 from the Russian Foundation for Basic Research and 2339.2003.2 from the Russian Ministry of Industry, Science, and Technologies.

## References

1. Wolfram, S.: *A New Kind of Science.* Wolfram Media, Inc., 2002
2. Gardner, M.: On cellular automata self-reproduction, the garden of Eden and the game of life. Sci. Am. **224** (1971) 112–117
3. http://psoup.math.wisc.edu/mcell/rullex_life.html
4. Harary, F.: *Graph Theory.* Addison-Wesley Publishing Company, 1969
5. Grunbaum, B., Sheppard, G. C.: *Tilings and Patterns.* W. H. Freeman, New York, 1987
6. Atiyah, M.F., Sutcliffe, P.M.: Polyhedra in Physics, Chemistry and Geometry. *Milan Journal of Mathematics* **71** (2003) 33–58; http://arXiv.org/abs/math-ph/0303071.
7. Kornyak, V.V.: On Compatibility of discrete relations. *Computer Algebra in Scientific Computing 2005,* LNCS 3718, V.G. Ganzha, E.W. Mayr, E.V. Vorozhtsov (Eds.), Springer-Verlag, Berlin, Heidelberg (2005) 272–284; http://arXiv.org/abs/math-ph/0504048

# Parallel Laplace Method with Assured Accuracy for Solutions of Differential Equations by Symbolic Computations

Natasha Malaschonok

Tambov State University,
Internatsionalnaya 33, 392622 Tambov, Russia
malaschonok@narod.ru

**Abstract.** We produce a parallel algorithm realizing the Laplace transform method for symbolic solution of differential equations. In this paper we consider systems of ordinary linear differential equations with constant coefficients, nonzero initial conditions, and the right-hand sides reduced to the sums of exponents with the polynomial coefficients.

## 1 Introduction

We produce a parallel algorithm applying the Laplace transform method to symbolic solution of differential equations.

An application of Laplace transform in differential equations theory in spite of its long history is topical. It has been very useful in classical or modified forms for solving ordinary or partial differential equations [2, 3, 8, 19, 21]. It is frequently applied for problems of fractional order equations [6, 20].

We consider systems of ordinary linear differential equations with constant coefficients, nonzero initial conditions, and right-hand sides as composite functions reducible to the sums of exponents with the polynomial coefficients. We place an emphasis on the symbolic character of computations. The efficient algorithmizataton of symbolic solution is achieved at several stages.

At the first stage, the preparation of data functions for the formal Laplace transform is performed (section 3). It is achieved by applying the Heaviside function and moving the obtained functions into the bounds of smoothness intervals. The parallelization of computations is realized as the multilevel tree, in the paper it is evident from the numeration of algorithm blocks.

The second stage is the parallel solution of the algebraic system with polynomial coefficients and the right-hand side obtained after the Laplace transform of the data system (section 4). There are parallel algorithms that are very efficient for solving this type of equations, and are different for various types of such systems.

At the third stage, the obtained solution of algebraic system is prepared to the inverse Laplace transform. It is reduced to the sum of partial fractions with exponential coefficients. One of the problems is calculation of roots of a polynomial. In [17, 18] the algorithm to determine the error of the roots sufficient for

V.G. Ganzha, E.W. Mayr, and E.V. Vorozhtsov (Eds.): CASC 2006, LNCS 4194, pp. 251–260, 2006.
© Springer-Verlag Berlin Heidelberg 2006

the required accuracy of the data system solution is obtained. The solution of the algebraic system for reducing into the sum of partial fractions is performed by means of parallel algorithms cited in the paper.

At the last stage, the solution of the data system is produced (section 5). It is obtained as the real part of the inverse Laplace transform image of the algebraic system solution prepared previously.

In the last section, an example is considered.

## 2    Input Data

Denote by $x_j, j = 1, \ldots, n$, unknown functions of argument $t, t \geq 0$, $x_j^k$ is the derivative of order $k$ of the function $x_j$, $k = 0, \ldots, N$. As the right-hand sides of equations we consider here composite functions $f_l, l = 1, \ldots, n$ whose components are represented as finite sums of exponents with polynomial coefficients. So we have to solve the system

$$\sum_{j=1}^{n} \sum_{k=0}^{N} a_{kj}^l x_j^k = f_l, \quad l = 1, \ldots, n, \quad a_{kj}^l \in \mathbb{R}, \tag{1}$$

of $n$ differential equations of order $N$ with initial conditions $x_j^k(0) = x_{0j}^k, k = 0, \ldots, N-1$ with functions $f_l$ reduced to the form

$$f_l(t) = f_l^i(t), \ t_l^i < t < t_l^{i+1}, \ i = 1, \ldots, I_l, t_l^1 = 0, t_l^{I_l+1} = \infty, \tag{2}$$

where

$$f_l^i(t) = \sum_{s=1}^{S_l^i} P_{ls}^i(t) e^{b_{ls}^i t}, \quad i = 1, \ldots, I_l, \quad l = 1, \ldots, n,$$

and $P_{ls}^i(t) = \sum_{m=0}^{M_{ls}^i} c_{sm}^{li} t^m$.

**Remark.** The algorithm tree is exposed by multi-index numbering of blocks. For example, **Block** $42jk$, $k = 1, \ldots, K_\Theta$, denotes the vertex $42jk$, which is an entrance of the $k$th tree-edge, outgoing from the vertex $42j$ – **Block** $42j$, and there are $K_\Theta$ such edges. As $k$ is the fourth index in the multi-index $42jk$, **Block** $42jk$ is the vertex of the fourth level. All the blocks **Block** $42jk$, $k = 1, \ldots, K_\Theta$ are performed independently and parallel.

**Block1: Block10, Block1$l$, $l = 1, \ldots, n$.** *Data file.*

Data file contains the coefficients $a_{kj}^l$, the initial conditions $x_{0j}^k$, $k = 0, \ldots, N-1$, $j = 1, \ldots, n$, and the right-hand sides $f_l, l = 1, \ldots, n$.

The data for functions $f_l$ consists of the polynomial coefficients $c_{sm}^{li}$, parameters $b_{ls}^i$ of exponents, the bounds $t_i$ of smoothness intervals. Here $m = 0, \ldots, M_{ls}^i$, $s = 1, \ldots, S_l^i$, $i = 1, \ldots, I_l$. The numbers $M_{ls}^i$ are degrees of corresponding polynomials, $S_l^i$ are amounts of exponents in the expressions for $f_l$.

## 3    Laplace Transform

Denote the Laplace images of the functions $x_j(t)$ and $f_l(t)$ by $X_j(p)$ and $F_l(p)$, respectively.

The Laplace transform of the left-hand side of system (1) with respect to the initial conditions is performed by formal writing of the expression

$$\sum_{j=1}^{n}\sum_{k=0}^{N} a_{kj}^l p^k X_j(p) - \sum_{j=1}^{n}\sum_{k=0}^{N-1} x_{0j}^k p^{N-1-k},$$

starting directly from input data.

**Block21l,  $l = 1,\ldots,n$.** *Preparation of right-hand functions $f_l(t)$ to the Laplace transform .*

The functions $f_l(t), l = 1,\ldots,n$ are composite and reduced to form (2). We use the Heaviside function $\eta(t)$ and present $f_l(t)$ as a sum

$$f_l(t) = \sum_{i=2}^{I_l-1} [f_l^i(t) - f_l^{i-1}(t)]\eta(t - t_l^i) + f_l^1(t)\eta(t).$$

**Block21li,  $i = 1,\ldots,I_l$.** *Transform into the function of $t - t_l^i$.*

Transform $f_l^i(t) - f_l^{i-1}(t)$ into the function of $t - t_l^i$:

$$f_l^i(t) - f_l^{i-1}(t) = \phi_l^i(t - t_l^i).$$

Generally, the functions $f_l^i(t) - f_l^{i-1}(t)$ are decomposed into power series at point $t_l^i$.

In our case the function $\phi_l^i(t - t_l^i)$ is represented as a finite sum

$$\phi_l^i(t - t_l^i) = \sum_{s=1}^{S_l^i} \psi_{ls}^i(t - t_l^i)e^{b_{ls}^i t_l^i}e^{b_{ls}^i(t-t_l^i)} - \sum_{s=1}^{S_l^{i-1}} \psi_{ls}^{i-1}(t - t_l^i)e^{b_{ls}^{i-1} t_l^i}e^{b_{ls}^{i-1}(t-t_l^i)}.$$

Here $\psi_{ls}^k(t - t_l^i) = P_{ls}^k(t)$ and $\psi_{ls}^k(t - t_l^i) = \sum_{m=0}^{M_{ls}^k} \gamma_{lsm}^{ki}(t - t_l^i)^m$. Coefficients $\gamma_{lsm}^{ki}$ are calculated by the formula

$$\gamma_{lsm}^{ki} = \sum_{j=0}^{M_{ls}^k - m} c_{s,m+j}^{lk}\binom{m + j}{j}(t_l^i)^j.$$

Finally the function $f_l(t)$ is reduced to the form

$$f_l(t) = \sum_{i=2}^{I_l-1} \phi_l^i(t - t_l^i)\eta(t - t_l^i) + \sum_{s=1}^{S_l^1} P_{ls}^1(t)e^{b_{ls}^1 t}\eta(t).$$

**Block22$l$,  $l = 1, \ldots, n$.** *The parallel Laplace transform of the functions $f_l(t)$.*

Since the Laplace image of $(t - t^*)^n e^{\alpha(t-t^*)} \eta(t - t^*)$ is $\frac{n!}{(p-\alpha)^{n+1}} e^{-t^*p}$ the Laplace transform of $\phi_l^i(t - t_l^i)\eta(t - t_l^i)$ equals

$$
\Phi_l^i(p) = \left[ \sum_{s=1}^{S_l^i} \sum_{m=0}^{M_{ls}^i} \gamma_{lsm}^{ii} e^{b_{ls}^i t_l^i} \frac{m!}{(p - b_{ls}^i)^{m+1}} \right.
$$
$$
\left. - \sum_{s=1}^{S^{i-1}{}_{il}} \sum_{m=0}^{M_{ls}^{i-1}} \gamma_{lsm}^{i,\,i-1} e^{b_{ls}^{i-1} t_l^i} \frac{m!}{(p - b_{ls}^{i-1})^{m+1}} \right] e^{-t_l^i p}.
$$

Finally, the Laplace transform of $f_l(t)$ is the following:

$$
F_l(p) = \sum_{i=2}^{I_l-1} \Phi_l^i(p) + \sum_{s=1}^{S_l^1} \sum_{m=0}^{M_{sl}^1} c_{sm}^{l1} \frac{m!}{(p - b_{ls}^1)^{m+1}}. \tag{3}
$$

In the case when the right-hand side of the given system is exposed in the form

$$
f_l(t) = \sum_{s=1}^{S_l} \sum_{m=0}^{M_{sl}} c_{sm}^l t^m e^{b_{ls}t}, \quad l = 1, \ldots, n,
$$

the Laplace transform is performed formally – according to input data we present the expression for $F_l(p)$:

$$
F_l(p) = \sum_{s=1}^{S_l} \sum_{m=0}^{M_{sl}} c_{sm}^l \frac{m!}{(p - b_{ls})^{m+1}}, l = 1, \ldots, n. \tag{4}
$$

For each $l = 1, \ldots, n$ we reduce (3)(or (4)) to the common denominator. The common denominator is left factorized. At that the nominator is the sum of exponents with polynomial coefficients.

In the case of a periodic function $f_l(t)$ with the period $T$, the respective denominator contains the expression $1 - e^{-pT}$. Then such fraction is expanded into power series.

## 4    Parallel Solution of Algebraic System

**Block 31.** *The construction of the algebraic system.*

As a result of the Laplace transform of system (1) we obtain an algebraic system relative to $X_j, j = 1, \ldots, n$:

$$
\sum_{j=1}^{n} \sum_{k=0}^{N} a_{kj}^l p^k X_j(p) = \sum_{j=1}^{n} \sum_{k=0}^{N-1} x_{0j}^k p^{N-1-k} + F_l(p), \quad l = 1, \ldots, n. \tag{5}
$$

For each $l = 1, \ldots, n$ the expressions on the right-hand side of (5) are reduced to the common denominator. The calculations are carried out in parallel. Denote

$$\Phi_l(p) = \sum_{j=1}^{n} \sum_{k=0}^{N-1} x_{0j}^k p^{N-1-k} + F_l(p).$$

We obtain the system

$$\sum_{j=1}^{n} \sum_{k=0}^{N} a_{kj}^l p^k X_j(p) = \Phi_l(p), \; l = 1, \ldots, n. \tag{6}$$

**Block 32.** *The parallel solution of the algebraic system.*

The system (6) may be solved by any possible classical method, for example, Cramer's method. But now there are new effective procedures for parallel computations, for example, p-adic method ([5], [9], [10]), modula method ([10] - [14]), the method based on determinant identities ([10] – [16]). The fastest method for solving such systems is p-adic method. But its code parallelization is not very effective. The best one for parallelization is the modula method based on Chinese Remainder Theorem.

# 5   Inverse Laplace Transform

**Block41j,** $\; j = 1, \ldots, n$. *Preparation of $X_j(p)$ to the inverse Laplace transform.*

Finally the solution of (6), i.e., each desired function $X_j(p)$, $j = 1, \ldots, n$, is represented as a fraction with polynomial denominator $D_j(p)$. Note that $D_j(p)$ is partially factorized – it contains the multipliers of $F_l(p)$ denominators and the determinant $D(p)$ of system (6). The nominator is the sum of exponents with polynomial coefficients.

We reduce the function $X_j(p)$, $j = 1, \ldots, n$, to the sum of exponents with fractional coefficients. The denominator of each fraction is $D_j(p)$, and the numerators are polynomials.

The next step is the decomposition of each fraction in the $X_j(p)$ expansion into the sum of partial fractions $A/(p - p^*)^v$, $p^* \in \mathbb{C}$. The first action here is the determination of the $D_j(p)$ roots.

**Block42j,** $\; j = 1, \ldots, n$. *Computation of the denominator roots.*

As it was pointed out the denominator $D_j(p)$ is already represented as a product of partial multipliers and the polynomial $D(p)$. So we have to find the roots of $D(p)$.

The accuracy of these calculations is determined first of all. Its value must be sufficient for the preassigned precision of system solution. An algorithm to compute such accuracy is written about in §5.

**Block42jk, $k = 1, \ldots, K_\Theta$.** *Decomposition into a sum of partial fractions.*

We decompose rational fractions or fractional coefficients of exponents into the sums of partial fractions $A/(p - p^*)^v$, $p^* \in \mathbb{C}$. The calculations for all fractions are in parallel, the number of blocks is formally denoted by $K_\Theta$. It depends upon the parameters that we do not describe in detail here.

One step of the algorithm is the solution of a system of linear equations with constant coefficients. Depending on the size of system matrix we use one or another fast parallel algorithm, for example, modular method ([4] - [7], [10] - [16]).

If the roots of $D(p)$ have been found exactly, then we obtain the exact solution of the system (6) – the functions $X_j(p)$. Each of them is represented as a sum

$$X_j(p) = \sum_m \sum_k \frac{A_{mk}}{(p - p_{ik})^{\beta_{mk}}} e^{-\alpha_m p}. \tag{7}$$

Denote by $\Xi_j(p)$ the expression that represents $X_j(p)$ after its reduction to the partial fractions form in the case when the roots of $D(p)$ are calculated not exactly. Each $\Xi_j(p)$ is also written in the form (7).

**Block43j, $j = 1, \ldots, n$.** *Inverse Laplace transform.*

The Laplace originals of functions $X_j(p)$ are obtained formally – by writing the expressions

$$x_j(t) = \sum_m \sum_k \frac{A_{mk}}{(\beta_{mk} - 1)!} (t - \alpha_m)^{\beta_{mk}-1} e^{p_{ik}(t-\alpha_m)} \eta(t - \alpha_m), j = 1, \ldots, n. \tag{6}$$

In the case when the roots of $D(p)$ are calculated not exactly denote by $\xi_j(t)$ the Laplace original of $\Xi_j(p)$. It is also written in the form (7). In general, the functions $\xi_j(t)$ are complex valued. We take the real part of $\xi_j(t)$ for each $j = 1, \ldots, n$. The functions $\mathrm{Re}\xi_j(t)$ may be taken as the solution of system (1), i.e., the required functions $x_j(t)$. It is easy to show that the error would not exceed the established precision assured by the calculated accuracy of roots of $D(p)$.

## 6    On Accuracy Estimation

We shall consider all functions and make calculations on the interval $[0, T]$, where $T$ is sufficiently high for the input problem. Denote by $\tilde{x}_i(t)$ the approximate value of the solution $x_i(t)$. We require the following accuracy for the solutions on the interval $T$:

$$\max_{t \in [0,T]} |x_i(t) - \tilde{x}_i(t)| < \varepsilon, \ i = 1, \ldots, n.$$

We must determine an error $\Delta$ of the $D(p)$ roots sufficient for the required accuracy $\varepsilon$ for $x_i(t)$. An algorithm for computation of $\Delta$ is produced in ([17] -

[18]). Notice only that $\Delta$ depends on the input parameters of the problem, $T$, the numbers $\mathcal{M}(f_i) = \max_{t \in [0,T]} |f_i(t)|, i = 1, \ldots, n$, the appreciation $\delta$ of minimal distance between the roots of $D(p)$, the number $\sigma$ such that the functions $X_j(p)$ are analytic in the domain $\mathrm{Re}\, p > \sigma$ for all $j = 1, \ldots, n$.

## 7  Example

**Block 10** $(a_{kj}^1) = \begin{pmatrix} 1 & 0 & -1 & -2 \\ 3 & 1 & -2 & 0 \end{pmatrix}; \quad (a_{kj}^2) = \begin{pmatrix} -1 & 0 & 0 & 1 \\ 1 & 0 & 0 & 1 \end{pmatrix};$

$x_{01}^0 = 5, \quad x_{01}^1 = 10, \quad x_{01}^2 = 30, \quad x_{02}^0 = 4, \quad x_{02}^1 = 14, \quad x_{02}^2 = 20;$

**Block 11** $f_1^1 = e^t, \qquad f_1^2 = t^2 e^{2t}, \qquad t_1^1 = 0, \quad t_1^2 = 1;$

**Block 12** $f_2^1 = te^t, \qquad f_2^2 = e^{2t}, \qquad t_2^1 = 0, \quad t_2^2 = 1;$

**Block 211** $f_1 = (f_1^2 - f_1^1)\eta(t - t_1^2) + f_1^1 \eta(t);$

**Block 212** $f_2 = (f_2^2 - f_2^1)\eta(t - t_2^2) + f_2^1 \eta(t);$

**Block 221** $F_1 = \dfrac{1}{-1+p} - \dfrac{e^{1-p}}{-1+p} + \dfrac{e^{2-p}(p^2 - 2p + 2)}{(-2+p)^3};$

**Block 222** $F_2 = \dfrac{e^{2-p}}{-2+p} + \dfrac{1}{(-1+p)^2} - \dfrac{e^{1-p}p}{(-1+p)^2};$

**Block 31**

$$
\begin{cases}
-2X_1 - pX_1 + p^3 X_1 + X_2 - p^3 = 10 - 4p - 4p^2 + 5(-1 + p^2) \\
\qquad\qquad + \dfrac{1}{-1+p} - \dfrac{e^{1-p}}{-1+p} + \dfrac{e^{2-p}(p^2 - 2p + 2)}{(-2+p)^3} \\
-2pX_1 + p^2 X_1 + 3p^3 X_1 + X_2 + p^3 X_2 = 110 + 14p + 4p^2 + 10(1 + 3p) + \\
\qquad\qquad + (-2 + p + 3p^2) \\
\qquad\qquad + \dfrac{e^{2-p}}{-2+p} + \dfrac{1}{(-1+p)^2} - \dfrac{e^{1-p}p}{(-1+p)^2};
\end{cases}
$$

$D(p) = -2 + p - p^2 - 4p^3 - 3p^4 + p^5 + 4p^6;$

**Block 411**

$$X_1(p) = e^{-p} \Big( \frac{8e + 2e^2 - 4ep - 4e^2 p + 2ep^2 + 2e^2 p^2 + 9ep^3 - 6e^2 p^3 - 19ep^4}{(-2+p)^3(-1+p)(-2+p-p^2-4p^3-3p^4+p^5+4p^6)}$$

$$+ \frac{12e^2 p^4 + 11ep^5 - 8e^2 p^5 - 2ep^6 + 2e^2 p^6}{(-2+p)^3(-1+p)(-2+p-p^2-4p^3-3p^4+p^5+4p^6)} \Big)$$

$$+ \frac{-856 + 1692p - 982p^2 + 1061p^3 - 1991p^4 + 1398p^5 - 412p^6 + 160p^7 - 95p^8 + 20p^9}{(-2+p)^3(-1+p)(-2+p-p^2-4p^3-3p^4+p^5+4p^6)};$$

## Block 412

$$X_2(p) = e^{-p}\left(\frac{-8e^2-32ep+24e^2p+64ep^2-28e^2p^2-32ep^3+17e^2p^3-24ep^4-5e^2p^4}{(-2+p)^3(-1+p)^2(-2+p-p^2-4p^3-3p^4+p^5+4p^6)}\right.+$$
$$+\left.\frac{34ep^5-3e^2p^5-14ep^6+5e^2p^6+2ep^7-2e^2p^7}{(-2+p)^3(-1+p)^2(-2+p-p^2-4p^3-3p^4+p^5+4p^6)}\right)+$$
$$+\frac{1776-4576p+3568p^2-1404p^3+2465p^4-2751p^5+841p^6+133p^7+2p^8-68p^9+16p^{10}}{(-2+p)^3(-1+p)^2(-2+p-p^2-4p^3-3p^4+p^5+4p^6)};$$

## Block 421

$$p_{x_1}{}^1 = 1, p_{x_1}{}^2 = -0.5949378 - 0.830714i,$$
$$p_{x_1}{}^3 = -0.5949378 + 0.830714i,$$
$$p_{x_1}{}^4 = 0.355937 - 0.513128i, p_{x_1}{}^5 = 0.355937 + 0.513128i, p_{x_1}{}^6 = 1,$$
$$p_{x_1}{}^7 = 1.228, p_{x_1}{}^8 = 2, p_{x_1}{}^9 = 2, p_{x_1}{}^{10} = 2,$$

## Block 422

$$p_{x_2}{}^1 = -1, p_{x_2}{}^2 = -0.5949378 - 0.830714i, p_{x_2}{}^3 = -0.5949378 + 0.830714i,$$
$$p_{x_2}{}^4 = 0.355937 - 0.513128i, p_{x_2}{}^5 = 0.355937 + 0.513128i, p_{x_2}{}^6 = 1,$$
$$p_{x_2}{}^7 = 1, p_{x_2}{}^8 = 1.228, p_{x_2}{}^9 = 2, p_{x_2}{}^{10} = 2, p_{x_2}{}^{11} = 2;$$

## Block 431

$$x_1(t) = \begin{cases} 10.031249e^{-t} - 1.25e^t + 5.538602e^{1.228001t}+ \\ + 2e^{0.355937t}(-3.735568\text{Cos}[0.513128t] + 15.529795\text{Sin}[0.513128t])+ \\ + 2e^{-0.594937t}(-0.924357\text{Cos}[0.830713t] + 0.061193\text{Sin}[0.830713t]), \\ 0 < t < 1; \\ 9.425260e^{-t} + 4.378584e^{1.228001t}+ \\ + 0.322878e^{2t} - 0.185554e^{2t}t + 0.0441176e^{2t}t^2+ \\ + 2e^{0.355937t}(-3.708886\text{Cos}[0.513128t] + 15.104078\text{Sin}[0.513128t])+ \\ + 2e^{-0.594938t}(-0.953417\text{Cos}[0.830713t] + 0.057591\text{Sin}[0.830713t]), \\ t > 1; \end{cases}$$

## Block 432

$$x_2(t) = \begin{cases} 10.03125e^{-t} + 0.5e^t - 8.948223e^{1.228001t} + 0.5e^t t+ \\ + 2e^{0.355937t}(-0.493116\text{Cos}[0.513128t] + 33.959275\text{Sin}[0.513128t])+ \\ + 2e^{-0.594938t}(1.701602\text{Cos}[0.830713t] + 0.929609\text{Sin}[0.830713t]), \\ 0 < t < 1; \\ 9.42526e^{-t} - 7.074087e^{1.228001t}- \\ - 0.577176e^{2t} + 0.435986e^{2t}t - 0.117647e^{2t}t^2+ \\ + 2e^{0.355937t}(-0.636666\text{Cos}[0.513128t] + 33.06373\text{Sin}[0.513128t])+ \\ + 2e^{-0.594938t}(1.748891\text{Cos}[0.830714t] + 0.968596\text{Sin}[0.830714t]), \\ t > 1; \end{cases}$$

The table gives the values of $\Delta$ for three values of $\varepsilon$ and three values of $T$.

| $T\backslash\varepsilon$ | $\varepsilon = 0.1$ | $\varepsilon = 0.01$ | $\varepsilon = 0.001$ |
|---|---|---|---|
| $T = 2$ | $1.37 \cdot 10^{-10}$ | $1.37 \cdot 10^{-11}$ | $1.37 \cdot 10^{-12}$ |
| $T = 3$ | $1.25 \cdot 10^{-12}$ | $1.25 \cdot 10^{-13}$ | $1.25 \cdot 10^{-14}$ |
| $T = 4$ | $5.93 \cdot 10^{-15}$ | $6.10 \cdot 10^{-16}$ | $5.55 \cdot 10^{-17}$ |

The work is partially supported by the grant RFBR 04-07-90268b.

# References

1. Akritas, A.G.: Elements of Computer Algebra with Applications. J. Wiley Interscience, New York, 1989
2. Burghelea, D., Haller, S.: Laplace transform, dynamics and spectral geometry. arXiv:math.DG/0405037v2, 17 Jan. (2005)
3. Dahiya, R.S., Jabar Saberi-Nadjafi: Theorems on n-dimensional Laplace transforms and their applications. In: 15th Annual Conf. of Applied Math., Univ. of Central Oklahoma, Electr. Journ. of Differential Equations, Conf.02 (1999) 61–74
4. Davenport, J., Siret, Y., Tournier, E.: Calcul formel. Systèmes et algorithmes de manipulations algébriques. MASSON, Paris, 1987
5. Dixon, J.: Exact solution of linear equations using p-adic expansions. Numer. Math. **40** (1982) 137–141
6. Felber, F.S.: New exact solutions of differential equations derived by fractional calculus. arXiv:math/0508157v1, 9 Aug. (2005)
7. Von zur Gathen, J., Gerhard, J.: Modern Computer Algebra. Cambridge University Press, 1999
8. Goursat E.: Sur les équations linéaires et la méthode de Laplace. Amer. J. Math. **18** (1896) 347–385
9. Malaschonok, G.I.: Solution of systems of linear equations by the p-adic method. Programming and Computer Software **29** (2) (2003) 59–71
10. Malaschonok, G.I.: Matrix Computational Methods in Commutative Rings. Tambov, TSU, 2002
11. Malaschonok, G.I.: Effective Matrix Methods in Commutative Domains. In: Formal Power Series and Algebraic Combinatorics, Springer, Berlin (2000) 506–517
12. Malaschonok, G.I.: Recursive method for the solution of systems of linear equations. In: Computational Mathematics (15th IMACS World Congress Vol. I, Berlin, August 1997), Wissenschaft und Technik Verlag, Berlin (1997) 475–480
13. Malaschonok, G.I.: Algorithms for computing determinants in commutative rings. Discrete Math. Appl. **5** (6) (1996) 557–566
14. Malaschonok, G.I.: Algorithms for the solution of systems of linear equations in commutative rings. In: Effective Methods in Algebraic Geometry. Progr. Math. V. 94, Birkhauser, Boston (1991) 289–298
15. Malaschonok, G.I.: Solution of a system of linear equations in an integral domain. USSR J. Comput. Math. and Math. Phys. **23** (6) (1983) 497–1500
16. Malaschonok, G.I.: The solution of a system of linear equations over a commutative ring. Math. Notes **42**, Nos. 3–4 (1987) 801–804
17. Malaschonok, N.: An algorithm to settle the necessary exactness in Laplace transform method. In: Computer Science and Information Technologies. Yerevan (2005) 453–456

18. Malaschonok, N.: Estimations in Laplace transform method for solutions of differential equations in symbolic computations. In: Differential Equations and Computer Algebra Systems. Minsk (2005) 195–199
19. Mizutani, N., Yamada H.: Modified Laplace transformation method and its applications to anharmonic oscillator. 12 Feb. (1997)
20. Podlubny, I.: The Laplace transform method for linear differential equations of the fractional order. arXiv:funct-an/9710005v1, 30 Oct. (1997)
21. Roux, J. Le.: Extensions de la méthodes de Laplace aux équatons linéaires aux deriveées partielles d'ordre supérieur du second. Bull. Soc. Math. de France **27** (1899) 237–262

# On Connection Between Constructive Involutive Divisions and Monomial Orderings

Alexander Semenov*

Moscow State University, Department of Mechanics and Mathematics

**Abstract.** This work considers the basic issues of the theory of involutive divisions, namely, the property of constructivity which assures the existence of minimal involutive basis. The work deals with class of $\succ$-divisions which possess many good properties of Janet division and can be considered as its analogs for orderings different from the lexicographic one. Various criteria of constructivity and non-constructivity are given in the paper for these divisions in terms of admissible monomial orderings $\succ$. It is proven that Janet division has the advantage in the minimal involutive basis size of the class of $\succ$-divisions for which $x_1 \succ x_2 \succ \ldots \succ x_n$ holds. Also examples of new involutive divisions which can be better than Janet division in minimal involutive basis size for some ideals are given.

## 1 Introduction

In papers [1, 3, 4], a general algorithmic approach to involutive bases, which are Gröbner bases of a special form, has been developed. This approach became one of the most efficient and widely used tools for computation of standard bases of polynomial ideals. In turn, the theory of involutive bases relies on the theory of involutive divisions on monomial sets.

In the process of involutive division investigation, it has been found that for the correctness and efficiency of the involutive basis algorithm an involutive division should possess some special properties, namely, noetherianity, and continuity [3, 4, 5], or admissibility [1]. There is a consensus among many specialists that Janet division is the best to use in the involutive basis computation algorithm, but there is almost no attempt to prove this fact rigorously. Also, in this situation the question of the Janet division generalization arises. This work presents some results in this direction by using pair property of involutive divisions.

In paper [5] a class of pairwise involutive divisions was introduced. Its study was continued in paper [7], where the connection between the pairwise property and the filter axiom was established. A procedure for construction of pairwise continuous involutive divisions by the pairwise closure method was proposed. This procedure defines explicitly the rule for specification of an involutive division only for sets consisting of no more than two different monomials. The

---

* The work was partially supported by the Russian Foundation for Basic Research, project no. 05-01-00671.

V.G. Ganzha, E.W. Mayr, and E.V. Vorozhtsov (Eds.): CASC 2006, LNCS 4194, pp. 261–278, 2006.

method of pairwise closure explicitly establishes the relation between the Janet division and the lexicographic ordering. A natural generalization of the Janet divisions is the class of $\succ$-divisions ($\succ$ is a admissible monomial ordering), which inherits many useful properties of Janet division. These divisions are defined by the way, which makes the sets of involutive multiples of each element as wide as possible. The concept of $\succ$-divisions encompasses the idea of pairwise property described in [5] and the connection between an involutive division and a monomial ordering, explicitly revealed in [1].

In papers [3, 4, 5], it is shown that a sufficient condition for the existence of the minimal involutive basis is the constructivity property of the involutive division. So, the problem arises to describe constructive involutive divisions in the $\succ$-division class, for which the minimal involutive basis exists for every polynomial set. These divisions can be considered as analogs of Janet division admissible for use in practice. Another question is to find sufficient conditions of non-constructivity if they exist.

The paper concerns all these problems, and the first results are the conditions of non-constructivity for a wide set of $\succ$-divisions, what can exclude it from practical use. The next result is constructivity of all $\succ$-division, when for $\succ$ the relations $x_1 \succ x_2 \succ \ldots \succ x_n$ hold. Nevertheless, the minimal Janet basis is contained in the minimal involutive bases for non-lexicographic orderings of this type, so Janet division can be considered optimal in this class.

The paper contains examples of constructive $\succ$-divisions, for which the relations $x_1 \succ x_2 \succ \ldots \succ x_n$ do not hold. These examples are more interesting as they can show better efficiency than Janet division in minimal involutive basis number for some ideals.

## 2    Basic Definitions

By $\mathbb{N}$ we denote the set of nonnegative integers. Then $\mathbb{M} = \{x_1^{d_1} \ldots x_n^{d_n} | d_i \in \mathbb{N}\}$ is the set of all possible monomials in $n$ variables.

By $\deg(u)$ and $\deg_i(u)$, we denote the total degree of monomial $u$ and the degree of $u$ with respect to variable $x_i$. For the least common multiple and the greatest common divisor of two monomials $u$ and $v$, we use the notation $\mathrm{lcm}(u, v)$ and $\gcd(u, v)$.

In this work $U$ always denotes a finite monomial set with distinct elements. The same is true for $V$.

We say that an *involutive division* $L$ is specified, if, for any $u \in U$, a submonoid $L(u, U)$ of $\mathbb{M}$ is defined such that the following axioms hold [3, 4]:

- if $w \in L(u, U)$ and $v | w$, then $v \in L(u, U)$,
- if $u, v \in U$ and $uL(u, U) \cap vL(v, U) \neq \emptyset$, then $u \in vL(v, U)$ or $v \in uL(u, U)$,
- if $v \in U$ and $v \in uL(u, U)$, then $L(v, U) \subseteq L(u, U)$,
- if $U \subseteq V$, then $L(u, V) \subseteq L(u, U)$ (filter axiom).

Elements of $L(u, U)$ are *multiplicative* for $u$. If $w \in uL(u, U)$, then $u$ is *involutive divisor* of $w$. This is denoted as $u |_L w$. The monomial $w$ is an *involutive multiple* of $u$. The monomial $v = w/u$ is *multiplicative* for $u$, and the equality $w = uv$ is written as $w = u \times v$. If $u$ is an ordinary divisor of $w$, but not involutive one, then this equality is written as $w = u \cdot v$. In this case, the monomial $v$ is *non-multiplicative* for $u$.

For any $u$ in $U$, there exists a partition of the set of all variables into two disjoint sets, namely, *multiplicative variables* $M_L(u, U) \subset L(u, U)$ and *non-multiplicative variables* $NM_L(u, U) \notin L(u, U)$.

The submonoids $L(u, U)$ may be naturally interpreted geometrically [2]. Consider a set $uL(u, U)$ for an involutive division $L$. Denote this set by $C_L(u, U)$. One can easily verify that the image of the set $C_L(u, U)$ under the bijective mapping from $\mathbb{M}$ onto $\mathbb{Z}^n$ is a discrete cone. The first three axioms are equivalent to the following two geometric facts:

- the set $C_L(u, U)$ is a discrete cone,
- $C_L(u, U) \cap C_L(v, U) \neq \emptyset \Longrightarrow C_L(u, U) \subseteq C_L(v, U) \vee C_L(v, U) \subseteq C_L(u, U)$

The notation $C_L(U) = \cup_{u \in U} C_L(u, U)$ is used further.

**Definition 1.** *An involutive division on a finite monomial set $U$ is disjoint, if $\not\exists u, v \in U$, $v \neq u$, $v \in uL(u, U)$.*

**Example 1.** *(Thomas division) Consider a finite monomial set $U$ with distinct elements. Variable $x_i$ is treated as multiplicative for $u \in U$, if $\deg_i(u) = \max\{\deg_i(v) \mid v \in U\}$ and non-multiplicative otherwise.*

**Example 2.** *(Janet division $J$) Consider a finite monomial set $U$ with distinct elements. For any $1 \leq i \leq n$, the set $U$ may be partitioned into subsets labeled with nonnegative integers $d_1, \ldots, d_i$ as follows:*

$$[d_1, \ldots, d_i] = \{u \in U \mid d_j = \deg_j(u), 1 \leq j \leq i\}.$$

*A variable $x_i$ is multiplicative for an element $u \in U$, if $i = 1$ and $\deg_1(u) = \max\{\deg_1(v) \mid v \in U\}$, or if $i > 1$, $u \in [d_1, \ldots, d_{i-1}]$, and $\deg_i(u) = \max\{\deg_i(v) \mid v \in [d_1, \ldots, d_{i-1}]\}$.*

Janet and Thomas divisions are disjoint.

The filter axiom can be reformulated the following way:

$$\forall u \in U \cap V \quad M_L(u, U \cup V) \subseteq M_L(u, U) \cap M_L(u, V).$$

Involutive divisions are mainly used for involutive basis determination of polynomial ideals (see [3, 4, 1]). Most algorithms for involutive basis computations have a similar structure. The core of the algorithm is a main loop (which finishes the computation if the division possesses some special characteristic properties: noetherianity, continuity, constructivity [3, 4], or admissibility [1]). Each iteration of the loop consists of taking the polynomial $g \cdot x$, where $x$ is non-multiplicative for $\mathrm{lm}(g)$ and $g$ belongs to the basis $G$ in construction, getting its involutive normal form, and adding it to $G$, if it is non-zero. The logic of any algorithm for involutive basis computation supposes that, the greater the size and dimension of the sets of involutive multiples, the less the number of involutive prolongations should be considered. Starting from this idea, it is worth to find optimal divisions among all of them. According to the filter axiom, these are the divisions with property $\forall u \in U \cap V$

$$M_L(u, U \cup V) = M_L(u, U) \cap M_L(u, V). \tag{1}$$

This class coincides with the class of **pairwise** involutive divisions, introduced in [5].

**Definition 2.** *[5] An involutive division $L$ is pairwise, if $\forall U$ , $\forall u \in U(U \setminus \{u\} \neq \emptyset)$ the following condition holds:*

$$M_L(u, U) = \bigcap_{v \neq u, v \in U} M_L(u, \{u, v\}) \tag{2}$$

The proof of the equivalence is given in [7]. In [5] it is proved that the Thomas and Janet divisions are pairwise.

So, a pairwise involutive divisions is entirely described by its definition on the sets consisting of no more than two distinct elements. The set which contains two distinct elements is called *basic set*.

**Definition 3.** *Consider a rule specifying a partition of variables into multiplicative and non-multiplicative on all basic sets, which satisfies the first three axioms of involutive divisions. Such a rule is denoted by $L_2$ and referred to as an involutive 2-partition. In other words, if $B$ is a basic set and an element $u \in B$, then $L_2$ uniquely determines the cone $uL_2(u, B)$. The 2-partition is disjoint, if $\forall U \in B$, $\nexists u, v \in U, v \neq u, v \in uL(u, U)$.*

Any involutive division being restricted onto basic sets specifies the corresponding 2-partition.

**Definition 4.** *(Pairwise closure) Consider a 2-partition $L_2$ on the basic sets. The rule, which defines multiplicative variables for any finite monomial set $U$ with distinct elements and $\forall u \in U$ as*

$$M_L(u, U) = \bigcap_{v \in U, v \neq u} M_{L_2}(u, \{u, v\}),$$

*is called pairwise closure of involutive 2-partition $L_2$.*

As shown in [7], it is not always true that the pairwise closure determines an involutive division. Nevertheless, in [7] it is proved that pairwise closure $L$ of the disjoint 2-partition $L_2$ is a disjoint involutive division.

If a set $U$ contains one element, namely, $U = \{u\}$, then $NM_L(u, \{u\}) = \{\}$ is the best issue for involutive division. So, for description of pairwise involutive divisions with the widest possible involutive cones, we should choose the optimal disjoint 2-partitions.

If a set $U$ contains two distinct elements, the optimal possibility is when one element has $n$-dimensional involutive cone and another element has $(n-1)$-dimensional cone or, in other words, one non-multiplicative variable. The sizes of involutive cones for sets $U$, which consist of more than two elements, are maximal possible for pariwise divisions, if 2-partitions on the basic sets are given.

The pairwise property for Janet involutive division is established in [5]. The Janet divisions has some other remarkable properties, which is worth to preserve for other involutive divisions considered as Janet analogs:

- efficient maximality i.e. existence of one $n$-dimensional cone for every finite set $U$,
- homotheticity, i.e. same structure on the set $U = \{u_1, \dots, u_n\}$ and on $mU = \{mu_1, \dots, mu_n\}$ ($m$ is an arbitrary monomial), namely, $NM_L(u, U) = NM_L(mu, mU)$,
- efficient variable separation.

The last property can be illustrated by following simple example.

**Example 3.** *(Inefficient variable separation) Consider $U = \{1, x_1 x_2, x_1^3 x_2^3\}$ and the case when $NM_L(1, \{1, x_1 x_2\}) = x_1$, $NM_L(1, \{1, x_1^3 x_2^3\}) = x_2$. Consequently, $NM_L(1, \{1, U\}) = \{x_1, x_2\}$. From the standpoint of the number of multiplicative variables for 1, both configurations $NM_L(1, \{1, x_1 x_2\}) = NM_L(1, \{1, x_1^3 x_2^3\}) = x_1$ and $NM_L(1, \{1, x_1 x_2\}) = NM_L(1, \{1, x_1^3 x_2^3\}) = x_2$ are preferable.*

Let $L$ be a pairwise involutive division. Consider the relation $\succ_L$: $u \succ_L v$ if $NM_L(v, \{u, v\}) \neq \emptyset$.

**Theorem 1.** *Let $L$ be a disjoint homothetic pairwise division for which efficient maximality takes place. Then $\succ_L$ is an admissible monomial ordering.*

*Proof.* As the division is disjoint the relation $1 \prec_L v \neq 1$ takes place. As the division is homothethic, then $v \prec_L u \Rightarrow vm \prec_L um$. The relation $\prec_L$ is transitive as in the opposite case $v \prec_L w \prec_L u \prec_L v$ every element would have a non-multiplicative element, and no $n$-dimensional cones exist.

For a variable $x_l$ which is a non-multiplicative for $v$ and pair $\{u, v\}$, $u \succ_L v$, the property $\deg_l(v) < \deg_l(u)$ should be held for division $L$ be disjoint on $\{v, u\}$. The choice of the variable $x_p$ as non-multiplicative for $v$, if $p = \min\{l | \deg_l(v) < \deg_l(u)\}$, excludes the cases of inefficient separations analogous to those in example 3.

On the other hand, every admissible monomial ordering $\succ$ defines an involutive $\succ$-division $L$ as pairwise closure of the $\succ$-partition $L_2$ on all basic sets. Involutive 2-partition $L_2$ is defined the following way for an $\{u, v\}(u \succ v)$:

1. $NM_{L_2}(u, \{u, v\}) = \{\}$,
2. $NM_{L_2}(v, \{u, v\}) = x_p$, where $p = \min\{l | \deg_l(v) < \deg_l(u)\}$

If the set $U$ contains one element, namely, $U = \{u\}$, then $NM_L(u, \{u\}) = \{\}$.

This way, it is shown that every admissible ordering defines an involutive division, which has all the properties listed above (as Janet division has), and can be considered as "optimal". Also, homotheticity and efficient maximality yield that $\succ_L$ is an admissible ordering. So, it is reasonable to search for new involutive divisions (analogs of Janet division) in the class of $\succ$-divisions. Janet division itself may be described as the lex-division, where symbol lex corresponds to the lexicographic ordering for which $x_1 >_{\text{lex}} x_2 >_{\text{lex}} \cdots >_{\text{lex}} x_n$.

For the sake of convenience, we introduce the notation, which is used further in the text. The assertion $\forall p < q, \deg_p(v) \geq \deg_p(u)$ is briefly written as $\deg_{q-}(v) \geq \deg_{q-}(u)$. If $\forall p < q$, $\deg_p(v) = \deg_p(u)$, then it is denoted as $\deg_{q-}(v) = \deg_{q-}(u)$. Moreover, we set

$$\text{ind}(v, \{v, u\}) = \min\{p | \deg_p(v) < \deg_p(u)\}.$$

Obviously, if $q = \text{ind}(v, \{v, u\})$, then we have $\deg_{q-}(v) \geq \deg_{q-}(u)$.

In case of involutive $\succ$-divisions ordering $\succ$ determines an associated permutation $\xi \in S_n$, namely, $x_{\xi(1)}, \ldots, x_{\xi(n)}$ is the permutation of variables such that $x_{\xi(1)} \succ \cdots \succ x_{\xi(n)}$. Lexicographic ordering with respect to the permutation of variables $\xi$ is called *induced lexicographic ordering* and is denoted by $\text{lex}(\prec)$.

**Theorem 2.** *Let $L$ be an involutive $\succ$-division, $U$ be an arbitrary finite monomial set with distinct elements, and $u_1, u_2$ be such elements, that $u_1 \cdot x \in u_2 L(u_2, U)$, where $x$ is a non-multiplicative variable for $u_1$ and $U$. Then $u_1 <_{\text{lex}(\prec)} u_2$.*

*Proof.* Consider monomials $u_1$, $u_2$ in $U$ and a variable $x \in NM_L(u, U)$ such that $u_1 \cdot x \in C_L(u_2, U)$. According to the filter axiom and pair property $u_1 \cdot x \in u_2 L(u_2, \{u_1, u_2\})$ and $NM_L(u_1, \{u_1, u_2\}) = x$. Then $u_1 \cdot x = u_2 \times w$, and $w$ does not contain $x$ as $L$ is disjoint. Then $u_1 = \frac{u_2 \times w}{x} \prec u_2$ according to pairwise property and $u_2 w \prec u_2 x$. Hence $w \prec x$. If $w = 1$ then, obviously, $1 <_{\text{lex}(\prec)} x$. Otherwise, let $z$ be a variable such that $z | w$, $w = z w_1$. It is clear that $z \prec x$, and $z <_{\text{lex}(\prec)} x$. That is true for every $z$ and that yields $w <_{\text{lex}(\prec)} x$.

As $w <_{\text{lex}(\prec)} x$ was obtained, expression $u_2 w <_{\text{lex}(\prec)} u_2 x$ is also valid. That leads to the fact that $u_1 = \frac{u_2 \times w}{x} <_{\text{lex}(\prec)} u_2$.

In papers [3, 4], a concept of *continuity* of involutive divisions is introduced. Let an involutive division $L$ be given. A sequence of monomials $\{u_t\}$ ($t \in \{1, 2 \ldots\}$) in a finite monomial set $U$ with distinct elements, such that exists $x_{k(t)} \in NM_L(u_t, U)$ and $[u_{t+1} |_L u_t \cdot x_{k(t)}]$ is called a *chain*.

**Definition 5.** *[3, 4] An involutive division $L$ is **continuous**, if, in every chain, the inequality $\forall s_1 \neq s_2 \ u_{s_1} \neq u_{s_2}$ holds, and, hence, every chain is finite.*

**Theorem 3.** *Let $L$ be an involutive $\succ$-division, $U$ be an arbitrary finite monomial set with distinct elements, and $\{u_t\}$ be a chain in $U$. Then the following is true:*

- *if $u_s$, $u_{s+1}$ are in $\{u_t\}$, then $u_s \prec u_{s+1}$,*
- *if $u_s$, $u_{s+1}$ are in $\{u_t\}$, then $u_s <_{\text{lex}(\prec)} u_{s+1}$,*
- *$L$ is continuous,*
- *if $m$ has an involutive divisor $u$ in $U$ $(u|_L m)$, then $u = \max_{\prec}\{v \in U, v|m\}$.*

*Proof.* According to the pairwise property $u_s \cdot x_{k(s)} \in u_{s+1} L(u_{s+1}, \{u_s, u_{s+1}\})$ and also the relation $NM_L(u_s, \{u_s, u_{s+1}\}) = x_{k(s)}$ is valid. Then $u_s \prec u_{s+1}$, and that proves the continuity. Theorem 2 shows that $u_s <_{\text{lex}(\prec)} u_{s+1}$.

The last item is proven from the opposite. Let $u|_L m$, $\overline{u} = \max_{\prec}\{v \in U, v|m\}$, and $u \neq \overline{u}$. Consider the pair $\{u, \overline{u}\}$. As $\overline{u} \succ u$, and $\overline{u}|m$ then $\text{lcm}(u, \overline{u})|m$ and the monomial $\text{lcm}(u, \overline{u})$ contains a non-multiplicative variable for $u$ which lies in $NM_L(u, \{u, \overline{u}\})$ according to the definintion. That leads to the contradiction.

## 3   Constructivity

In the framework of the involutive approach, an analog of the minimal autore-duced Gröbner basis is the minimal involutive basis [4]. A necessary condition for existence of the minimal involutive basis of an arbitrary ideal for division $L$ is the existence of such a basis in monomial ideals. Below, we present the theoretical background for the monomial case.

**Definition 6.** *An $L$-autoreduced monomial set $V$ is an $L-$involutive basis of monomial ideal $Id(U)$, if*

$$(\forall u \in U)(\forall w \in \mathbb{M})[\exists v \in V v|_L uw].$$

**Definition 7.** *[4] Let $L$ be an involutive division, and $Id(U)$ be a monomial ideal. Its $L$-involutive basis $\overline{U}$ is minimal, if for an involutive basis $V$ of the same ideal inclusion $\overline{U} \subseteq V$ takes place.*

In paper [4], a condition is formulated under which the minimal involutive basis exists and can be obtained by the algorithm **Minimal Monomial Involutive Basis**.

The sufficient condition of the existence of minimal involutive basis is constructivity. In [4], it is proven that, in the polynomial case constructivity also guarantees the existence of the minimal involutive basis.

**Definition 8.** *[3, 4] A continuous involutive division $L$ is constructive on a finite set $U \subset \mathbb{M}$ with distinct elements, if for any $u \in U$, $x_i \in NM_L(u, U)$ such that $u \cdot x_i$ has no involutive divisors in $U$ and*

$$(\forall v \in U)(\forall x_j \in NM_L(v, U))(v \cdot x_j | u \cdot x_i, v \cdot x_j \neq u \cdot x_i) \Rightarrow$$

$$\Rightarrow [v \cdot x_j \in C_L(U)]$$

*the following condition holds:*

$$\forall w \in C_L(U)[u \cdot x_i \notin C_L(U \cup \{w\})].$$

*A continuous involutive division $L$ is constructive if it is constructive on every finite set $U$ with distinct elements.*

**Theorem 4.** *Let a continuous involutive division $L$ be non-constructive on a finite set $U$ with distinct elements. Then $U$ is not autoreduced in the sense of conventional division.*

*Proof.* Consider the opposite, namely $U$ is autoreduced, and $u, u_1, u_1 \times v = w$ are the monomials which deliver non-constructivity. In other words, $\exists x_i, u \cdot x_i \notin C_L(U), u \cdot x_i \in C_L(U \cup \{w\}), (\forall v \in U) \, (\forall x_j \in NM_L(v, U)) \, (v \cdot x_j | u \cdot x_i, v \cdot x_j \neq u \cdot x_i) \Rightarrow [v \cdot x_j \in C_L(U)]$. Then $u \cdot x_i \in wL(w, U \cup \{w\})$.

There exists a variable $x_{s1} \in NM_L(u_1, U)$ such that $\frac{u \cdot x_i}{u_1}$ contains $x_{s1}$. The monomial $u_1 \cdot x_{s1}$ properly divides $u \cdot x_i$ and $u_1 \cdot x_{j1} = u_2 \times v_2$ according to the definition 8.

As $u \cdot x_i$ has no involutive divisors in $U$, the monomial $\frac{u \cdot x_i}{u_2}$ is not multiplicative for $u_2$ and contains a variable $x_{s2}$, non-multiplicative for $u_2$. The element $u_3 |_L u_2 \cdot x_{s2}$ (if $u_2 \cdot x_{s2} \neq u \cdot x_i$) is constructed analogously. So, there is a chain $u_1, u_2, u_3, \ldots$. As the division is continous, this chain breaks on an element $u_t \in U : \exists x_{st}, u_t \cdot x_{st} = u \cdot x_i$ ( it may occur that $u_t = u_2$).

Then we have $u_{t-1} \cdot x_{s(t-1)} = u_t \times q, u_t \times q | u_t \cdot x_{st}$. Hence, $q = 1$, $u_{t-1} | u_t$, and $U$ is not autoreduced in the sence of conventional division.

The next example shows that there are non-constructive divisions among $\succ$-divisions.

**Theorem 5.** *Consider the $\succ$-division $L$. If for the ordering $\succ$ satisfy the condition: $\exists i < j < k < l$ and $\exists s \in \mathbb{N}, s > 0$ s.t. $x_j x_l \succ x_k^s \succ x_i x_l$ then $L$ is non-constructive.*

*Proof.* The relation $x_j x_l \succ x_k^s \succ x_i x_l$ implies $x_i^2 x_j x_l \succ x_i^2 x_k^s$ and $x_i x_j x_k^s \succ x_i^2 x_j x_l$. Also, $x_i^2 x_j x_l \succ x_i x_k^s$ is valid.

Then consider $U = \{x_i x_k^s, x_i^2 x_k^s, x_i^2 x_j x_l\}$, $w = x_i x_k^s \times x_j$. The main relation is $x_i^2 x_k^s \cdot x_j = x_i x_j x_k^s \times x_i$.

| $U$ | $NM_L(U)$ | $U \cup \{x_i x_j x_k^s\}$ | $NM_L(U \cup \{x_i x_j x_k^s\})$ |
|---|---|---|---|
| $x_i x_k^s$ | $x_i$ | $x_i x_k^s$ | $x_i, x_j$ |
| $x_i^2 x_k^s$ | $x_j$ | $x_i^2 x_k^s$ | $x_j$ |
| $x_i^2 x_j x_l$ | $-$ | $x_i^2 x_j x_l$ | $x_k$ |
| | | $x_i x_j x_k^s$ | $-$ |

That proves the theorem.

This example explicitly shows non-existence of minimal involutive basis. Consider the case $s = 1$ (for example, ordering $\text{lex}_{\{j,k,i,l\}}$). Consider $U = \{x_i^2 x_j x_l, x_i x_k\}$, which is autoreduced. The sets $U_1 = \{x_i x_j x_k, x_i^2 x_j x_l, x_i^2 x_k, x_i x_k\}$ and $U_2 = \{x_i^2 x_j x_k, x_i^2 x_j x_l, x_i^2 x_k, x_i x_k\}$ are both involutive monomial bases. Their intersection is $U_1 \cap U_2 = \{x_i x_k, x_i^2 x_k, x_i^2 x_j x_l\}$. Neither $U_1 \cap U_2$ nor its subsets cannot be involutive bases of $U$. As the minimal involutive basis is contained in every monic involutive basis, that shows non-existence of it in this particular case.

The following examples show that there exist many non-constructive divisions in three-variable case.

**Theorem 6.** *Consider the $\succ$-division $L$. If for the ordering $\succ$ one of four conditions*

1. $\exists i < j < k$ *s.t.* $x_j \prec x_i \prec x_k$,
2. $\exists i < j < k$ *s.t.* $x_i \prec x_j \prec x_k$,
3. $\exists i < j < k, p \in \mathbb{N}$ *s.t.* $x_i \prec x_k \prec x_j \prec x_k^{p-1}$,
4. $\exists i < j < k < l$ *s.t.* $x_j \succ x_k \succ x_i, \; x_j \succ x_k \succ x_l$

*is satisfied, $L$ is non-constructive.*

*Proof.* The proof is done according to the following scheme. Firstly, the set $U$ is constructed, and $L$ is defined on $U$ according to pairwise property. Then, for each case, the example of non-constructivity is given by specifying the set $U$, the non-multiplicative prolongation $\bar{u}$ of the element $u \in U$, and the multiplicative prolongation $w$ of the element $v \in U$. All the conditions on these elements can be checked from tables of variable partitioning for $U$ and $U \cup \{w\}$ given below for each case.

In case 1, relation $x_j \prec x_i \prec x_k$ implies $x_i \prec x_i x_j \prec x_j x_k \prec x_i x_k \prec x_i^2 x_j x_k$. Then $U = \{x_i, x_i x_j, x_j x_k, x_i^2 x_j x_k\}$, and $w = x_i \times x_k$. The main relation is $x_i x_j \cdot x_k = x_j x_k \cdot x_i = x_i x_k \times x_j$.

| $U$ | $NM_L(U)$ | $U \cup \{x_i x_k\}$ | $NM_L(U \cup \{x_i x_k\})$ |
|---|---|---|---|
| $x_i$ | $x_i, x_j$ | $x_i$ | $x_i, x_j, x_k$ |
| $x_i x_j$ | $x_i, x_k$ | $x_i x_j$ | $x_i, x_k$ |
| $x_j x_k$ | $x_i$ | $x_j x_k$ | $x_i$ |
| | | $x_i x_k$ | $x_i$ |
| $x_i^2 x_j x_k$ | $-$ | $x_i^2 x_j x_k$ | $-$ |

In the case 2, relation $x_i \prec x_j \prec x_k$ implies $x_j \prec x_i x_j \prec x_i x_k \prec x_j x_k \prec x_j^2 x_k$. Then $U = \{x_j, x_i x_j, x_i x_k, x_j^2 x_k\}$, and $w = x_j \times x_k$. The main relation is $x_i x_k \cdot x_j = x_i x_j \cdot x_k = x_j x_k \times x_i$.

| $U$ | $NM_L(U)$ | $U \cup \{x_j x_k\}$ | $NM_L(U \cup \{x_j x_k\})$ |
|---|---|---|---|
| $x_j$ | $x_i, x_j$ | $x_j$ | $x_i, x_j, x_k$ |
| $x_i x_j$ | $x_j, x_k$ | $x_i x_j$ | $x_j, x_k$ |
| $x_i x_k$ | $x_j$ | $x_i x_k$ | $x_j$ |
| | | $x_j x_k$ | $x_j$ |
| $x_j^2 x_k$ | $-$ | $x_j^2 x_k$ | $-$ |

In case 3, relation $x_i \prec x_k \prec x_j \prec x_k^{p-1}$ implies $x_k \prec x_i x_k \prec x_i x_j \prec x_j x_k \prec x_k^p$. Then $U = \{x_k, x_i x_k, x_i x_j, x_k^p\}$, $w = x_k \times x_j$. The main relation is $x_i x_k \cdot x_j = x_i x_j \cdot x_k = x_j x_k \times x_i$.

| $U$ | $NM_L(U)$ | $U \cup \{x_j x_k\}$ | $NM_L(U \cup \{x_j x_k\})$ |
|---|---|---|---|
| $x_k$ | $x_i, x_k$ | $x_k$ | $x_i, x_j, x_k$ |
| $x_i x_k$ | $x_j, x_k$ | $x_i x_k$ | $x_j, x_k$ |
| $x_i x_j$ | $x_k$ | $x_i x_j$ | $x_k$ |
| | | $x_j x_k$ | $x_k$ |
| $x_k^p$ | $-$ | $x_k^p$ | $-$ |

The fourth situation is proven by such a way. Consider the case $x_j \succ x_k \succ x_l \succ x_i$. If $x_i x_l \succ x_k$ then $x_l^2 \succ x_k$ and the variables $x_l^2 \succ x_k \succ x_l \succ x_i$ satisfy the case 3. Then $x_i x_l \prec x_k$, and it can be obviously seen that $x_j x_l \succ x_k$, what also yield non-constructivity by the theorem 5.

Consider the case $x_j \succ x_k \succ x_i \succ x_l$. If $x_j x_l \prec x_k^2$ then $x_j \prec x_k^2$, and the relation $x_k^2 \succ x_j \succ x_k \succ x_i$ satisfies case 3. Then $x_j x_l \succ x_k^2$ and $x_k^2 \succ x_i x_l$ what also yield non-constructivity by the theorem 5. Thus, all four cases are considered, and that proves the theorem.

This theorem gives many examples of non-constructive $\succ$-divisions among which are divisions $(\mathrm{deglex}_{\{3,1,2\}}, \{1,2,3\})$ (case 1), $(\mathrm{deglex}_{\{3,2,1\}}, \{1,2,3\})$ (case 2), $(\mathrm{deglex}_{\{2,3,1\}}, \{1,2,3\})$ (case 3), $(\mathrm{deglex}_{\{2,3,1,4\}}, \{1,2,3,4\})$ (case 4).

Below, we will identify new constructive $\succ$-divisions. Further, the special notation is developed for this task.

**Definition 9.** *Let $L$ be an involutive $\succ$-division. A set of distinct monomials $\{u_1, u, w, \widehat{u}\}$ is $\gamma$-configuration, if it satisfies the following conditions:*

1. $u \prec \widehat{u}$,
2. $w = u_1 \times v$, $v \in L(u_1, \{u_1, u, \widehat{u}\})$,
3. $u \cdot x_i \in wL(w, \{\widehat{u}, u, u_1, w\})$, where $x_i = NM_L(u, \{u, \widehat{u}\})$.

Further the index $i$ and notation $x_i$ is always used in the previous sense $(x_i = NM_L(u, \{u, \widehat{u}\}))$. Also in case $u_1 \prec u$ the notation $x_j$ and $x_k$ will be used for $x_j = NM_L(u_1, \{u_1, u\})$ and $x_k = NM_L(u_1, \{u_1, \widehat{u}\})$

**Lemma 1.** *Let an involutive division $L$ be given. Suppose that there exist admissible monomial orderings $\lhd_1, \ldots, \lhd_t$ and the following two conditions hold ($U$ is a finite monomial set with distinct elements):*

- *$\forall U, \forall u \in U, x \in NM_L(u, U)$ such that $u \cdot x \notin C_L(U), u \cdot x \in C_L(U \cup \{u_1 \times v\}), u_1 \in U, v \in L(u_1, U)$ $\exists p$ $u \lhd_p u_1$ is satisfied*
- *$\forall U, \forall u, \overline{u} \in U, x \in NM_L(u, U)$ such that $u \cdot x \in C_L(\overline{u}, U)$ $\forall p$ $\overline{u} \rhd_p u$ is satisfied.*

*Then, the division $L$ is constructive.*

*Proof.* Let $u, u_1, u_1 \times v = w$ be monomials, which provide the non-constructivity; i.e., $\exists x, u \cdot x \notin C_L(U), u \cdot x \in C_L(U \cup \{w\})$. Then $u \cdot x \in wL(w, U \cup \{w\})$. There exists a variable $x_{s1} \in NM_L(u_1, U)$ such that $\frac{u \cdot x}{u_1}$ contains $x_{s1}$.

Consider the monomial $u_1 \cdot x_{s1}$. By the definition of constructivity, we have $u_1 \cdot x_{s1} = u_2 \times v_2, \forall p$ $u_2 \rhd_p u_1$ (by the second condition).

Since $u \cdot x$ has no involutive divisors in $U$, the monomial $\frac{u \cdot x}{u_2}$ is non-multiplicative for $u_2$ and contains a variable $x_{s2}$ which is non-multiplicative for $u_2$. So we have the relation $\forall p$ $u_1 \lhd_p u_2$. If $u_2 \cdot x_{s2} \neq u \cdot x$, we construct an element $u_3 |_L u_2 \cdot x_{s2}$, $\forall p$ $u_1 \lhd_p u_2 \lhd_p u_3$ and so on. This chain must finish only on element $u_k \cdot x_{sk} = u \cdot x$ due to the continuity of the division. By the initial condition there exists $q$, $u_k \lhd_q u_1$. So, we have $u_k \lhd_q u_1 \lhd_q u_2 \lhd_q u_3 \ldots \lhd_q u_k$, what is impossible.

**Lemma 2.** *Let an involutive $\succ$-division $L$ be given. Suppose that there exist admissible monomial orderings $\lhd_1, \ldots, \lhd_t$ and the following two conditions hold ($U$ is a finite monomial set with distinct elements).*

- *For all $\gamma$-configurations $\{\widehat{u}, u, u_1, w\}$ $\exists p$  $u_1 \rhd_p u$ is satisfied.*
- *$\forall U, \forall u, \overline{u} \in U, x \in NM_L(u, U)$ such that $u \cdot x \in C_L(\overline{u}, U)$ $\forall p$  $\overline{u} \rhd_p u$ is satisfied.*

*Then division $L$ is constructive.*

*Proof.* For the proof it is sufficient to verify the conditions of Theorem 1 for the division $L$ and orderings $\rhd_1, \ldots, \rhd_t$.

The second conditions of this theorem and lemma 1 coincide and, thence, are satisfied.

Suppose the opposite for the the first condition of lemma 1. Namely, there exists a monomial set $U$ and monomials $u_1, u \in U$ such that for $u, u_1, u_1 \times v = w$ $\exists x, u \cdot x \notin C_L(U), u \cdot x \in wL(w, U \cup \{w\})$, and $\forall p, u_1 \lhd_p u$ is also satisfied.

As $x$ is non-multiplicative for $u$, then there exists $\widehat{u} \in U$ such that $NM_L(u, \{u, \widehat{u}\}) = x$ and $u \prec \widehat{u}$.

Consider the set $\{u_1, u, w, \widehat{u}\}$. From $v \in L(u_1, U)$, $u \cdot x \in wL(w, U \cup \{w\})$ and from filter axiom it can be seen that $v \in L(u_1, \{u_1, u, \widehat{u}\})$ and $u \cdot x \in wL(w, \{\widehat{u}, u, u_1, w\})$. So, the set $\{u_1, u, w, \widehat{u}\}$ is $\gamma$-configuration, for which $\forall p$ $u_1 \lhd_p u$. That contradicts the first condition of the current lemma.

*Remark 1.* For a $\gamma$-configuration $\{u_1, u, w, \widehat{u}\}$ the relation $u \cdot x_i \in \widehat{u}L(\widehat{u}, \{u, \widehat{u}\})$ can be valid, so $u \cdot x_i \in C_L(\{u_1, u, \widehat{u}\})$

**Lemma 3.** *For every $\gamma$-configuration $\{u_1, u, w, \widehat{u}\}$ corresponding to $\succ$-division, where $u_1 \prec u$, the relation $x_i \succ x_j$ is valid.*

*Proof.* Firstly, note that $x_i \neq x_j$. As $\deg_j(u_1) < \deg_j(u)$ then $\deg_j(u_1 v) < \deg_j(u)$ so $u \cdot x_i = u_1 v \times x_j p$. If $x_i = x_j$ then $u = u_1 v \times p$ which is impossible since $\succ$-division $L$ is disjoint.

Due to Theorem 2, $u_1 v >_{\text{lex}(\succ)} u$. As $u_1 \cdot x_i = (u_1 v) \times p$, where $p$ is a monomial, $\forall x_l \succ x_i \deg_l(u_1 v) \preceq \deg_l(u)$. In this case inequality $u_1 v >_{\text{lex}(\succ)} u$ leads to the relation $\forall x_l \succ x_i \deg_l(u_1 v) = \deg_l(u)$. Then $x_j \prec x_i$ what proves the theorem.

**Lemma 4.** *(Sufficient conditions of constructivity) Let $L$ be an involutive $\succ$-division. If there exists no $\gamma$-configuration $\{u_1, u, w, \widehat{u}\}$ for which relations $u_1 \prec u$, and $u_1 <_{\text{lex}(\prec)} u$ are simultaneously valid, then $L$ is constructive.*

*Proof.* The proof can be done by the lemma 2, supposing that $\prec$ and $\text{lex}(\prec)$ are used as orderings $\lhd_1, \lhd_2$.

**Theorem 7.** *Let $\succ$ be an admissible monomial ordering, such that $x_1 \succ x_2 \succ \ldots \succ x_n$. There is no $\gamma$-configurations with $u_1 \prec u$ and $u_1 <_{\text{lex}(\prec)} u$ and involutive $\succ$-division $L$ is constructive.*

*Proof.* The proof is performed with the help of lemma 4. Consider the opposite, namely, that $L$ is non-constructive and so there exists a $\gamma$-configuration $\{u_1, u, w, \widehat{u}\}$ for which relations $u_1 \prec u$ and $u_1 <_{\mathrm{lex}(\prec)} u$ hold simultaneously.

Due to Theorem 2 and the pairwise property we have:

$$u_1 \prec u \prec u_1 v,$$

$$u_1 <_{\mathrm{lex}(\prec)} u <_{\mathrm{lex}(\prec)} u_1 v.$$

Let $NM_L(u_1, \{u_1, u\}) = x_j$, and, hence, $\deg_{j-}(u_1) \geq \deg_{j-}(u)$. Due to $u_1 <_{\mathrm{lex}(\prec)} u$ and the initial conditions $x_1 \succ x_2 \succ \ldots \succ x_n$ we have $\deg_{j-}(u_1) = \deg_{j-}(u)$.

Due to lemma 3, $x_i \succ x_j$ and $i < j$. Also, $\deg_i(u_1 v) = \deg_i(u) + 1$. As $\deg_{j-}(u_1) = \deg_{j-}(u)$, then $x_i \in M_L(u_1, \{u_1, u, \widehat{u}\})$ and $x_i \in M_L(u_1, \{u_1, \widehat{u}\})$. That is impossible, as $\deg_{j-}(u_1) = \deg_{j-}(u)$, and $x_i \in NM_L(u, \{u, \widehat{u}\})$. That proves the theorem.

The immediate consequence of the theorem is the constructivity of the Janet lex-division, deglex-division, and degrevlex-division, where for corresponding orderings relation $x_1 \succ x_2 \succ \ldots \succ x_n$ takes place.

Nevertheless, the further investigation shows that Janet division can be considered optimal among those $\succ$-divisions for which $x_1 \succ x_2 \succ \ldots \succ x_n$ is valid.

**Definition 10.** *A finite monomial set with distinct elements is involutive with respect to the involutive division* $L$, *if* $\forall U$ *and* $\forall u_i \in U, w \in \mathbb{M} \ \exists u_j \in U, w_1 \in L(u_j, U)$ *such that* $u_i w = u_j \times w_1$.

**Theorem 8.** *Consider an involutive* $\succ$-*division* $L$ *and a finite monomial set* $U$ *with distinct elements, which is involutive with respect to* $L$. *Then* $\forall u \in U$

$$uL(u, U) = \{m \in \mathbb{M}, u = \max_{\mathrm{lex}(\prec)} \{v \in U, v|m\}\}.$$

*Proof.* Firstly, it can be proven that $\forall U, u \in U \ \{m \in \mathbb{M}, u = \max_{\mathrm{lex}(\prec)}\{v \in U, v|m\}\} \subseteq uL(u, U)$. Consider that there exists $U$ such that $\exists m \in \mathbb{M}, u \in U$, $u = \max_{\mathrm{lex}(\prec)}\{v \in U, v|m\}, m \notin uL(u, U)$. Then $m$ contains a non-multiplicative variable $x \in NM_L(u, U)$. As $U$ is involutive $u \cdot x$ has an involutive divisor $u_1 \in U$, $u \cdot x = u_1 \times w$. According to Theorem 2, $u <_{\mathrm{lex}(\prec)} u_1$ what contradicts the initial supposition.

Secondly, it is proven that $\forall U, u \in U \ \{m \in \mathbb{M}, u = \max_{\mathrm{lex}(\prec)}\{v \in U, v|m\}\} \supseteq uL(u, U)$. Consider that there exists $U$ such that $\exists m \in \mathbb{M}, u \in U$, and $u_1 >_{\mathrm{lex}(\prec)} u$ for which $u \in U, m \in uL(u, U)$, and $u_1|m$ take place. We can consider an element $u_1$ as the maximal element, which conventionally divides $m$. As the division is pairwise, $\exists x \in NM_L(u_1, \{u_1, U\}), u_1 \cdot x|m$. Due to the involutivity of $U$ with respect to $L$, $\exists u_2|_L u_1 \cdot x$, $u_2 \succ u_1$, and $u_2 >_{\mathrm{lex}(\prec)} u_1$, according to Theorem 2. As $u_2|m$ we obtain the contradiction. That proves the theorem.

**Theorem 9.** *Consider an involutive $\succ$-division $L$ and a finite monomial set $U$ with distinct elements, which is involutive with respect to $L$, and for $\succ$ the relation $x_1 \succ x_2 \succ \ldots \succ x_n$ is valid. Then $U$ is involutive with respect to lex($\prec$)-division, namely, Janet division.*

*Proof.* In the proof, the Janet division will be denoted as $J$. For the proof it is sufficient to show that $U$ is involutive with respect to J. If $\forall u \in U \ NM_J(u, U) \subseteq NM_L(u, U)$ the proof is done.

Consider the opposite, i.e., there exists an element $u \in U$ and a variable $x_s$, $x_s \in NM_J(u, U), x_s \notin NM_L(u, U)$. Then there exists $\widehat{u} \in U, NM_J(u, \{u, \widehat{u}\}) = x_s$. The case $\widehat{u} \succ u$ is not possible as it would yield $NM_L(u, \{u, \widehat{u}\}) = x_s$.

Thus, $\widehat{u} \prec u, u <_{\text{lex}(\prec)} \widehat{u}$. As the variable $x_s$ is non-multiplicative for $u$ and Janet division, then

$$\deg_{s-}(u) = \deg_{s-}(\widehat{u}), \deg_s(u) < \deg_s(\widehat{u}).$$

Consider the monomial $\text{lcm}(u, \widehat{u})$. The following relations are valid for it:

$$\deg_{s-}(u) = \deg_{s-}(\text{lcm}(u, \widehat{u})),$$

$$\deg_s(u) < \deg_s(\text{lcm}(u, \widehat{u})).$$

Let $\text{lcm}(u, \widehat{u}) \in u_0 L(u_0, U)$. Due to Theorem 8 $u <_{\text{lex}(\prec)} \widehat{u} \leq_{\text{lex}(\prec)} u_0$ and also $\widehat{u} \prec u \preceq u_0$. Then $u_0 \neq u, u_0 \neq \widehat{u}$. As $u_0 | \text{lcm}(u, \widehat{u})$ and $\widehat{u} <_{\text{lex}(\prec)} u_0$ we have $\deg_{s-}(u_0) = \deg_{s-}(\widehat{u}), \deg_s(u_0) > \deg_s(u)$. So $NM_L(u, \{u, u_0\}) = x_s$. That contradicts the initial conditions and proves the theorem.

If symbols $MB_L(U)$ and $MB_J(U)$ denote the minimal involutive bases for the $Id(U)$, $\succ$-division $L$ and Janet lex($\prec$)-division $J$ then, if $x_1 \succ x_2 \succ \ldots \succ x_n$ takes place, we have

$$MB_J(U) \subseteq MB_L(U).$$

This fact shows that that the minimal monomial involutive basis with respect to $\succ$-division with $x_1 \succ x_2 \succ \ldots \succ x_n$ is Janet involutive set and hence a Janet basis. But this basis can be not minimal for lex($\prec$)-order. So, the computing of minimal Janet basis can seem preferable over the use of other $\succ$-divisions with property $x_1 \succ x_2 \succ \ldots \succ x_n$.

So this reasoning shows that potentially useful constructive $\succ$-divisions which could compete with Janet division lie outside the group which preserves ordering $x_1 \succ x_2 \succ \ldots \succ x_n$. The next theorem shows that there exist some constructive $\succ$-divisions of that type.

**Theorem 10.** *Consider the two-variable case, i.e., when $n = 2$ and all the monomials are formed by variables $x_1, x_2$. Then for $\succ$-division $L$ no $\gamma$-configurations with $u_1 \prec u$ and $u_1 <_{\text{lex}(\prec)} u$ exist, and involutive $\succ$-division $L$ is constructive.*

*Proof.* The proof of the theorem is performed with the help of lemmas 3 and 4. Consider the opposite, namely, that $L$ is non-constructive and so there exists a $\gamma$-configuration $\{u_1, u, w, \widehat{u}\}$ for which relations $u_1 \prec u$ and $u_1 <_{\text{lex}(\prec)} u$ hold.

The case $x_1 \succ x_2$ was considered in Theorem 7. So, it is sufficient to suppose $x_2 \succ x_1$. Due to the lemma 3 $j = 1$ and $i = 2$.

As $v \neq 1$ we have $NM_L(u_1, \{u_1, u\}) = NM_L(u_1, \{u_1, \widehat{u}\}) = x_j = x_1$. So, $\deg_1(u_1) = \deg_1(u_1 v) < \deg_1(\widehat{u})$. Accodring to the definition of $i$ $\deg_2(u) + 1 = \deg_2(u_1 v) \leq \deg_2(\widehat{u})(i = 2)$.

These relations immediately yield $u_1 v | \widehat{u}$. Thus $NM_L(u_1 v, \{u_1 v, \widehat{u}\}) = x_1$ what is impossible. That proves the theorem.

The next theorem establishes constructivity for a wide amount of divisions. Consider natural numbers $r_1, \ldots, r_q$ such that $r_1 + \ldots + r_q = n$ and $1 \leq r_i \leq n$. The set of variables $x_1, \ldots, x_n$ is divided into $q$ groups, each consisting of $r_i$ variables. The numbering of variables can be redefined in such a way:

$$x_{1,1}, \ldots, x_{1,r_1}, x_{2,1}, \ldots, x_{2,r_2}, \ldots x_{s,1}, \ldots, x_{s,r_i}, \ldots, x_{q,1}, \ldots, x_{q,r_q}$$

The degree of variable $x_{s,p}$ in a monomial $m$ is denoted as $\deg_{s,p}(m)$. The value of $\mathrm{ind}(v, \{v, u\})$ is also a variable index in double indexing.

The monomial $[u]_s, 1 \leq s \leq q$ is defined the following way:

$$\deg_{h,p}([u]_s) = \begin{bmatrix} 0, & h \neq s \\ \deg_{h,p}(u), & h = s \end{bmatrix}$$

Let $\prec_s$ $(1 \prec s \prec q)$ be admissible orderings on sets of monomials of type $\{x_{s,1}^{n_{s,1}}, \ldots, x_{s,r_s}^{n_{s,r_s}}\}$. On the set $\mathbb{M} = \{x_1, \ldots, x_n\} = \{x_{1,1} \ldots x_{1,r_1}, \ldots, x_{q,1} \ldots x_{q,r_q}\}$ the admissible ordering $\prec_{1,\ldots,q}$ is defined by such a way:

$$u \prec_{1,\ldots,q} v \Longleftrightarrow \exists 1 \leq s \leq q, \forall p < s, [u]_p = [v]_p, [u]_s \prec_s [v]_s.$$

**Theorem 11.** *Let $L_k$ — be a finite set of $\succ_k$-divisions $(1 \leq k \leq q)$, where $\succ_k$ are admissible orderings. For every division $L_k$ no $\gamma$-configurations with $u_1 \prec_k u$, $u_1 <_{\mathrm{lex}(\prec_k)} u$ exist. Then $\succ = (\succ_1, \ldots, \succ_q)$ is an admissible ordering, for $\succ$-division $L$ no $\gamma$-configurations with $u_1 \prec u$, $u_1 <_{\mathrm{lex}(\prec)} u$ exist, and $L$ is constructive.*

*Proof.* For simplicity, consider the case $q = 2$. Suppose that there exists a $\gamma$-configuration $\{\widehat{u}, u, u_1, w = u_1 v\}$ with $u_1 \prec u$, $u_1 <_{\mathrm{lex}(\prec)} u$ for the division $L$.

The case $[u]_1 = [u_1]_1 = [u_1 \times v]_1 = [\widehat{u}]_1$ is impossible as $\{[\widehat{u}]_2, [u]_2, [u_1]_2, [u_1 v]_2\}$ yield a $\gamma$-configuration with corresponding properties of $u_1$ for $\succ_2$-division $L_2$.

Let the variable $x_i$ lie in the $s$th group, namely $x_i \in \{x_{s,1}, \ldots, x_{s,r_s}\}$. Within the theorem, $x_i$ will be denoted by $x_{s,i}$.

If $s = 2$ then from the properties of the ordering $[\widehat{u}]_1 = [u]_1 = [u \times v]_1 = [u_1]_1$. First two relations are valid since $\deg_{2,i-}(u) \geq \deg_{2,i-}(\widehat{u})$ $\deg_{2,i-}(u) \geq \deg_{2,i-}(u_1 \times v)$, and the last is valid since in the opposite case $v$ would contain some non-multiplicative variables for $u_1$. That cannot be true so $s = 1$ and $[\widehat{u}]_1 \prec_1 [u]_1, [u]_1 \prec_1 [u_1 \times v]_1$.

Then it is necessary to show that monomials $[\widehat{u}]_1, [u]_1, [u_1 \times v]_1, [u_1]_1$ are distinct. The case $[u_1]_1 = [u]_1$ is impossible, as it leads to relations $\deg_{1,i}([u_1 \times v]_1) = \deg_{1,i}([u]_1) + 1 = \deg_{1,i}([u_1]_1) + 1$, which implie that $v$ contains $x_{1,i}$, what is impossible ( $[u_1]_1 = [u]_1 \Longrightarrow \mathrm{ind}(u, \{u, \widehat{u}\}) = \mathrm{ind}(u_1, \{u_1, \widehat{u}\}) = s, i)$. So, $[u_1]_1 \prec_1 [u]_1 \prec_1 [\widehat{u}]_1$, $[u_1]_1 \prec_1 [u]_1 \prec_1 [u_1 \times v]_1$, and $[u_1]_1 <_{\mathrm{lex}(\prec_1)} [u]_1 <_{\mathrm{lex}(\prec_1)} [u_1 \times v]_1$.

The last step is to check that $[u_1 \times v]_1 \neq [\widehat{u}]_1$. The relation $[u_1]_1 \prec [\widehat{u}]_1$ yields $\mathrm{ind}([u_1]_1, \{[u_1]_1, [\widehat{u}]_1\}) = \mathrm{ind}(u_1, \{u_1, \widehat{u}\}) = 1, k$. The assumption $[u_1 \times v]_1 = [\widehat{u}]_1$ implies that $x_{1,k}$ contains in $v$ and $[v]_s$, what is impossible.

Hence, the set $\{[\widehat{u}]_1, [u]_1, [u_1 \times v]_1, [u_1]_1\}$ is distinct. Then it can be trivially proven that it is a $\gamma$-configuration for $\succ_1$, and $[u_1]_1 \prec_1 [u]_1$ and $[u_1]_1 <_{\mathrm{lex}(\prec_1)} [u]_1$.

The proof for $q > 2$ can be done by induction as $(\succ_1, \succ_2, \ldots, \succ_q)$-division can be represented as the division for ordering $((\succ_1, \succ_2, \ldots, \succ_{q-1}), \succ_q)$.

As the examples of new constructive involutive divisions given by theorem 11 $\mathrm{lex}_{\{1,3,2\}}$-division and $(\mathrm{degrevlex}_{\{1,2,3\}}, \mathrm{lex}_{\{5,4\}})$-division may be mentioned.

**Theorem 12.** *Consider the case when $x_2 \succ x_3 \succ x_1$, and for ordering $\succ$ the following holds:*

$$\forall p, q \in \mathbb{M} \quad \deg_2(p) > \deg_2(q) \Rightarrow p \succ q.$$

*Then $\succ$-division $L$ in variables $x_1, x_2, x_3$ is constructive.*

*Proof.* Consider the opposite. Then there exists a set $U$ for which constructivity does not hold on elements $u_1, u, u_1 v \in U$ and corresponding $\gamma$-configuration $\{u_1, u, w, \widehat{u}\}$ for which relations $u_1 \prec u$ and $u_1 <_{\mathrm{lex}(\prec)} u$ hold simultaneously. In the proof of the theorem we will consider that $\widehat{u}$ is maximal possible with respect to $\succ$, which satisfies corresponding properties of $\gamma$-configuration.

Due to lemma 3, $i \neq 1$ so

$$\deg_1(\widehat{u}) \leq \deg_1(u).$$

According to the definition of the ordering $\succ$ we have

$$\deg_2(u_1) \leq \deg_2(u) \leq \deg_2(\widehat{u}).$$

If $\deg_2(u) < \deg_2(\widehat{u})$ then $i = 2$, and otherwise $i = 3$.

The case $\deg_2(u_1) = \deg_2(u) = \deg_2(\widehat{u})$ is impossible, as it yields that $2 \notin \{i, j, k\}$, and the proof can be done analogously to that of Theorem 10. So,

$$\deg_2(u_1) < \deg_2(\widehat{u}).$$

As $u_1 v >_{\mathrm{lex}(\prec)} u$ $\deg_2(u) \leq \deg_2(u_1 v)$, and, if $i = 2$, $\deg_2(u_1 v) = \deg_2(u) + 1$.

The next step is to prove that $\deg_2(u_1) < \deg_2(u_1 v)$. If $i = 2$ it is obvious. If $i = 3$ then $\deg_2(u) = \deg_2(\widehat{u})$, and $\deg_2(u_1) < \deg_2(u) = \deg_2(u_1 v)$.

Hence, $x_2 \in M_L(u_1, \{u_1, u\})$ and $x_2 \in M_L(u_1, \{u_1, \widehat{u}\})$. Due to $\deg_2(u_1) < \deg_2(\widehat{u})$, $k = 1$, what yields

$$\deg_1(u_1) < \deg_1(\widehat{u}) \le \deg_1(u),$$

and $j = 1$.

The last relation implies that if $u_1 v \prec \widehat{u}$ then $NM_L(u_1 v, \{u_1 v, \widehat{u}\}) = x_1$ what is impossible as $\deg_1(u_1 v) < \deg_1(u)$. So, $\widehat{u} \prec u_1 v$, hence

$$\deg_2(u_1 v) = \deg_2(\widehat{u}).$$

For $i = 3$ it is checked below that $u_1 v | \widehat{u}$ and $NM_L(u_1 v, \{u_1 v, \widehat{u}\}) = x_j$. That leads to the contradiction and proves the theorem.

$$\deg_1(u_1 v) < \deg_1(\widehat{u}) \le \deg_1(u) \qquad (i = 3, j = 1, k = 1),$$

$$\deg_2(u_1 v) = \deg_2(u) = \deg_2(\widehat{u}) \qquad (u_1 v >_{\text{lex}(\prec)} u, \quad u_1 v | u \cdot x_3),$$

$$\deg_3(u_1 v) = \deg_3(u) + 1 \le \deg_3(\widehat{u}) \qquad (i = 3).$$

Consider the case $i = 2$. The following relations are valid for it:

$$\deg_2(u_1) \le \deg_2(u) < \deg_2(\widehat{u}) = \deg_2(u_1 v) = \deg_2(u) + 1,$$

$$\deg_1(u_1 v) < \deg_1(\widehat{u}) \le \deg_1(u).$$

Due to $u_1 v \succ \widehat{u}$,

$$\deg_3(\widehat{u}) < \deg_3(u_1 v) \le \deg_3(u \cdot x_2),$$

and

$$\widehat{u} | u \cdot x_2.$$

Without any loss of generality we can suppose that $\widehat{u}$ is the maximal among those monomials in $U$ which divide $u \cdot x_2$, with respect to $\succ$. (if there exists $\widehat{w} \succ \widehat{u}$, $\widehat{w} | u \cdot x_2$ then $\{u_1, u, u_1 v, \widehat{w}\}$ is $\gamma$-configuration).

As $u \cdot x_2$ has no involutive divisors in $U$, then $\frac{u \cdot x_2}{\widehat{u}}$ contains a non-multiplicative variable $x_l$ for $\widehat{u}$. The case $\widehat{u} \cdot x_l \ne u \cdot x_2$ is impossible as $\widehat{u} \cdot x_l$ would have an involutive divisor greater than $\widehat{u}$, which also divides $u \cdot x_2$ and contradicts the maximality of $\widehat{u}$.

If $l \ne 3$ then $\deg_3(\widehat{u}) = \deg_3(u) \ge \deg_3(u_1 v)$ what is impossible. So, $u \cdot x_2 = \widehat{u} \cdot x_3$.

As $m$ we will take the monomial s.t. $x_3 = NM_L(\widehat{u}, \{\widehat{u}, m\})$. Due to $m \succ \widehat{u}$ and $m >_{\text{lex}(\prec)} \widehat{u}$

$$\deg_1(m) \le \deg_1(\widehat{u}) < \deg_1(u),$$

$$\deg_2(m) = \deg_2(\widehat{u}),$$

$$\deg_3(m) \ge \deg_3(\widehat{u}) + 1.$$

Consider the set $\{u_1, u, u_1 v, m\}$. By the filter axiom it can be shown that it is a $\gamma$-configuration. As $m \succ \widehat{u}$, we obtained a contradiction.

The interesting fact concerning this division is that $\text{lex}_{\{2,3,1\}}$-division is constructive, but $\text{lex}_{\{2,3,1,4\}}$ is non-constructive. The explanation is natural as the divisions from the assertion of Theorem 12 do not satisfy the conditions of Theorem 11 because they allow $\gamma$-configurations with $u_1 \prec u$ and $u_1 <_{\text{lex}(\prec)} u$, but only when other conditions of non-constructivity fail ($\widehat{u}$ involutively divides $u \cdot x_i$).

Also it should be stressed that the relation $\succ$ on variables does not completely define whether $\succ$-division is constructive. Consider the case when $x_2 \succ x_1 \succ x_3 \succ x_4$. The $\text{lex}_{\{2,1,3,4\}}$-division is constructive due to Theorem 11. But if $\succ$ is such an ordering for which $x_2 x_4 \succ x_3^2 \succ x_1 x_4$, then division is non-constructive according to the theorem 5.

For the constructive involutive divisions, for which relations $x_1 \succ x_2 \succ \ldots \succ x_n$ do not hold, the minimal involutive basis can contain less elements than minimal Janet basis.

**Example 4.** *Consider the set $U = \{x_1^2, x_2\}$. This set is autoreduced. Let $J$ be a Janet involutive division, and $L$ be a $\text{lex}_{\{2,1\}}$-division. Both divisions are constructive, so the Minimal Involutive Basis algorithm gives correct results. In is clear to see, that $MB_J(U) = \{x_2, x_1 x_2, x_1^2\}$, and $MB_L(U) = \{x_1^2, x_2\}$. So, the size of $MB_J(U)$ exceeds that of $MB_L(U)$.*

## 4 Conclusions

The result of the paper is the description of a wide group of involutive $\succ$-divisions, which are constructive and so have a unique minimal involutive basis. They can be considered as the generalization of Janet division because they are constructed according to the same principles and allow the existence of minimal involutive basis.

Also, the sufficient conditions of non-constructivity for $\succ$-divisions are presented, which allow description of a wide class of non-constructive involutive divisions in terms of $\succ$, what is still the problem of theoretic interest. For example, the divisions for which consistence is twice violated (i.e., $i < j < k$, $x_k \succ x_i, x_k \succ x_j$) are not constructive.

The last group of results establish that Janet division has the superiority in minimal involutive basis size than divisions from a wide subset of all $\succ$-divisions, namely, those divisions for which $x_1 \succ x_2 \succ \ldots \succ x_n$.

## Acknowledgements

The author is grateful to his scientific advisor E.V. Pankratiev, as well as to V.P. Gerdt, V.A. Mityunin, and A.I. Zobnin for the help, remarks, and useful ideas which influenced the work.

The work was partly supported by the Russian Foundation for Basic Research, project no. 05-01-00671.

# References

1. Apel, J.: The theory of involutive divisions and an application to Hilbert function computations. Journal of Symbolic Computation **25**(6) (1998) 683–704
2. Calmet, J., Hausdorf, M., Seiler, W. M.: A constructive introduction to involution. In: Proc. Int. Symp. Applications of Computer Algebra – ISACA 2000 (2001) 33–50
3. Gerdt, V. P., Blinkov, Yu. A.: Involutive bases of polynomial ideals. Mathematics and Computers in Simulation **45** (1998) 519–542
4. Gerdt, V. P., Blinkov, Yu. A.: Minimal involutive bases. Mathematics and Computers in Simulation **45** (1998) 543–560
5. Gerdt, V. P.: Involutive division technique: some generalizations and optimizations. Journal of Mathematical Sciences **108**(6) (2002) 1034–1051
6. Hemmecke, R.: Involutive Bases for Polynomial Ideals. Institut für Symbolisches Rechnen, Linz, 2003
7. Semenov, A.S.: Pair analysis of involutive divisions. Fundamental and Applied Mathematics **9**(3) (2003) 199–212
8. Gerdt, V. P., Blinkov., Yu. A., Yanovich, D. A.: Quick search of Janet divisor. Programmirovanie **1** (2001) 32–36 (in Russian)
9. Zharkov, A. Yu., Blinkov, Yu. A.: Involutive systems of algebraic equations. Programmirovanie (1994) (in Russian)
10. Zharkov, A. Yu., Blinkov, Yu. A.: Involutive Bases of Zero-Dimensional Ideals. Preprint No. E5-94-318, Joint Institute for Nuclear Research, Dubna, 1994

# A Symbolic-Numeric Approach to Tube Modeling in CAD Systems

Gerrit Sobottka and Andreas Weber

Institut für Informatik II, Universität Bonn, Römerstr. 164, Bonn, Germany
{sobottka, weber}@cs.uni-bonn.de

**Abstract.** In this note we present a symbolic-numeric method to the problem of tube modeling in CAD systems. Our approach is based on the Kirchhoff kinetic analogy which allows us to find analytic solutions to the static Kirchhoff equations for rods under given boundary conditions.

**Keywords:** Kirchhoff rod, boundary value problems, automatic differentiation.

## 1 Introduction

In this short note we address the problem of physics based tube modeling which frequently occurs in *computer aided design* (CAD) applications. The task at hand is to connect a tube of given length and with certain material properties to connectors fixed in space. In particular, we aim to predict the internal forces and torques along the tube for dimensioning purposes as well as the final configuration. The equilibrium shape of the tube is governed by the well known static *Kirchhoff equations*. Along with the boundary conditions at both ends defined by the position and orientation of the connectors we have to solve a two-point boundary value problem (BVP). Such BVPs can be solved employing standard shooting techniques which usually perform at slow convergence rates. Our approach is based on the analytic solution to the static Kirchhoff equations and is a continuation of our work described in [1], to which we refer for more details on the basis of the method. In addition to adding a new case of boundary conditions we will also show how the "symbolic-numeric method" introduced in [1] can be made "more symbolic" by using the Jacobian Matrix in symbolic form within the numerical part.

## 2 Related Work

While the number of publications on solution methods for Kirchhoff equations is large, the task of tube modeling based on these equations has rarely been addressed before. Grégoire and Schömer [2] use an extended spring-mass system that is solved with an energy minimizing algorithm. In [3] Healey and Metha present a method to solve associated BVPs by augmenting the system of boundary conditions by a constraint on the magnitude of the quaternions used for the parameterization of the rotations. In particular, they show that if these constraints are met at the end points they are also met on the

V.G. Ganzha, E.W. Mayr, and E.V. Vorozhtsov (Eds.): CASC 2006, LNCS 4194, pp. 279–283, 2006.

whole domain. In [4] a geometrically exact approach is proposed, which is based on the explicit solution of the kinematic relation based on Rodriguez' formula. Henderson and Neukirch [5] study spatial equilibria of clamped elastica based on Kirchhoff rods. In contrast to their work we do not restrict ourselves to the case where the tangents at the end points are collinear. In [6] Nizette and Goriely study explicit solution of the static Kirchhoff equations in terms of Euler-Kirchhoff filaments.

## 3   Physics Based Tube Modeling

### 3.1   The Symbolic Part

Let $\mathbf{r}(s) : [a_1, a_2] \in \mathbb{R} \mapsto \mathbb{R}^3$ be the centerline of the tube. The centerline is furnished with a set of right-handed orthonormal triads $\{\mathbf{d}_1, \mathbf{d}_2, \mathbf{d}_3\}$, such that $\mathbf{d}_3 = \mathbf{r}'$ is the tangent of the centerline and $\mathbf{d}_1, \mathbf{d}_2$ span the cross section plane at each point of the rod. Further, we assume the tube to be inextensible, unshearable, and initially straight. The equilibrium state is given by the static Kirchhoff equations:

$$\mathbf{F}' = \mathbf{0}, \tag{1}$$

$$\mathbf{M}' + \mathbf{d}_3 \times \mathbf{F} = \mathbf{0}, \tag{2}$$

$$\mathbf{M} = u_1 \cdot \mathbf{d}_1 + u_2 \cdot \mathbf{d}_2 + b \cdot u_3 \cdot \mathbf{d}_3, \tag{3}$$

where $\mathbf{F}$ and $\mathbf{M}$ are the internal force and the torque of the rod. Note that since we assume the tube to have a circular cross section we use the scaled form of the constitutive equation for $\mathbf{M}$, where $b = 1/(1 + \nu)$ with $\nu$ being Poisson's ratio.

Further, we have the kinematic relation $\mathbf{d}_i = \mathbf{u} \times \mathbf{d}_i$, where $\mathbf{u} = \{u_1, u_2, u_3\}$ is the twist vector. The components of the twist vector as well as the local directors $\{\mathbf{d}_i\}$ are conveniently expressed in terms of Euler angles $(\varphi, \theta, \psi)$ w.r.t. to the global frame $\{\mathbf{e}_x, \mathbf{e}_y, \mathbf{e}_z\}$. With

$$\mathbf{F} = F \cdot \mathbf{e}_z, \tag{4}$$

$$M_z' = \mathbf{M}_z' \cdot \mathbf{e}_z = 0, \tag{5}$$

$$M_3' = \mathbf{M}_3' \cdot \mathbf{d}_3 = 0, \tag{6}$$

$$H = \frac{1}{2} \cdot \mathbf{M} \cdot \mathbf{u} + \mathbf{F} \cdot \mathbf{d}_3, \tag{7}$$

being first integrals, which do not depend on the arc length parameter, we obtain the following equations for the Euler angles:

$$\varphi' = \frac{M_z - M_3 \cdot z}{1 - z^2}, \tag{8}$$

$$\psi' = \frac{M_3 - M_z \cdot z}{1 - z^2} + M_3 \cdot \left( \frac{1}{b} - 1 \right), \tag{9}$$

$$z'^2 = 2F \cdot (h - z) \cdot (1 - z^2) - (M_z - M_3 \cdot z)^2, \tag{10}$$

where $z = \cos\theta$ and $h = 1/F \cdot (H - M_3^2/2b)$. The right hand side of $z'^2$ is a cubic polynomial in $z$ with the roots $z_1, z_2, z_3$, such that $-1 \leq z_1 \leq z_2 \leq 1 \leq z_3$. The solution is given by

$$z = z_1 + (z_2 - z_1) \cdot \text{sn}^2 \left[ \lambda(s + s_0), k \right], \tag{11}$$

where $\lambda = \sqrt{1/2 \cdot F \cdot (z_3 - z_1)}$, $k = \sqrt{(z_2 - z_1)/(z_3 - z_1)}$, and sn is one of the Jacobi elliptic function. With the function $z$ at hand the solutions for $\varphi$ and $\psi$ are obtained by directly integrating the above equations [1]:

$$\varphi = \int_0^s \frac{M_z - M_3 \cdot z(\sigma)}{1 - z^2(\sigma)} d\sigma + \varphi_0, \tag{12}$$

$$\psi = \int_0^s \frac{M_3 - M_z \cdot z(\sigma)}{1 - z^2(\sigma)} d\sigma + \psi_0 + M_3 \cdot s \cdot \left( \frac{1}{b} - 1 \right). \tag{13}$$

Since the system integral $H$ is a function of $\theta$ and $\theta'$ [1] the set of quantities determining the configurations of the centerline of the tube is given by $\eta = \{F, M_z, M_3, \theta(0), \theta(0)'\}$.

Since the tube is clamped at $s = 0$ and can freely rotate at the other end ($s = L$) the boundary conditions imposed by the underlying problem are thus given as $r(0) = x_0$, $d_1(0) = d_{10}$, $d_2(0) = d_{20}$, $r(L) = x_L$ and $\langle d_3(L), t_L \rangle = 1$, where $x_L$ and $t_L$ are the coordinates of the point and the tangent to be matched at $s = L$.

### 3.2 Numerical Computations

Thus the solution of this two point boundary value problem is reduced to the solution of the following set $\mathcal{F}(\eta) = 0$ of non-linear equations:

$$\begin{aligned} r(L) - x_L &= 0, \\ 1 - \langle d_3, t_L \rangle &= 0, \\ \theta_0 - \theta_0' &= 0. \end{aligned} \tag{14}$$

Standard numeric solution techniques [7, 8] require that the Jacobian matrix is known numerically at every iteration point. These methods use numeric approximations to the partial derivatives at the iteration points, if those are not given as program code. As we have derived a symbolic expression for the function $\mathcal{F}$, we will show how we can come up with rather efficient code for the Jacobian, too.

## 4   Using the Jacobian Matrix in Symbolic Form

Using the common subexpression elimination algorithm of Maple the function $\mathcal{F}$ can be described by a computation sequence involving the following number of commands:

$$43 \text{ assignments} + 29 \text{ additions} + 65 \text{ multiplications} + 5 \text{ divisions}$$
$$+19 \text{ functions} + 5 \text{ integrals} \tag{15}$$

The 19 function evaluations consist of 9 trigonometric and square root functions and 10 instances of the Jacobi elliptic functions sn. The 5 integrals come from the necessity to have the centerline of the space curve available in a form that allows for boundary conditions at two distinct points, i.e. to explicitly carry the integration of the tangent vector [1, 6].

Standard tools for automatic differentiation and also the automatic differentiation procedure available in Maple cannot handle integral operators in their inputs. Thus we could not use a straight-forward automatic differentiation approach on the computation sequence of $\mathcal{F}$ to obtain a computation sequence for the Jacobian matrix.

Whereas the symbolic differentiation algorithm of Maple can handle occurring integrals, the symbolic expression representing $\mathcal{F}$ was too large for a straight-forward symbolic differentiation.

However, with the following method we successfully derived symbolic computation sequences for the Jacobian matrix.

– We used auxiliary symbolic functions for the roots $z_1$, $z_2$, and $z_3$ of the cubic polynomial occurring on the right-hand-side of (10) and its partial derivatives.
– Using these auxiliary functions in the expression of $\mathcal{F}$ we could successfully compute the Jacobian matrix in symbolic form using Maple.
  This computation required several minutes of computation time.
– A computation sequence could be obtained by Maple generated from the expressions of the Jacobian and assignment of the expressions of the roots $z_1$, $z_2$, and $z_3$ and its partial derivatives to the auxiliary symbolic functions.
  Notice that all symbolic partial derivatives of the expressions of the roots $z_1$, $z_2$, and $z_3$ could be obtained by Maple easily.

Using the optimize function on computational sequences the result for computing the Jacobian required the following number of commands:

$$260 \text{ assignments} + 174 \text{ additions} + 419 \text{ multiplications}$$
$$+31 \text{ divisions} + 76 \text{ functions} + 24 \text{ integrals} \tag{16}$$

Notice that because of the symbolic differentiation rules for "special functions" used by Maple the computation sequence contains calls to various of Jacobi's elliptic function and also to incomplete elliptic integrals of the second kind.

*Remark.* If equation (14) is solved via a corresponding minimization problem of a real valued function $m_{\mathcal{F}}$, then the gradient function of $m_{\mathcal{F}}$ can be expressed in symbolic form similarly. The computational costs for this gradient after common subexpression elimination is almost identical to the one for the Jacobian after common subexpression evaluation, i.e. is also given by the number of commands stated in (16).

# References

1. Liu, S., Weber, A.: A symbolic-numeric method for solving boundary value problems of Kirchhoff rods. In Ganzha, V.G., Mayr, E.W., Vorozhtsov, E.V., eds.: Computer Algebra in Scientific Computing (CASC '05). Volume 3718 of Lecture Notes in Computer Science., Kalamata, Greece, Springer-Verlag (2005) 387–398
2. Grégoire, M., Schömer, E.: Interactive simulation of one-dimensional flexible parts. In: SPM '06: Proceedings of the 2006 ACM symposium on Solid and physical modeling, Cardiff, Wales, United Kingdom, ACM Press (2006) 95–103
3. Healey, T.J., Mehta, P.G.: Straightforward computation of spatial equilibria of geometrically exact Cosserat rods (2003) http://tam.cornell.edu/Healey.html.

4. Simo, J.C., Vu-Quoc, L.: On the dynamics in space of rods undergoing large motions – a geometrically exact approach. Computer Methods in Applied Mechanics and Engineering **66** (1988) 125–161
5. Henderson, M.E., Neukirch, S.: Classification of the spatial equilibria of the clamped elastica: numerical continuation of the solution set. International Journal of Bifurcation and Chaos **14** (2004) 1223–1239
6. Nizette, M., Goriely, A.: Towards a classification of Euler-Kirchhoff filaments. Journal of Mathematical Physics **40** (1999) 2830–2866
7. Hopkins, T.R., Phillips, C.: Numerical Methods in Practice – A Guide to the NAG Library. Addison-Wesley, Reading, MA, USA (1988)
8. Press, W.H., Teukolsky, S.A., Vetterling, W.T., Flannery, B.P.: Numerical Recipes in C++, Second Edition. Cambridge University Press (2002)

# Inequalities on Upper Bounds for Real Polynomial Roots

Doru Ştefănescu

University of Bucharest, Romania
stef@rms.unibuc.ro

**Abstract.** In this paper we propose two methods for the computation of upper bounds of the real roots of univariate polynomials with real coefficients. Our results apply to polynomials having at least one negative coefficient. The upper bounds of the real roots are expressed as functions of the first positive coefficients and of the two largest absolute values of the negative ones.

## 1 Introduction

In this paper we derive new upper bounds for the real roots of univariate polynomials with real coefficients. Such bounds are useful for the location of real roots, and for polynomial real root isolation (cf. A. G. Akritas– A. W. Strzeboński [1], I. Emiris–E. Tsigaridas [2], F. Rouiller–P. Zimmermann [10], C. K. Yap [12]). The computation of positive roots of univariate polynomials with real coefficients is also relevant to iterative numerical processes (J. Herzberger [5], N. Kjurkchiev [7]).

Our results imply tighter upper bounds for the positive roots than those of Lagrange, Longchamp and Kioustelidis [6].

## 2 Upper Bounds for the Roots

In this section we derive new bounds for positive roots greater than one. Our results extend a well known upper bound of Lagrange and other related results (see L. S. Grinstein [4]). Note that for bounds on positive roots smaller than one there exists another approach (see D. Ştefănescu [11]) which gives efficient bounds in many cases.

The following theorem gives upper bounds for positive roots larger than one as functions of the degree of the polynomial, the index of the first positive coefficients, and the largest absolute value of the negative ones.

**Theorem 1.** *Let* $P(X) = a_0 X^d + \cdots + a_m X^{d-m} - a_{m+1} X^{d-m-1} \pm \cdots \pm a_d \in \mathbb{R}[X]$, *with all* $a_i \geq 0$, $a_0, a_{m+1} > 0$. *Denote*

$$A = \max\{a_i \, ; \, i = m+1, \ldots, d\} \, .$$

V.G. Ganzha, E.W. Mayr, and E.V. Vorozhtsov (Eds.): CASC 2006, LNCS 4194, pp. 284–294, 2006.

*If $m \geq 2$, the number*

$$1 + \max \left\{ \left( \frac{2A}{3\big(s(s-1)a_0 + (s-1)(s-2)a_1 + \cdots + 2a_{s-2}\big)} \right)^{1/(m-s+3)} , \right.$$

$$\left. \left( \frac{A}{3\big(sa_0 + \cdots + +2a_{s-2} + a_{s-1}\big)} \right)^{1/(m-s+2)} , \left( \frac{A}{3\big(a_0 + a_1 + \cdots + a_s\big)} \right)^{1/(m-s+1)} \right\}$$

*is an upper bound for the positive roots of $P$ for any $s \in \{2, 3, \ldots, m\}$.*

*Proof.* We consider $x \in \mathbb{R}$, $x > 1$. We have

$$|P(x)| \geq |a_0 x^d + \cdots + a_m x^{d-m}| - |a_{m+1} x^{d-m-1} \mp + \ldots \mp a_d|$$

$$\geq a_0 x^d + \cdots + a_s x^{d-s} - A(x^{d-m-1} + \cdots + 1)$$

$$\geq (a_0 x^s + \cdots + a_s) x^{d-s} - A \frac{x^{d-m} - 1}{x - 1} \tag{1}$$

$$= \frac{(a_0 x^s + \cdots + a_s)(x-1) x^{m-s} - A}{x-1} \cdot x^{d-m} + \frac{A}{x-1}.$$

The last line in the right hand side of (1) is strictly positive provided that

$$(a_0 x^s + \cdots + a_s)(x-1) x^{m-s} \geq A. \tag{2}$$

Now let $x = 1 + y$ and note that $x^j \geq 1 + jy + \dfrac{j(j-1)}{2} y^2$ for all $j \geq 2$.
We observe that

$$(a_0 x^s + \cdots + a_s)(x-1) x^{m-s} \geq \left( a_0 \left( 1 + sy + \frac{s(s-1)}{2} y^2 \right) \cdots \right.$$

$$\left. + a_{s-2} \left( 1 + 2y + y^2 \right) + a_{s-1} \left( 1 + y \right) + a_s \right) y^{m-s+1}$$

$$= \frac{1}{2} \left( s(s-1)a_0 + (s-1)(s-2)a_1 + \cdots + 2a_{s-2} \right) y^{m-s+3}$$

$$+ \left( sa_0 + \cdots + 2a_{s-2} + a_{s-1} \right) y^{m-s+2}$$

$$+ \left( a_0 + \cdots + a_s \right) y^{m-s+1}.$$

It follows that (2) is satisfied if

$$\left( s(s-1)a_0 + (s-1)(s-2)a_1 + \cdots + 2a_{s-2} \right) y^{m-s+3} \geq 2A/3 \,,$$

$$\left( sa_0 + \cdots + 2a_{s-2} + a_{s-1} \right) y^{m-s+2} \qquad\qquad \geq A/3 \,,$$

$$\left( a_0 + \cdots + a_{s-1} + a_s \right) y^{m-s+1} \qquad\qquad \geq A/3 \,.$$

These inequalities are satisfied as soon as

$$y \geq \max \left\{ \left( \frac{2A}{3\left(s(s-1)a_0 + (s-1)(s-2)a_1 + \cdots + 2a_{s-2}\right)} \right)^{1/(m-s+3)} \,, \right.$$

$$\left. \left( \frac{A}{3(sa_0 + \cdots + +2a_{s-2} + a_{s-1})} \right)^{1/(m-s+2)} , \left( \frac{A}{3(a_0 + a_1 + \cdots + a_s)} \right)^{1/(m-s+1)} \right\} .$$

This proves that

$$1 + \max \left\{ \left( \frac{2A}{3\left(s(s-1)a_0 + (s-1)(s-2)a_1 + \cdots + 2a_{s-2}\right)} \right)^{1/(m-s+3)} \,, \right.$$

$$\left. \left( \frac{A}{3(sa_0 + \cdots + +2a_{s-2} + a_{s-1})} \right)^{1/(m-s+2)} , \left( \frac{A}{3(a_0 + a_1 + \cdots + a_s)} \right)^{1/(m-s+1)} \right\}$$

is an upper bound for the positive roots of the polynomial $P$. $\qquad\square$

**Corollary 2.** *The following numbers are upper bounds for the positive real roots of the polynomial $P$ :*

$$B_2(P) = 1 + \max \left\{ \left( \frac{2A}{3(2a_0 + a_1)} \right)^{1/(m+1)} , \left( \frac{A}{3(2a_0 + a_1)} \right)^{1/m} , \left( \frac{A}{3(a_0 + a_1 + a_2)} \right)^{1/(m-1)} \right\} ,$$

$$B_m(P) = \left\{ \begin{array}{l} 1 + \max \left\{ \left( \dfrac{2A}{3\left(m(m-1)a_0 + (m-1)(m-2)a_1 + \cdots + 2a_{m-2}\right)} \right)^{1/3} \,, \right. \\[2em] \left. \left( \dfrac{A}{3(ma_0 + \cdots + +2a_{m-2} + a_{m-1})} \right)^{1/2} , \left( \dfrac{A}{3(a_0 + a_1 + \cdots + a_m)} \right) \right\} . \end{array} \right.$$

*Proof.* For the first bound we consider $s = 2$ in Theorem 1. For the second, we let $s = m$. $\qquad\square$

*Remark.* A key step in the proof of Theorem 1 is to consider the inequality (2), with $s \in \{2, 3, \ldots, m\}$. Using $x^j \geq 1 + jy + \dfrac{j(j-1)}{2} y^2$, we then obtain an upper bound for the positive roots.

Similar techniques as in the proof of Theorem 1 can be used to derive different bounds. For this purpose one can also consider the second largest absolute value of the negative coefficients, then employ inequalities similar to (1). We use this approach in Theorem 4.

## Comparisons with Other Bounds

Many estimates exist for all roots of a univariate polynomial with real coefficients, however, there are only few known upper bounds for the real roots (cf. [2], [6], [11]).

We compare our results with the following upper bounds of J.–L. Lagrange, M. Longchamp, J. B. Kioustelidis and our bound $S(P)$ (see [11]):

$$L_1(P) = 1 + \left( \frac{A}{a_0} \right)^{1/(m+1)}, \qquad\qquad \text{(Lagrange)}$$

$$L_2(P) = 1 + \frac{A}{a_0 + \cdots + +a_{m-1}}, \qquad\qquad \text{(Longchamp)}$$

$$S(P) = 1 + \max \left\{ \left( \frac{A}{2(2a_0 + a_1)} \right)^{1/2}, \frac{A}{2(a_0 + a_1 + a_2)} \right\}, \qquad \text{(Ştefănescu)}$$

$$K(P) = 2 \cdot \max\{b_1^{1/m_1}, \ldots, b_k^{1/m_k}\}. \qquad\qquad \text{(Kioustelidis)}.$$

Denote by $B_s(P)$ the bound given in Theorem 1. In the bound $K(P)$ the polynomial $P$ is represented as

$$P(X) = X^d - b_1 X^{d-m_1} - \cdots - b_k X^{d-m_k} + g(X),$$

with all $b_j > 0$ and $g(X)$ a polynomial of degree $< d$ and with positive coefficients.

We consider the following polynomials:

$$P_1(X) = X^7 + X^6 + X^5 + X^4 - 3X^3 - 2X^2 + X - 3,$$

$$P_2(X) = X^9 + X^8 + X^7 + X^6 - 4X^5 - 5X^4 - 5,$$

$$P_3(X) = X^{10} + X^9 + 2X^8 + X^6 + 2X^5 - 11X^4 + X^3 - 4X^2 - 5.$$

| | $m$ | $L_1$ | $L_2$ | $K$ | $S$ | $B_2$ | $B_3$ | $B_4$ | $B_5$ | largest pos root |
|---|---|---|---|---|---|---|---|---|---|---|
| $P_1$ | 3 | 2.312 | 2.0 | 2.632 | 1.5 | 2.0 | 1.629 | — | — | 1.079 |
| $P_2$ | 3 | 2.495 | 2.666 | 2.828 | 1.912 | 2.136 | 1.746 | — | — | 1.345 |
| $P_3$ | 5 | 2.491 | 3.2 | 2.98 | 2.375 | 2.241 | 1.982 | 1.856 | 1.55 | 1.233 |

*Remark.* We observe that the bounds $B_m(P)$ give in many cases better estimates. However, in some cases $S(P)$ is better. For example, if we take $P(X) = X^7 + 2X^6 + X^5 - 3X^3 - 2X^2 - 2X + 1$, we have

$$m = 2, \quad S(P) = 1.612, \quad B_2(P) = 1.63.$$

**How to Choose $s$.** We have several possibilities for choosing $s$ among the bounds $B_s$, but the optimal choice is $s = m$.

For the polynomials

$$P_4(X) = X^{10} + X^9 + 2X^8 + X^7 + X^6 + 2X^5 - 11X^4 + X^3 - 4X^2 - 5,$$

$$P_5(X) = X^{11} + X^{10} + X^9 + 2X^8 + X^7 + X^6 + 2X^5 - 11X^4 + X^3 - 4X^2 - 5$$

we have

| | $m$ | $B_2$ | $B_3$ | $B_4$ | $B_5$ | $B_6$ | largest positive root |
|---|---|---|---|---|---|---|---|
| $P_4$ | 5 | 2.241 | 1.982 | 1.781 | 1.542 | — | 1.202 |
| $P_5$ | 6 | 2.203 | 1.985 | 1.848 | 1.7232 | 1.458 | 1.155 |

## 3    Iterative Computation of Upper Bounds

In this section we describe an iterative method for the computation of upper bounds for positive real roots. We consider first the largest absolute value of the negative coefficients. Upper bounds which depend on the two largest absolute values of the negative coefficients are examined in Theorem 4.

**Theorem 3.** *Let* $P(X) = a_0 X^d + \cdots + a_m X^{d-m} - a_{m+1} X^{d-m-1} \pm \cdots \pm a_d \in \mathbb{R}[X]$, *with all* $a_i \geq 0$, $a_0, a_{m+1} > 0$. *Denote*

$$A = \max\{a_i \, ; \, i = m+1, \ldots, d\}.$$

*For any positive root* $\alpha$ *of* $P$, *we have*

$$|\alpha| < g_s^n(\beta) \quad \text{for all integers} \quad n \geq 0, s \in \{0, 1, \ldots, m\},$$

*where*

$$g_s(x) = 1 + b_s \left(1 - x^{-d+m}\right)^{\dfrac{1}{m-s+1}}, \quad \text{with} \quad b_s = \left(\frac{A}{a_0+\cdots+a_s}\right)^{1/(m-s+1)},$$

$$\beta = \beta_s = 1 + \left(\frac{A}{a_0+\cdots+a_s}\right)^{1/(m-s+1)}.$$

*Proof.* Let $x \in \mathbb{R}$, $x > 1$. We have

$$|P(x)| \geq |a_0 x^d + \cdots + a_m x^{d-m}| - |a_{m+1} x^{d-m-1} \mp + \ldots \mp a_d|$$

$$\geq a_0 x^d + \cdots + a_s x^{d-s} - A(x^{d-m-1} + \cdots + 1)$$

$$\geq (a_0 + \cdots + a_s)x^{d-s} - A \frac{x^{d-m}-1}{x-1}$$

$$= \frac{(a_0 + \cdots + a_s)(x-1)x^{d-s} - A x^{d-m} + A}{x-1}.$$

It follows that $|P(x)| > 0$ provided that

$$(a_0 + \cdots + a_s)(x-1)x^{d-s} > A x^{d-m} - A,$$

which is equivalent to

$$(a_0 + \cdots + a_s)(x-1)x^{m-s} > A - A x^{-d+m}. \tag{3}$$

But (3) is satisfied if

$$(a_0 + \cdots + a_s)(x-1)^{m-s+1} > A - A x^{-d+m}$$

i.e.

$$x > 1 + \left(\frac{A}{a_0 + \cdots + a_s} \cdot \left(1 - x^{-d+m}\right)\right)^{1/(m-s+1)}. \tag{4}$$

Let

$$g_s(x) = 1 + b_s \left(1 - x^{-d+m}\right)^{\dfrac{1}{m-s+1}}, \quad \text{with} \quad b_s = \left(\frac{A}{a_0 + \cdots + a_s}\right)^{1/(m-s+1)}$$

and the previous inequality becomes

$$x > g_s(x). \tag{5}$$

Let

$$\beta = 1 + \left(\frac{A}{a_0 + \cdots + a_s}\right)^{1/(m-s+1)}.$$

We observe that

$$(\beta - 1)^{m-s+1} > \frac{A}{a_0 + \cdots + a_s},$$

therefore,

$$(a_0 + \cdots + a_s)(\beta - 1)^{m-s+1} > A > A(1 - \beta^{-d+m}),$$

which proves that inequality (5) is satisfied by $x = \beta$.

Note also that the function $g_s$ is increasing on $(1, \infty)$, hence it follows that the inequality (5) is satisfied by $g_s^n(\beta)$ for all $n \in \mathbb{N}$. We have

$$\beta > g_s(\beta) > g_s^2(\beta) > \cdots g_s^n(\beta) > g_s^{n+1}(\beta) > \cdots.$$

By (5) all the numbers $g_s^n(\beta)$ are upper bounds for the positive roots of the polynomial $P$.    □

More precise bounds can be computed if we consider not only the largest absolute value of the negative coefficients. In the next result we consider the first two largest absolute values of the negative coefficients.

**Theorem 4.** Let $P(X) = a_0 X^d + \cdots + a_m X^{d-m} - a_{m+1} X^{d-m-1} \pm \cdots \pm a_d \in \mathbb{R}[X]$, with all $a_i \geq 0$, $a_0, a_{m+1} > 0$. We assume that there exists $j \geq 1$ such that $0 < a_{m+j+1} < a_{m+1}$.

We denote by $t$ the smallest integer such that

$$0 < a_{m+t+1} < a_{m+1}.$$

*Let*

$$A = \max\{a_i \,;\, i = m+1, \ldots, d\},$$

*and*

$$B = \max\{a_i \,;\, i = m+t+1, \ldots, d\}.$$

*For any positive root $\alpha$ of $P$, we have*

$$|\alpha| < h_{st}^n(\beta) \quad \text{for all integers} \quad n \geq 0, s \in \{0, 1, \ldots, m\},$$

*where*

$$h_{st}(x) = 1 + b_s \left(1 - \frac{A-B}{A} x^{-t} - \frac{B}{A} x^{-d+m}\right)^{\frac{1}{m-s+1}},$$

$$b_s = \left(\frac{A}{a_0 + \cdots + a_s}\right)^{1/(m-s+1)}, \qquad \beta = \beta_s = 1 + \left(\frac{A}{a_0 + \cdots + a_s}\right)^{1/(m-s+1)}.$$

*Proof.* Since $t \geq 1$ is the smallest integer such that $0 < a_{m+t+1} < a_{m+1}$, we have

$$B = \max\{a_i \; ; \; i = m+t+1, \ldots, d\} \, .$$

Let $x \in \mathbb{R}$, $x > 1$. We have

$$|P(x)| \geq |a_0 x^d + \cdots + a_m x^{d-m}|$$

$$-|a_{m+1} x^{d-m-1} \mp + \ldots \mp a_{m+t} x^{d-m-t}| - |a_{m+t+1} x^{d-m-t-1} \mp + \ldots \mp a_d|$$

$$\geq (a_0 + \cdots + a_s) \, x^{d-s}$$

$$-A \left( x^{d-m-1} + \ldots + x^{d-m-t} \right) - B \left( x^{d-m-t-1} + \ldots + 1 \right)$$

$$= (a_0 + \cdots + a_s) \, x^{d-s} - A \frac{x^t - 1}{x - 1} x^{d-m-t} - B \frac{x^{d-m-t} - 1}{x - 1}$$

$$= \frac{(a_0 + \cdots + a_s)(x - 1) \, x^{d-s} - A(x^t - 1)x^{d-m-t} - B(x^{d-m-t} - 1)}{x - 1}$$

$$> \frac{(a_0 + \cdots + a_s)(x - 1)^{d-s+1} - A \, x^{d-m} + (A - B) \, x^{d-m-t} + B}{x - 1} \, .$$

It follows that $|P(x)| > 0$ provided that

$$(a_0 + \cdots + a_s)(x - 1)x^{d-s+1} > A x^{d-m} - (A - B) \, x^{d-m-t} - B \, ,$$

i. e.

$$(a_0 + \cdots + a_s)(x - 1)^{m-s+1} > A - (A - B)x^{-t} - Bx^{-d+m} \, . \tag{6}$$

But the previous inequality is satisfied if

$$(x - 1)^{m-s+1} > \frac{A}{a_0 + \cdots + a_s} \cdot \left( 1 - \frac{A - B}{A} x^{-t} - \frac{B}{A} x^{-d+m} \right) ,$$

which is satisfied if

$$x > 1 + \left( \frac{A}{a_0 + \cdots + a_s} \cdot \left( 1 - \frac{A - B}{A} x^{-t} - \frac{B}{A} x^{-d+m} \right) \right)^{1/(m-s+1)} . \tag{7}$$

Let

$$h_{st}(x) = 1 + b_s \left( 1 - \frac{A - B}{A} x^{-t} - \frac{B}{A} x^{-d+m} \right)^{\frac{1}{m - s + 1}} ,$$

with $b_s = \left( \dfrac{A}{a_0 + \cdots + a_s} \right)^{1/(m-s+1)} .$

The inequality (7) becomes

$$x > h_{st}(x).  \tag{8}$$

Let

$$\beta = 1 + \left( \frac{A}{a_0 + \cdots + a_s} \right)^{1/(m-s+1)}.$$

We observe that

$$(\beta - 1)^{m-s+1} > \frac{A}{a_0 + \cdots + a_s},$$

therefore,

$$(a_0 + \cdots + a_s)(\beta - 1)^{m-s+1} > A > A \left( 1 - \frac{A-B}{A} \beta^{-t} - \frac{B}{A} \beta^{-d+m} \right),$$

which proves that (8) is satisfied by $x = \beta$.

Since $h := h_{st}$ is increasing on $(1, \infty)$, we have

$$\beta > h(\beta) > h\big(h(\beta)\big) = h^2(\beta) > \cdots h^n(\beta) > h^{n+1}(\beta) > \cdots.$$

By (8) all the numbers $h^n(\beta)$ are upper bounds for the positive roots of the polynomial $P$.                    □

*Remark.* In Theorem 4 we considered polynomials $P \in \mathbb{R}[X]$ such that $0 < a_{m+t+1} < a_{m+1}$. Similar results can be obtained on upper bounds for the positive real roots also in the case $0 < a_{m+1} < a_{m+t+1}$.

## Comparisons

Theorems 3 and 4 give other upper bounds for the positive roots. We observe that Theorem 4 gives better estimates than Theorem 3 because $h_{st}(x) < g_s(x)$ for all $x > 1$. This happens because the largest two absolute values of the coefficients are considered in Theorem 4.

*Example 1.* Let $P(X) = X^7 + 2X^6 + X^5 - 3X^3 - 2X^2 - 2X + 1$. We have $m = 3$, $t = 1$ and we obtain $\beta_1 = 2$, $\beta_2 == \beta_3 = 1.866$. The largest positive root is $1.121$.
*We obtain*

| $n$ | $g_1^n(\beta_1)$ | $g_2^n(\beta_2)$ | $h_{11}^n(\beta_1)$ | $h_{21}^n(\beta_2)$ | $h_{31}^n(\beta_3)$ |
|---|---|---|---|---|---|
| 1 | 1.97871 | 1.82954 | 1.92508 | 1.75814 | 1.66370 |
| 2 | 1.97776 | 1.82647 | 1.91984 | 1.74530 | 1.61715 |
| 3 | 1.97772 | 1.82620 | 1.91945 | 1.74530 | 1.60309 |
| 4 | 1.97771 | 1.82617 | 1.91942 | 1.74331 | 1.59853 |
| 5 | 1.97771 | 1.82617 | 1.91942 | 1.74327 | 1.59701 |

### 3.1  Polynomials with Frequent Changes of Signs

We obtained in [11] alternative bounds for polynomials with real coefficients represented as

$$P(X) = c_1 X^{d_1} - b_1 X^{m_1} + c_2 X^{d_2} - b_2 X^{m_2} + \cdots + c_k X^{d_k} - b_k X^{m_k} + g(X),$$

with $g(X)$ with positive coefficients, $c_i > 0$, $b_i > 0$, $d_i > m_i$ for all $i$.

These bounds can be used also for positive roots smaller than one, and they give better estimates than $K(P)$.

For polynomials with frequent changes of sign the following bound gives better limits for the largest positive roots

$$S_2(P) = \max\left\{ \left(\frac{b_1}{c_1}\right)^{1/(d_1-m_1)}, \ldots, \left(\frac{b_k}{c_k}\right)^{1/(d_k-m_k)} \right\}.$$

If we consider

$$Q_1(X) = X^7 + 2X^6 + X^5 - 3X^3 - 2X^2 - 2X + 1,$$

$$Q_2(X) = X^{11} + 10X^9 - 17X^5 + 10X^4 - 13X + 1,$$

we use the representations

$$Q_1(X) = (X^7 - 2X^2) + (2X^6 - 3X^3) + (X^5 - 2X) + 1,$$

$$Q_2(X) = (10X^9 - 17X^5) + (10X^4 - 13X) + X^{11} + 1,$$

and obtain

|       | $L_1$ | $L_2$ | $S$   | $B_2$ | $S_2$ | largest positive root |
|-------|-------|-------|-------|-------|-------|-----------------------|
| $Q_1$ | 2.42  | 2.0   | 1.612 | 1.793 | 1.189 | 1.121                 |
| $Q_2$ | 2.571 | 2.545 | 2.683 | 2.782 | 1.141 | 1.097                 |

## References

1. Akritas, A.G., Strzeboński, A.W.: A comparative study of two real root isolation methods. Nonlin. Anal: Modell. Control **10** (2005) 297–304
2. Emiris, I.Z., Tsigaridas, E.P.: Univariate polynomial real root isolation: Continued fractions revisited. http://arxiv.org/pdf/cs.SC/0604066 (2006)
3. Eve, J.: The evaluation of polynomials. Numer. Math. **6** (1964) 17–21
4. Grinstein, L.S.: Upper limits to the real roots of polynomial equations, Amer. Math. Monthly **60** (1953) 608–615

5. Herzberger, J.: Construction of bounds for the positive root of a general class of polynomials with applications. In: Inclusion Methods for Nonlinear Problems with Applications in Engineering, Economics and Physiscs (Munich, 2000), Comput. Suppl. **16**, Springer, Vienna (2003)

6. Kioustelidis, J.B.: Bounds for positive roots of polynomials. J. Comput. Appl. Math. **16** (1986) 241–244

7. Kjurkchiev, N.: Note on the estimation of the order of convergence of some iterative methods. BIT **32** (1992) 525–528

8. Mignotte, M., Ştefănescu, D.: Polynomials – An Algorithmic Approach, Springer Verlag, 1999

9. Nuij, W.: A note on hyperbolic polynomials. Math. Scand. **23** (1968) 69–72

10. Rouiller, F., Zimmermann, P.: Efficient isolation of polynomials real roots. J. Comput. Appl. Math. **162** (2004) 33–50

11. Ştefănescu, D.: New bounds for the positive roots of polynomials. J. Univ. Comp. Sc. **11** (2005) 2125–2131

12. Yap, C.K.: Fundamental Problems of Algorithmic Algebra. Oxford University Press, 2000

# New Domains for Applied Quantifier Elimination
## (Plenary Talk)

Thomas Sturm

Fakultät für Mathematik und Informatik, Universität Passau, Germany

**Abstract.** We address various aspects of our computer algebra-based computer logic system REDLOG. There are numerous examples in the literature for successful applications of REDLOG to practical problems. This includes work by the group around the REDLOG developers as well as by many others. REDLOG is, however, not at all restricted to the real numbers but comprises a variety of other domains. We particularly point at the immense potential of quantifier elimination techniques for the integers. We also address another new REDLOG domain, which is queues over arbitrary basic domains. Both have most promising applications in practical computer science, viz. automatic loop parallelization and software security.

## 1 REDLOG

The REDLOG package of the computer algebra system REDUCE extends the idea of symbolic computation from computer algebra to first-order logic [1, 2, 3, 4]. It provides an integrated interactive environment for computing with first-oder formulas over temporarily fixed languages and domains within a computer algebra system. Its development started around 1992 with a focus on real quantifier elimination algorithms. The current version REDLOG 2.0 is included in the current REDUCE distribution 3.8, which has appeared in 2004.

Over the years the system has been continuously extended by further domains including complex numbers, p-adic numbers [5, 6], quantified propositional calculus [7], term algebras [8], uniform Presburger Arithmetic (linear theory of the integers), and queues over arbitrary domains.

On the algorithmic side, practical applications have triggered the development of several interesting variants of the quantifier elimination paradigm, comprising extended quantifier elimination [9], generic quantifier elimination [10, 11, 12], local quantifier elimination [13], and weak quantifier elimination [14].

## 2 Applications

There are numerous successful applications of the real elimination methods of REDLOG by the group around the developers:

- parametric and nonlinear optimization [15]
- transportation problems [16, 17]

V.G. Ganzha, E.W. Mayr, and E.V. Vorozhtsov (Eds.): CASC 2006, LNCS 4194, pp. 295–301, 2006.

- circuit analysis, design, diagnosis [18]
- generalized scheduling problems [15]
- real implicitation [15]
- automated theorem proving [10, 19]
- computational geometry [20]
- solid modeling [21, 22]
- robot motion planning [15, 23, 24].

Even more interesting, there are just as many such applications documented in the literature by people outside the REDLOG group:

- stability of differential equations [25]
- bifurcation analysis [26, 27, 28, 29]
- theoretical mechanics [30, 31]
- hydraulic network diagnosis
- hybrid control theory [32, 33]
- automatic loop parallelization [34]
- software security [35, 36].

As for applications in other domains, the list is not that exciting. There is one interesting application of REDLOG's (extended) quantifier elimination for discretely valued fields:

- solution of systems of integer congruences with parametric moduli [37].

Furthermore, it has been clearly demonstrated that the quantified propositional calculus has some straightforward applications in

- digital circuit design and testing [7].

This has, however, to our knowledge not been applied within real-world design processes so far. Taking into consideration the remarkable success of SAT-checking methods there is, however, a huge potential in the further development of the propositional domain, which from the SAT-checking point of view provides parametric QSAT-checking.

It might appear surprising that there is not a list of physical applications for the domain of complex numbers. The reason is that for this domain, there is the language of rings used without any additional functions. This is a very clean approach from the algebraic point of view. In practice one would, however, also use absolute values or functions for the real part and the imaginary part. At that point one introduces also real numbers, and in fact such problems can be generally coded as real quantifier elimination problems by using pairs of variables for the real and imaginary parts.

Finally, the REDLOG domain of term algebras is rather new, and the complexity of our elimination algorithm is in the first class of the Gzegorczyk hierarchy, which is rather hard but provably optimal [8].

# 3   Towards New Domains and Applications

We now turn to two new REDLOG domains, which have been mentioned at the beginning but have not been discussed in the previous section: uniform Presburger Arithmetic and queues. The development of both these domains has been inspired by possible applications in practical computer science [34, 35], at the first place in software security [35, 36]: Both integers and queues frequently occur as fundamental data types in computer programs.

## 3.1   Weak Quantifier Elimination for the Integers

For the integers, we speak of *uniform* Presburger Arithmetic or, alternatively, of the *full* linear theory of the integers [14].

Recall that Presburger Arithmetic is the *additive* theory of the integers with ordering and congruences. There is no multiplication. In a Presburger formula like $\exists x(3x + b = 0)$ the notation $3x$ is just a handy way of writing $x + x + x$. A quantifier-free equivalent for our formula is $b \equiv_3 0$, which illustrates the necessity for congruences in the Presburger framework.

Uniform Presburger Arithmetic, in contrast, uses the language of ordered rings including multiplication plus one additional ternary relation symbol for congruences. Over that language, there are *linear* formulas admitted in the meanwhile classical sense known from real quantifier elimination [38, 16]: Without loss of generality, all terms are polynomials. Within these polynomials the total degree with respect to the quantified variables must not exceed 1. That is, quantified variables must not be multiplied with one another, in particular not with themselves. In addition there must not occur any quantified variables within the moduli of the congruences. Note that in uniform Presburger Arithmetic there are not more sentences than in regular Presburger arithmetic. So for decision problems there is no difference. For general quantifier elimination there is, however, a huge gain in expressiveness. Consider, e.g, the following formula asking for necessary and sufficient conditions in parameters $a, \ldots, d$ for the solvability of a system of two parametric constraints:

$$\varphi := \exists x(ax + b \leq 0 \wedge cx + d \geq 0).$$

It is well-known that our uniform Presburger Arithmetic does not admit regular quantifier elimination [39]. Instead, we use *weak quantifier* elimination, which admits *bounded quantifiers* in the elimination result. Consider, e.g., REDLOG's elimination result for $\varphi$ above:

$$(b \leq 0 \wedge d \geq 0)$$
$$\vee \bigsqcup_{k: \, -|a| \leq k \leq |a|} (b - k \equiv_a 0 \wedge a \neq 0 \wedge a^2 k \leq 0 \wedge a^2 d - abc + ack \geq 0)$$
$$\vee \bigsqcup_{k: \, -|c| \leq k \leq |c|} (d - k \equiv_c 0 \wedge c \neq 0 \wedge c^2 k \geq 0 \wedge acd - ack - bc^2 \geq 0).$$

The bounded quantifiers $\bigsqcup_k$ are essentially disjunctions, where the range for $k$ is finite for all possible choices of values for the parameters.

Recent research has shown that the idea of weak quantifier elimination for the integers can be even extended to higher degrees subject to two restrictions: First, in each atomic formula there must occur at most one quantified variable with a degree greater than one. Second, there must not occur any quantified variables with a degree greater than one within congruences. Products between different quantified variables are still prohibited. Also, there still must not occur any quantified variables within the moduli of the congruences [40].

The particular importance of integer methods in practical computer science origins from the fact that they serve as array indices in programs. For the automatic parallelization of for-loops one wishes to automatically check for assignments affecting the same array cell in different executions of the body [34]. Similarly, in software security there is the question for data dependencies between array cells and other data [35, 36].

### 3.2   Queues over Arbitrary Basic Types

The queue domain is two-sorted. That is there are two sorts of variables, one for the basic type and one for the queue type. Accordingly, there is first-order quantification possible for both sorts.

A queue is a finite sequence of elements of the basic type. There are functions available for adding elements and for forming the tail on both sides independently. Our framework thus comprises regular queues, two-sided queues, and stacks. As an example, consider the following formula, which states that $q$ is a 2-periodic queue of odd length with prefix 00 and postfix 11:

$$\exists q'(q = \mathrm{ladd}(0, \mathrm{ladd}(0, \mathrm{radd}(1, \mathrm{radd}(1, q')))) \wedge$$
$$\exists x \exists y (x \neq y \wedge \mathrm{ladd}(y, (\mathrm{ladd}(x, q')) = \mathrm{radd}(y, \mathrm{radd}(x, q'))))).$$

The quantifier elimination for queues works in such a way that queue quantifiers are eliminated independently from basic type quantifiers. Thus, if the basic type admits quantifier elimination, then so does the corresponding queue domain. If the basic type does not admit quantifier elimination, then it is still possible to eliminate all the queue quantifiers.

The work on the queue domain in REDLOG is still at quite an early stage. So far, the implementation is restricted to the reals as basic type [41].

Similarly to the integers there is a most promising application area for the queue domain in software security: Queue-like data structures are systematically used for the implementation of protocols, where there is a particular need for verification and security [35, 36].

## References

1. Dolzmann, A., Sturm, T.: Redlog: Computer algebra meets computer logic. ACM SIGSAM Bulletin **31**(2) (1997) 2–9
2. Dolzmann, A., Sturm, T.: Redlog user manual. Technical Report MIP-9905, FMI, Universität Passau, D-94030 Passau, Germany (1999) Edition 2.0 for Version 2.0.

3. Hearn, A.C.: Reduce User's Manual for Version 3.8, Santa Monica, CA. (2004) http://reduce-algebra.com/.

4. Hearn, A.C.: Reduce: The first forty years. In Dolzmann, A., Seidl, A., Sturm, T., eds.: Algorithmic Algebra and Logic. BoD, Norderstedt (2005) 19–24

5. Sturm, T.: Linear problems in valued fields. Journal of Symbolic Computation **30**(2) (2000) 207–219

6. Dolzmann, A., Sturm, T.: P-adic constraint solving. In Dooley, S., ed.: Proceedings of the 1999 International Symposium on Symbolic and Algebraic Computation (ISSAC 99), Vancouver, BC. ACM Press, New York, NY (1999) 151–158

7. Seidl, A.M., Sturm, T.: Boolean quantification in a first-order context. In Ganzha, V.G., Mayr, E.W., Vorozhtsov, E.V., eds.: Computer Algebra in Scientific Comput-ing. Proceedings of the CASC 2003. Institut für Informatik, Technische Universität München, Passau (2003) 329–345

8. Sturm, T., Weispfenning, V.: Quantifier elimination in term algebras. The case of finite languages. In Ganzha, V.G., Mayr, E.W., Vorozhtsov, E.V., eds.: Computer Algebra in Scientific Computing. Proceedings of the CASC 2002. TUM München (2002) 285–300

9. Weispfenning, V.: Simulation and optimization by quantifier elimination. Journal of Symbolic Computation **24**(2) (1997) 189–208 Special issue on applications of quantifier elimination.

10. Dolzmann, A., Sturm, T., Weispfenning, V.: A new approach for automatic the-orem proving in real geometry. Journal of Automated Reasoning **21**(3) (1998) 357–380

11. Seidl, A., Sturm, T.: A generic projection operator for partial cylindrical alge-braic decomposition. In Sendra, R., ed.: Proceedings of the 2003 International Symposium on Symbolic and Algebraic Computation (ISSAC 03), Philadelphia, Pennsylvania. ACM Press, New York, NY (2003) 240–247

12. Dolzmann, A., Gilch, L.A.: Generic Hermitian quantifier elimination. In Bruno Buchberger, J.A.C., ed.: Artificial Intelligence and Symbolic Computation: 7th International Conference, AISC 2004, Linz, Austria. Volume 3249 of Lecture Notes in Computer Science. Springer-Verlag, Berlin, Heidelberg (2004) 80–93

13. Dolzmann, A., Weispfenning, V.: Local quantifier elimination. In Traverso, C., ed.: Proceedings of the 2000 International Symposium on Symbolic and Algebraic Computation (ISSAC 2000), St Andrews, Scotland. ACM Press, New York, NY (2000) 86–94

14. Lasaruk, A., Sturm, T.: Weak quantifier elimination for the full linear theory of the integers. a uniform generalization of presburger arithmetic". Technical Report MIP-0604, FMI, Universität Passau, D-94030 Passau, Germany (2006)

15. Dolzmann, A., Sturm, T., Weispfenning, V.: Real quantifier elimination in practice. In Matzat, B.H., Greuel, G.M., Hiss, G., eds.: Algorithmic Algebra and Number Theory. Springer, Berlin (1998) 221–247

16. Loos, R., Weispfenning, V.: Applying linear quantifier elimination. THE Computer Journal **36**(5) (1993) 450–462 Special issue on computational quantifier elimina-tion.

17. Dolzmann, A.: Algorithmic Strategies for Applicable Real Quantifier Elimination. Doctoral dissertation, Department of Mathematics and Computer Science. Univer-sity of Passau, Germany, D-94030 Passau, Germany (2000)

18. Sturm, T.: Reasoning over networks by symbolic methods. Applicable Algebra in Engineering, Communication and Computing **10**(1) (1999) 79–96

19. Dolzmann, A., Sturm, T., Weispfenning, V.: Automatic theorem proving in geometry. In Grabmeier, J., Kaltofen, E., Weispfenning, V., eds.: Computer Algebra Handbook. Springer, Berlin (2003) 201–207

20. Sturm, T., Weispfenning, V.: Computational geometry problems in Redlog. In Wang, D., ed.: Automated Deduction in Geometry. Volume 1360 of Lecture Notes in Artificial Intelligence (Subseries of LNCS). Springer-Verlag, Berlin Heidelberg (1998) 58–86

21. Sturm, T., Weispfenning, V.: Rounding and blending of solids by a real elimination method. In Sydow, A., ed.: Proceedings of the 15th IMACS World Congress on Scientific Computation, Modelling, and Applied Mathematics (IMACS 97). Volume 2., Berlin, IMACS, Wissenschaft & Technik Verlag, Berlin, 1997 (1997) 727–732

22. Sturm, T.: An algebraic approach to offsetting and blending of solids. In Ganzha, V.G., Mayr, E.W., Vorozhtsov, E.V., eds.: Computer Algebra in Scientific Computing. Proceedings of the CASC 2000. Springer, Berlin (2000) 367–382

23. Weispfenning, V.: Semilinear motion planning among moving objects in REDLOG. In Ganzha, V.G., Mayr, E.W., eds.: Computer Algebra in Scientific Computing. Proceedings of the CASC 2001. Springer, Berlin (2001) 541–553

24. Weispfenning, V.: Semilinear motion planning in REDLOG. Applicable Algebra in Engineering, Communication and Computing **12** (2001) 455–475

25. Hong, H., Liska, R., Steinberg, S.: Testing stability by quantifier elimination. Journal of Symbolic Computation **24**(2) (1997) 161–187 Special issue on applications of quantifier elimination.

26. Kahoui, M., Weber, A.: Deciding hopf bifurcations by quantifier elimination in a software component architecture. Journal of Symbolic Computation **30**(2) (2000) 161–179

27. Kahoui, M., Weber, A.: Symbolic equilibrium point analysis in parameterized polynomial vector fields. In Ganzha, V.G., Mayr, E.W., Vorozhtsov, E.V., eds.: Computer Algebra in Scientific Computing. Proceedings of the CASC 2002. TUM München (2002) 71–83

28. Seiler, W.M., Weber, A.: Deciding ellipticity by quantifier elimination. In Ganzha, V.G., Mayr, E.W., Vorozhtsov, E.V., eds.: Computer Algebra in Scientific Computing. Proceedings of the CASC 2003. Institut für Informatik, Technische Universität München, Passau (2003) 347–355

29. Brown, C.W., Kahoui, M., Novotni, D., Weber, A.: Algorithmic methods for investigating equilibria in epidemic modeling. (To appear in the Journal of Symbolic Computation.)

30. Ioakimidis, N.I.: Automatic derivation of positivity conditions inside boundary elements with the help of the REDLOG computer logic package. Engineering Analysis with Boundary Elements **23**(10) (1999) 847–856

31. Ioakimidis, N.I.: REDLOG-aided derivation of feasibility conditions in applied mechanics and engineering problems under simple inequality constraints. Journal of Mechanical Engineering (Strojnícky Časopis) **50**(1) (1999) 58–69

32. Jirstrand, M.: Nonlinear control system design by quantifier elimination. Journal of Symbolic Computation **24**(2) (1997) 137–152 Special issue on applications of quantifier elimination.

33. Lafferriere, G., Pappas, G.J., Yovine, S.: A new class of decidable hybrid systems. In Vaandrager, F.W., van Schuppen, J.H., eds.: Hybrid Systems and Control. Proceedings of the Second International Workshop, HSCC'99, Berg en Dal, The Netherlands, March 1999. Volume 1569 of Lecture Notes in Computer Science. Springer, Berlin, Germany (1999) 137–151

34. Größlinger, A., Griebl, M., Lengauer, C.: Quantifier elimination in automatic loop parallelization. In Dolzmann, A., Seidl, A., Sturm, T., eds.: Algorithmic Algebra and Logic. Proceedings of the A3L 2005. BoD, Germany, Norderstedt (2005) 123–128

35. Snelting, G.: Quantifier elimination and information flow control for software security. In Dolzmann, A., Seidl, A., Sturm, T., eds.: Algorithmic Algebra and Logic. Proceedings of the A3L 2005. BoD, Germany, Norderstedt (2005) 237–242

36. Snelting, G., Robschink, T., Krinke, J.: Efficient path conditions in dependence graphs for software safety analysis. ACM Transactions on Software Engineering and Methodolody (2006) To appear.

37. Dolzmann, A., Sturm, T.: Parametric systems of linear congruences. In Ganzha, V.G., Mayr, E.W., Vorozhtsov, E.V., eds.: Computer Algebra in Scientific Computing. Proceedings of the CASC 2001. Springer, Berlin (2001) 149–166

38. Weispfenning, V.: The complexity of linear problems in fields. Journal of Symbolic Computation **5**(1&2) (1988) 3–27

39. Weispfenning, V.: Complexity and uniformity of elimination in Presburger Arithmetic. In Küchlin, W.W., ed.: Proceedings of the 1997 International Symposium on Symbolic and Algebraic Computation, Maui, HI (ISSAC 97). ACM Press, New York, NY (1997) 48–53

40. Lasaruk, A.: Univariate weak quantifier elimination for the integers. In Gerdt, V., Spiridonova, M., Nisheva-Pavlova, M., eds.: ACA 2006. 12th International Conference on Applications of Computer Algebra. June 26–29, 2006, Varna, Bulgaria. Abstracts of Presentations. Institute of Mathematics and Informatics, Bulgarian Academy of Sciences, Sofia, Bulgaria (2006) 8

41. Straßer, C.: Quantifier elimination for queues. In: Rhine Workshop on Computer Algebra. Proceedings of the RWCA 2006. Universität Basel, Basel (2006)

# Algorithms for Symbolic Polynomials
## (Plenary Talk)

Stephen M. Watt

Ontario Research Centre for Computer Algebra
University of Western Ontario London
Canada
Watt@orcca.on.ca

**Abstract.** We wish to work with polynomials where the exponents are not known in advance, such as $x^{2n} - 1$. There are various operations we will want to be able to do, such as squaring the value to get $x^{4n} - 2x^{2n} + 1$, or differentiating it to get $2nx^{2n-1}$. Expressions of this sort arise frequently in practice, for example in the analysis of algorithms, and it is very difficult to work with them effectively in current computer algebra systems.

We consider the case where multivariate polynomials can have exponents which are themselves integer-valued multivariate polynomials, and we present algorithms to compute their GCD and factorization. The algorithms fall into two families: algebraic extension methods and interpolation methods. The first family of algorithms uses the algebraic independence of $x$, $x^n$, $x^{n^2}$, $x^{nm}$, etc, to solve related problems with more indeterminates. Some subtlety is needed to avoid problems with fixed divisors of the exponent polynomials. The second family of algorithms uses evaluation and interpolation of the exponent polynomials. While these methods can run into unlucky evaluation points, in many cases they can be more appealing. Additionally, we also treat the case of symbolic exponents on rational coefficients (e.g. $4^{n^2+n} - 81$) and show how to avoid integer factorization.

V.G. Ganzha, E.W. Mayr, and E.V. Vorozhtsov (Eds.): CASC 2006, LNCS 4194, p. 302, 2006.
© Springer-Verlag Berlin Heidelberg 2006

# Testing Mersenne Primes with Elliptic Curves[*]

Song Y. Yan and Glyn James

Faculty of Engineering & Computing
Coventry University
Coventry CV1 5FB, UK
s.yan@coventry.ac.uk

**Abstract.** The current primality test in use for Mersenne primes continues to be the Lucas-Lehmer test, invented by Lucas in 1876 and proved by Lehmer in 1935. In this paper, a practical approach to an elliptic curve test of Gross for Mersenne primes, is discussed and analyzed. The most important advantage of the test is that, unlike the Lucas-Lehmer test which requires $\mathcal{O}(p)$ arithmetic operations and $\mathcal{O}(p^3)$ bit operations in order to determine whether or not $M_p = 2^p - 1$ is prime, it only needs $\mathcal{O}(\lambda)$ arithmetic operations and $\mathcal{O}(\lambda^3)$ bit operations, with $\lambda \ll p$. Hence it is more efficient than the Lucas-Lehmer test, but is still as simple, elegant and practical.

**Keywords:** Mersenne numbers, Mersenne primes, Lucas-Lehmer test, elliptic curve test.

## 1 Introduction

A number is called *Mersenne number* if it is of the form:

$$M_p = 2^p - 1, \quad \text{with } p \text{ prime.} \tag{1}$$

If a Mersenne number is also a prime, then it is called a *Mersenne prime*. For example, $M_{11} = 2^{11} - 1$ is a Mersenne number but not a Mersenne prime since $M_{11} = 2^{11} - 1 = 23 \times 89$, whereas $M_{13} = 2^{13} - 1$ is not only a Mersenne number but also a Mersenne prime, and in fact it is the 5th Mersenne prime found in 1456. However, most of the Mersenne numbers are not prime, this makes it very difficult to actually find a new Mersenne prime. To date, only forty-three Mersenne primes have been found, although the first four were known to the ancient Greek mathematicians 2000 years ago. Table 1 gives all the known 43 Mersenne primes, together with some information about the digits in $M_p$ and the discovery time (year). In spite of its long history, there are still many open problems regarding the Mersenne primes; for example, it is still does not

---

[*] The authors would like to thank Prof Benedict Gross of the Department of Mathematics at Harvard University for his advice to work on the computational aspects of the elliptic curve test for Mersenne primes. This paper was written while the first author visited the Dept of Computer Science at the University of Toronto hosted by Prof Stephen Cook and supported by the Royal Society London.

V.G. Ganzha, E.W. Mayr, and E.V. Vorozhtsov (Eds.): CASC 2006, LNCS 4194, pp. 303–312, 2006.
© Springer-Verlag Berlin Heidelberg 2006

know whether or not there are infinitely many Mersenne primes although there are infinitely many primes. Note that the 43rd Mersenne primes is the largest known prime to date and it is over nine million digits. The Electronic Frontier Foundation in USA is offering a \$100,000 award to the first person or group to discover a prime number with more than ten million digits.

**Table 1.** The 43 Known Mersenne Primes of the Form $2^p - 1$ with $p$ Prime

| No. | $p$ | Digits | Discovery Time | No. | $p$ | Digits | Discovery Time |
|-----|-----|--------|----------------|-----|-----|--------|----------------|
| 1 | 2 | 1 | Ancient | 2 | 3 | 1 | Ancient |
| 3 | 5 | 2 | Ancient | 4 | 7 | 3 | Ancient |
| 5 | 13 | 4 | 1456 | 6 | 17 | 6 | 1588 |
| 7 | 19 | 6 | 1588 | 8 | 31 | 10 | 1772 |
| 9 | 61 | 19 | 1883 | 10 | 89 | 27 | 1911 |
| 11 | 107 | 33 | 1914 | 12 | 127 | 39 | 1876 |
| 13 | 521 | 157 | 1952 | 14 | 607 | 183 | 1952 |
| 15 | 1279 | 386 | 1952 | 16 | 2203 | 664 | 1952 |
| 17 | 2281 | 687 | 1952 | 18 | 3217 | 969 | 1957 |
| 19 | 4253 | 1281 | 1961 | 20 | 4423 | 1332 | 1961 |
| 21 | 9689 | 2917 | 1963 | 22 | 9941 | 2993 | 1963 |
| 23 | 11213 | 3376 | 1963 | 24 | 19937 | 6002 | 1971 |
| 25 | 21701 | 6533 | 1978 | 26 | 23209 | 6987 | 1979 |
| 27 | 44497 | 13395 | 1979 | 28 | 86243 | 25962 | 1982 |
| 29 | 110503 | 33265 | 1988 | 30 | 132049 | 39751 | 1983 |
| 31 | 216091 | 65050 | 1985 | 32 | 756839 | 227832 | 1992 |
| 33 | 859433 | 258716 | 1994 | 34 | 1257787 | 378632 | 1996 |
| 35 | 1398269 | 420921 | 1996 | 36 | 2976221 | 895932 | 1997 |
| 37 | 3021377 | 909526 | 1998 | 38 | 6972593 | 2098960 | 1999 |
| 39 | 13466917 | 4053946 | 2001 | 40 | 20996011 | 6320430 | 2003 |
| 41 | 24036583 | 7235733 | 2004 | 42 | 25964951 | 7816230 | 2005 |
| 43 | 30402457 | 9152052 | 2005 | 44 | ? | ? | ? |

One of most important properties of the Mersenne primes is that they have a one-to-one correspondence to perfect numbers. A number $n$ is called a *perfect number* if $\sigma(n) = 2n$, where $\sigma(n) = \sum_{d|n} d$. For example, 8128 is a perfect number (in fact, it is the 4th perfect number), since $\sigma(8128) = 1 + 2 + 4 + 8 + 16 + 32 + 64 + 127 + 254 + 508 + 1016 + 2032 + 4064 + 8128 = 162562 = 2 \times 8128$. The famous Euclid-Euler Theorem states that $n$ is an even perfect number if and only if $n = 2^{p-1}(2^p - 1)$, where $2^p - 1$ is a Mersenne prime. The sufficient condition of this theorem was established in Euclid's *Elements* (Book IX, Proposition 36) 2000 years ago, but the fact that is is also necessary was established by Euler in work published posthumously. Readers who are interested in the history of Mersenne primes and perfect numbers could consult [14] for more information.

The Euclid-Euler Theorem implies that if one finds a new Mersenne prime, one finds a new perfect number. Note however that all known perfect numbers are even and that there are no odd perfect numbers up to $10^{300}$ [2]. Whether or not there exists an odd perfect number is one of the most important open problems in mathematics. In this paper, we concentrate on tests particularly those tests based on elliptic curves for Mersenne primes from a computational point of view.

## 2    The Lucas-Lehmer Test

Define recursively the Lucas-Lehmer sequence (of integers) $L_k$ as follows [14]:

$$\begin{cases} L_0 = 4 \\ L_{k+1} = L_k^2 - 2, k \geq 0. \end{cases} \tag{2}$$

Then the Lucas-Lehmer sequence begins with integers:

4, 14, 194, 37634, 1416317954, 2005956546822746114,
40238616677410360228256356556102100994,
1619146272111567178177755907012051366495859012549915851432930874097 5788034,
$\cdots$.

The famous Lucas-Lehmer test for Mersenne primes is stated as follows:

**Theorem 1.** *Let $M_p = 2^p - 1$ be a Mersenne number. Then $M_p$ is prime if and only if $M_p \mid L_{p-2}$. That is, $M_p$ is prime if and only if*

$$L_{p-2} \equiv 0 \pmod{M_p}. \tag{3}$$

This theorem was proposed by Lucas in 1876 [8] and proved by Lehmer in 1935 [6]. Since then, there have been many interesting proofs for this theorem, some of which, say, [11], being based on some deep ideas from algebraic number theory. Clearly, if $M_p$ is composite, then $M_p \nmid L_k$ for $0 \leq k < p-2$. The following efficient algorithm can be derived directly from this theorem.

**Algorithm 1 (Lucas-Lehmer Test).** Given a Mersenne number $M_p$, the following algorithm will determine whether or not $M_p$ is prime.

```
Initialize the value for p
L ← 4
for i from 1 to p − 2 do
    L ← L² − 2 (mod (2ᵖ − 1))
If L = 0 then 2ᵖ − 1 is prime
    else 2ᵖ − 1 is composite
```

Algorithm 1 can be efficiently performed in $\mathcal{O}(p^3)$ bit operations, since the test only requires $\mathcal{O}(p)$ squaring modulo $M_p$. However, when $p$ becomes very large, even with the help of some special computing techniques, such as FFT (Fast Fourier Transform), it is still very hard to find a new Mersenne prime. This is the reason that there are only 43 Mersenne primes have been found so far. This is also one of the main motivations to find a more powerful test for Mersenne primes.

**Example 1.** For $M_{31} = 2^{31} - 1$ we have the following Lucas-Lehmer sequence:

14, 194, 37634, 1416317954, 669670838, 1937259419, 425413602, 842014276, 12692426, 2044502122, 1119438707, 1190075270, 1450757861, 877666528, 630853853, 940321271, 512995887, 692931217, 1883625615, 1992425718, 721929267, 27220594, 1570086542, 1676390412, 1159251674, 211987665, 1181536708, 65536, 0.

Thus, by Theorem 1, $M_{31} = 2^{31} - 1$ is prime since the last number in the sequence is zero. On the other hand, for $M_{37} = 2^{37} - 1$, we have the following sequence:

14, 194, 37634, 1416317954, 111419319480, 75212031451, 42117743384, 134212256520, 54923239684, 61369726979, 100682126153, 46790825955, 120336432403, 15532303443, 43487582705, 63215664337, 24881968247, 36378170995, 23347868395, 34319987212, 27325339261, 67024860468, 67821607698, 45433743622, 32514699513, 51489094388, 44855569738, 31479590378, 32455804440, 54840899833, 71222372297, 35230286592, 24416019713, 80429963578, 117093979072.

Thus, by Theorem 1, $M_{37} = 2^{37} - 1$ is not a prime since the last term in the sequence is not zero.

An equivalent form of Theorem 1 is as follows:

**Theorem 2.** Let $M_p = 2^p - 1$ be a Mersenne number. Then $M_p$ is prime if and only if $\gcd(L_k, M_p) = 1$ for $0 \le k \le p - 3$ and $\gcd(L_{p-2}, M_p) > 1$.

**Example 2.** Let $M_7 = 2^7 - 1$ be a Mersenne number. Then we have

$$\gcd(\{L_0, L_1, L_2, L_3, L_4\}, M_7) = \gcd(\{4, 14, 194, 37634, 1416317954\}, 2^7 - 1) = 1,$$

but $\gcd(L_5, M_7) = \gcd(2005956546822746114, 2^7 - 1) = 127 > 1$, thus, by Theorem 2, $M_7$ is prime. On the other hand, for $M_{11} = 2^{11} - 1$, we have

$$\gcd(\{L_0, L_1, L_2, L_3, L_4, L_5, L_6, L_7, L_8, L_9\}, M_{11})$$
$$= \gcd(\{4, 14, 194, 788, 701, 119, 1877, 240, 282, 1736\}, 2^{11} - 1)$$
$$= 1,$$

thus, by Theorem 2, $M_{11}$ is not prime.

From an algebraic point of view, the Lucas-Lehmer test is based on the successive squaring of a point on the one dimensional algebraic torus over $\mathbb{Q}$, associate to the real quadratic field $k = \mathbb{Q}(\sqrt{3})$. This suggests, as Gross noted[3], some other tests for Mersenne numbers, using different algebraic groups, e.g., elliptic curve groups, based on successive squaring of a point on an elliptic curve. This is the idea underpinning the elliptic curve test for Mersenne numbers considered in this paper.

## 3    The Elliptic Curve Test

Define a sequence of rational numbers $G_k$ (see [3] and [12]) as follows:

$$\begin{cases} G_0 = -2 \\ G_{k+1} = \dfrac{(G_k^2 + 12)^2}{4 \cdot G_k \cdot (G_k^2 - 12)} \end{cases} \quad k \ge 0. \tag{4}$$

Then, the sequence begins with rational numbers:

$$-2, 4, \frac{49}{4}, \frac{6723649}{1731856}, \frac{6593335793533896979873913089}{42923806093478210389068294 4}, \dots$$

Remarkably enough, this sequence, as in the case of the Lucas-Lehmer sequence, is intimately connected to the primality testing of Mersenne numbers. The Mersenne number $M_p = 2^p - 1$ is prime if and only if $\gcd(G_k(G_k^2 - 12), M_p) = 1$ for $0 \le k \le p - 2$, and $\gcd(G_{p-1}, M_p) > 1$. This test, in fact, involves the successive squaring of a point on an elliptic curve $E$ over $\mathbb{Q}$:

$$E/\mathbb{Q}: \ y^2 = x^3 - 12x, \tag{5}$$

with discriminant $\Delta = 2^{12} \cdot 3^3$ and conductor $N = 2^5 \cdot 3^2 = 288$ [3].

**Theorem 3.** *Let $M_p = 2^p - 1$ be a Mersenne number. Then $M_p$ is prime if and only if $\gcd(G_k(G_k^2 - 12), M_p) = 1$ for $0 \le k \le p - 2$, and $\gcd(G_{p-1}, M_p) > 1$.*

**Remark 1.** Theorem 3 is just an elliptic curve analog of Theorem 1.

We note that Theorem 2 and Algorithm 2 can be modified and simplified to produce a practical primality test for Mersenne primes, that is as *simple, elegant* and *efficient* as the Lucas-Lehmer test.

**Theorem 4.** *Let $M_p = 2^p - 1$ be a Mersenne number. Then $M_p$ is prime if and only if $M_p \mid G_{p-1}$, or equivalently,*

$$G_{p-1} \equiv 0 \ (\mathrm{mod} \ M_p). \tag{6}$$

**Remark 2.** Theorem 4 is an elliptic curve analog of Theorem 2. The test based on Theorem 4 is competitive to the test based on Theorem 2. More importantly and more interestingly, in the case that $M_p$ is a composite, the test is much better and quicker than the test based on Theorem 2.

## 4    Algorithms and Complexities

The test based on (and derived from) Theorem 3 for Mersenne primes can be easily converted to an algorithm (and implemented by a computer algebra system such as Maple) as follows:

**Algorithm 2.** This algorithm provides a test for primality of Mersenne numbers $M_p$ based on elliptic curves.

Initialize the value for $p$
$G \leftarrow -2$
for $i$ from 1 to $p - 2$ do
    $G \leftarrow (G^2 + 12)^2 / (4 \cdot G \cdot (G^2 - 12)) \ (\mathrm{mod} \ (M_p))$
    $\gcd \leftarrow \gcd(G(G^2 - 12), M_p)$
    $G \leftarrow (G^2 + 12)^2 / (4 \cdot G \cdot (G^2 - 12)) \ (\mathrm{mod} \ (M_p))$

$$\gcd \leftarrow \gcd(G, M_p)$$

If all gcd $= 1$ (except the last one) and the last gcd $> 1$
    then $M_p$ is prime
    else $M_p$ is not prime

The complexity of the elliptic curve is almost the same as that of Lucas-Lehmer test, since it also requires $\mathcal{O}(p)$ successive squaring of a point on an elliptic curve modulo $M_p$ and the gcd computation.

**Example 3.** For $M_{13} = 2^{13} - 1$ we have:

$G_1 = 4$ and $\gcd(G_1(G_1^2 - 12), 2^p - 1) = \gcd(16, 8191) = 1$
$G_2 = 2060$ and $\gcd(G_2(G_2^2 - 12), 2^p - 1) = \gcd(8741791280, 8191) = 1$
$G_3 = 4647$ and $\gcd(G_3(G_3^2 - 12), 2^p - 1) = \gcd(100350092259, 8191) = 1$
$G_4 = 6472$ and $\gcd(G_4(G_4^2 - 12), 2^p - 1) = \gcd(271091188384, 8191) = 1$

$$\vdots$$

$G_{10} = 3036$ and $\gcd(G_{10}(G_{10}^2 - 12), 2^p - 1) = \gcd(27983674224, 8191) = 1$
$G_{11} = 362$ and $\gcd(G_{11}(G_{11}^2 - 12), 2^p - 1) = \gcd(47433584, 8191) = 1$
$G_{12} = 0$ and $\gcd(G_{12}, 2^p - 1) = \gcd(0, 8191) = 8191 > 1$

Thus, by Theorem 2, $M_{13}$ is prime.

Theorem 4, however, can be converted to a very efficient algorithm as follows:

**Algorithm 3 (Elliptic Curve Test).** This algorithm provides a test for primality of Mersenne numbers $M_p$ based on elliptic curves. The procedure is similar to that of Algorithm 1 for the Lucas-Lehmer test.

Initialize the value for $p$
$G \leftarrow -2$
for $i$ from 1 to $p - 1$ do
    $G \leftarrow (G^2 + 12)^2/(4 \cdot G \cdot (G^2 - 12)) \ (\text{mod } M_p)$
    If $G$ does not exist, then stop the Algorithm, $M_p$ is not prime
    end_do
If $G = 0$ then $M_p$ is prime
        else $M_p$ is not prime

The most important feature of Algorithm 3 is that it can stop anytime whenever $G$ does not exist (the multiplication inverse modulo $M_p$ does not exist which implies that $M_p$ is not prime). Unlike Algorithm 1 and even Algorithm 2, Algorithm 3 does not need to perform the loops from 1 up to $p - 1$. Thus, if the complexity, measured by bit operations, for Algorithm 1 and even Algorithm 2 is $\mathcal{O}(p^3)$, then the complexity for Algorithm 3 is $\mathcal{O}(\lambda^3)$, with $\lambda \ll p$. It is this feature that makes it very suitable to test the primality for Mersenne primes, since most of the Mersenne numbers are not prime, and in fact very very few of the Mersenne numbers are prime. Algorithm 3 can stop the useless loops at a very earlier stage of the test if the Mersenne number is not prime. On the other hand, no matter $M_p$ is prime or not, the Lucas-Lehmer test will need to perform the test for $p-1$ time. Thus, for the Lucas-Lehmer test, most of the computation times are spent on the useless loops on the uninteresting Mersenne composite numbers. Therefore, the elliptic curve test of Algorithm 3 is very practical and efficient for finding new Mersenne primes than any other existing methods; these can be seen from the following two examples.

**Example 4.** Let $M_{61} = 2^{61} - 1$. Then by Algorithm 3, we have the following sequence of $G$:

$-2, 4, 576460752303423500, 2273229660002968910, 61942924192900875,$
$227188641361726013, 584955155938028078, 1625992443540546788,$
$2281402858238123895, 96833079332604917, 190667854843477861, \cdots,$
$754634924515350806, 580226091912597714, 485337391446314291,$
$523773010142196837, 2066643542538005360, 0.$

Hence, by Theorem 3, $M_{61} = 2^{61} - 1$ is prime. On the other hand, for $M_{59} = 2^{59} - 1$ with 59 prime, then by Algorithm 3, we have

$-2, 4, 144115188075855884, 525264983758271552, 359803122487490816,$
$307956120154006017, 125273927552899636, 455734762531475987,$
$443763109181002475, 516877268683260273, 459543861857489578, \cdots,$
$118819180406808830, 343866746792758738, 379530138081340611,$
$158582427978263517, 89062835136920118.$

Hence, by Theorem 3, $M_{59} = 2^{59} - 1$ is not prime since the last term in the squence is not zero.

**Example 5.** For comparison purpose, we tabulate the results for primality testing of $M_{31}$ and $M_{23}$, by using the two methods, in Table 2. As can be seen from Table 2 the computational costs of the two tests for $M_{31}$ are the same. However for $M_{23}$ a big saving has been made using the elliptic curve test over the Lucas-Lehmer test, since, for a composite Mersenne number, the Lucas-Lehmer test will still need to go through all the loops from 1 to $p - 2$, whereas the elliptic curve test will stop whenever $G_k$ does not exist. This is because the computation of $G_k$ is based on fractions modulo $M_p$ which may not exist (because $M_p$ is composite). Whenever a modular inverse does not exist the computation stops and hence a saving is made in computation time; the earlier the computation stops the bigger the saving can be made. This is a significant advantage of adopting the elliptic curve test over the Lucas-Lehmer test. As the Mersenne primes become parser and sparser, the elliptic curve test can skip those composite $M_p$ and concentrate on the prime $M_p$. For example, for $M_{23}$, the Lucas-Lehmer test will need 21 tests whereas the Elliptic curve test will only need 3 tests to reach an answer that $M_{23}$ is not a prime.

**Remark 3.** As can be seen from Table 2 that if $M_p = 2^p - 1$ is prime, we do need to perform $p - 2$ recursions/loops in order to decide whether or not $M_p$ is prime. However, if $M_p$ is composite, then we do not need to perform all the $p - 2$ loops; we can conclude that $M_p$ is not a prime at a very earlier stage of the loops, hence a big saving in time can be made. For example, for $p = 23$,

**Table 2.** Comparisons of the Lucas-Lehmer and Elliptic Curve Tests for $M_{31}$ and $M_{23}$

| Recursion | LL Test | EC Test |
|---|---|---|
| $k$ | $L_k \bmod M_{31}$ | $G_k \bmod M_{31}$ |
| 0 | 4 | $-2$ |
| 1 | 14 | 4 |
| 2 | 194 | 536870924 |
| 3 | 37634 | 242940031 |
| 4 | 1416317954 | 1997781005 |
| 5 | 669670838 | 1070166402 |
| 6 | 1937259419 | 1316556811 |
| 7 | 425413602 | 1539940455 |
| 8 | 842014276 | 2000813575 |
| 9 | 12692426 | 361728374 |
| 10 | 2044502122 | 602038520 |
| 11 | 1119438707 | 1031405401 |
| 12 | 1190075270 | 553331261 |
| 13 | 1450757861 | 2131672851 |
| 14 | 877666528 | 1777654456 |
| 15 | 630853853 | 321910276 |
| 16 | 940321271 | 1025466243 |
| 17 | 512995887 | 102277241 |
| 18 | 692931217 | 1545817800 |
| 19 | 1883625615 | 1322945035 |
| 20 | 1992425718 | 1314104653 |
| 21 | 721929267 | 427090555 |
| 22 | 27220594 | 326614488 |
| 23 | 1570086542 | 2116120380 |
| 24 | 1676390412 | 1193217808 |
| 25 | 1159251674 | 47494329 |
| 26 | 211987665 | 73295707 |
| 27 | 1181536708 | 1114213282 |
| 28 | 65536 | 40029428 |
| $29 = p - 2$ | 0 | 388539999 |
| $30 = p - 1$ | $M_{31}$ is prime STOP | 0 |
| | | $M_{31}$ is prime STOP |

| Recursion | LL Test | EC Test |
|---|---|---|
| $k$ | $L_k \bmod M_{23}$ | $G_k \bmod M_{23}$ |
| 0 | 4 | $-2$ |
| 1 | 14 | 4 |
| 2 | 194 | 2097164 |
| 3 | 37634 | $G_3 = \perp$ |
| 4 | 7031978 | $M_{23}$ is not prime STOP |
| | | - saving |
| 5 | 7033660 | - saving |
| 6 | 1176429 | - saving |
| 7 | 7643358 | - saving |
| 8 | 3179743 | - saving |
| 9 | 2694768 | - saving |
| 10 | 763525 | - saving |
| 11 | 4182158 | - saving |
| 12 | 7004001 | - saving |
| 13 | 1531454 | - saving |
| 14 | 5888805 | - saving |
| 15 | 1140622 | - saving |
| 16 | 4321431 | - saving |
| 17 | 7041324 | - saving |
| 18 | 2756392 | - saving |
| 19 | 1280050 | - saving |
| 20 | 6563009 | - saving |
| $21 = p - 2$ | $6107895 \neq 0$ | - saving |
| | $M_{23}$ is not prime STOP | - |

only three loops (not $p - 2 = 21$ loops) are needed in order to decide that $M_{23} = 2^{23} - 1$ is not prime. As most of the Mersenne numbers are not prime but composite, and in fact for $p = 2, 3, 5, 7, \cdots, 25964951$, there are 1622441 primes (i.e., $\pi(25964951) = 1622441$), but only 43 such $p$ (primes) that can make $M_p = 2^p - 1$ to be prime. This is why we say that the complexity of the elliptic curve test is $\mathcal{O}(\lambda^3)$ with $\lambda \ll p$, because it can quickly skip the Mersenne composite numbers and only concentrate on the verification of the Mersenne primes. Although at the beginning of the test we do not know which $M_p$ is indeed prime, we can quickly know it is not prime by the elliptic curve test; whenever the current $M_p$ is not prime, we just simply throw it away and choose a next $M_p$ to test. Our complexity analysis and computing experiments show that the elliptic curve test is more efficient than the Lucas-Lehmer test and is a good candidate as a replacement to the Lucas-Lehmer test for Mersenne primes. Of course, much needs to be done in order to make the elliptic curve test more practical and more useful.

## 5   Conclusion

Elliptic curves have been studied by number theorists for about a century; not for applications in either mathematics or computing science but because of their intrinsic mathematical beauty and interest. In recent years, elliptic curves have found applications in many areas of mathematics and computer science. For example, by using the theory of elliptic curves, Lenstra [7] invented the powerful factoring method ECM, Morain (jointly with Atkin, see [1] and [10]) designed the practical and fast elliptic curve primality proving algorithm ECPP, Koblitz [4] and Miller [9] proposed the idea of elliptic public-key cryptosystems and, more interestingly, Wiles proved the famous Fermat's Last Theorem [13]. However, little attention has been paid to the use of elliptic curves in primality testing of Mersenne numbers.

In this paper a practical elliptic curve test for Mersenne primes, based on Gross's original idea [3] is proposed and evaluated. The computational evidence shows that the new elliptic curve approach can be efficiently used as an alternative to the Lucas-Lehmer test for Mersenne primes. More importantly, The most important advantage of the test is that, unlike the Lucas-Lehmer test which requires $\mathcal{O}(p)$ arithmetic operations and $\mathcal{O}(p^3)$ bit operations in order to verify whether or not $M_p = 2^p - 1$ is prime, it only needs $\mathcal{O}(\lambda)$ arithmetic operations and $\mathcal{O}(\lambda^3)$ bit operations, with $\lambda \ll p$. Hence it is more efficient than the Lucas-Lehmer test, whilst remaining to be still simple and elegant.

With respect to the Electronic Frontier Foundation's $100,000 prize, we suggest that interested readers can simply try to use the elliptic curve test and verify the Mersenn numbers starting from $M_p = 2^p - 1$ for $p = 33375371$, 33375383, 33375389, 33375409, 10046988, $\cdots$. All these numbers are over 10 million digits, and if you are lucky enough, you may be able to get the EFF $100,000 prize!

**Acknowledgments.** The author would like to thank Dr Werner Meixner and the anonymous referees for their very helpful comments and suggestions.

# References

1. A. O. L. Atkin and F. Morain, Elliptic Curves and Primality Proving, *Mathematics of Computation*, **61**, 1993, pp 29–68.
2. R. P. Brent, G. L. Cohen and H. J. J. te Riele, Improved Techniques for Lower Bounds for Odd Perfect Numbers, *Mathematics of Computation*, **57**, 1991, pp 857–868.
3. B. H. Gross, An Elliptic Curve Test for Mersenne Primes, *Journal of Number Theory*, **110**, 2005, pp 114-119.
4. N. Koblitz, Elliptic Curve Cryptography, *Mathematics of Computation*, **48**, 1987, pp 203–209.
5. S. Lang, *Fundamentals of Dionphtine Geometry*, Springer, 1983.
6. D. H. Lehmer, On Lucas's Test for the Primality of Mersenne's Numbers, Journal of London Mathematical Society, **10**, 1935, pp 162–165.
7. H. W. Lenstra, Jr., Factoring Integers with Elliptic Curves, *Annals of Mathematics*, **126**, 1987, pp 649–673.
8. E. Lucas, Noveaux Théorèmes d'Arithmétique Supèrieure, C. R. Acad. Sci. Paris, **83**, 1876, pp 1286–1288.
9. V. Miller, Uses of Elliptic Curves in Cryptography, *Advances in Cryptology*, CRYPTO '85, Proceedings, Lecture Notes in Computer Science **218**, Springer-Verlag, 1986, pp 417–426.
10. F. Morain, Implementing the Asymptotically Fast Version of the Elliptic Curve Primality Proving Algorithm, 21 October 2005.
    See http://www.lix.polytechnique.fr/ morain/Articles/fastecpp-final.pdf
11. M. Rosen, A Proof of the Lucase-Lehmer Test, *American Mathematics Monthly*, **95**, 1988, pp 855–856.
12. J. H. Silverman, *The Arithmetic of Elliptic Curves*, Graduate Texts in Mathematics **106**, Springer-Verlag, 1994.
13. A. Wiles, Modular Elliptic Curves and Fermat's Last Theorem, *Annals of Mathematics*, **141**, 1995, pp 443–551.
14. S. Y. Yan, Number Theory for Computing, 2nd Edition, Springer, 2002.

# Author Index

# Lecture Notes in Computer Science

For information about Vols. 1–4097

please contact your bookseller or Springer